# World Climatology

## An environmental approach

# World Climatology

## An environmental approach

## John G Lockwood
Lecturer in Geography, University of Leeds

Edward Arnold

© John G Lockwood 1974
First published 1974 by
Edward Arnold (Publishers) Ltd
25 Hill Street, London W1X 8LL

ISBN: 0 7131 5701 1

Printed in Great Britain by
Fletcher & Son Ltd, Norwich

# Contents

Contents

Contents

Contents

# Preface

In recent years there has been a growing interest in the natural environment and this book is an attempt to give an up-to-date review of the physical basis of climatology and to describe those aspects of climatology which are environmentally important. As most climatological text books concentrate on the mathematics and physics of the atmosphere and neglect the environment created by the atmosphere, it endeavours to fill this particular gap in the available literature. Early chapters consider both microclimatology and the general circulation of the atmosphere while later chapters study the environment of the major climatic zones. In particular, the climates of southern and southeast Asia, North America, western Europe, and the USSR are described in detail. The book also contains a collection of climatological maps and data which are not readily obtainable elsewhere under one cover.

This book is intended for students taking university courses on climatology, particularly those at a more advanced level. It should also prove useful to scientists who are interested in environmental aspects of climatology, such as geographers, agriculturists, botanists, civil engineers, as well as meteorologists, since it examines climatological factors which influence plant growth, water and energy balance, extreme rainfall and extreme winds, and many other climatological parameters.

It is assumed that the reader is familiar with the basic concepts of mathematics and physics, and acquainted with elementary meteorology; therefore, mathematical equations and concepts are used when they are considered to be necessary. At the time of writing, units created an extremely difficult problem, since this is a transitional period from non-metric to SI units. Where possible all units used are metric, but they do not necessarily belong to the SI system because it is sometimes impossible to convert them without making nonsense of a particular illustration or table. The radiation units, especially, though metric, often do not conform to the SI system.

Various people have helped in the production of the manuscript, and I would in particular like to thank Mr D. H. Johnson, Principal of the Meteorological Office Training College, Reading, for reading the original manuscript and making some very useful and interesting comments on it. Naturally, however, I am responsible for any errors or deficiencies which may be found in it. The maps were drawn by Mr G. Hodgson and Mr T. Hadwin, and the diagrams by Bucken Ltd. I am indebted to the Meteorological

*Preface*

Office Library at Bracknell for their assistance in obtaining books, journals, and photographs. The National Oceanic and Atmospheric Administration, USA, have also been very helpful in obtaining photographs. Lastly I thank my wife for typing most of the manuscript and also for giving me great help and encouragement during its preparation.

John G. Lockwood

Leeds
January 1973

# Acknowledgements

The author and publishers gratefully acknowledge permission received from the following to modify and make use of copyright material in diagrams and maps:

The American Geographical Society for figures 7.12B, 10.6, 10.7, 10.10[1]; the Canadian Department of Northern Affairs and National Resources, Water Research Branch for figure 1.3; the University of Chicago Press for figures 1.4, 1.22B, 1.24, 2.17; the Clarendon Press and Dr J. C. Kendrew for figure 10.2; the Elsevier Publishing Company for figures 1.9, 2.3, 2.11, 4.10, 9.2, 9.11, 10.9, 11.1, 11.2, 11.3, 11.4; the Geophysical Society of Finland for figures 2.18, 3.8; the University Press of Hawaii (East-West Center Press) for figures 3.9A, 3.9B, 6.4, 6.16; Hutchinson Publishing Group Ltd for figures 1.19, 1.20; the Indian Meteorological Department for figures 3.2, 6.15; the Indian Meteorological Department, Office of the Deputy Director General of Observatories (Forecasting) for figure 6.12; the International Association of Hydrological Sciences for figure 10.11; the International Institute for Land Reclamation and Improvement, Wageningen, for figures 9.9, 9.13; the International Society for Tropical Ecology, Banaras Hindu University for figure 6.14; the Johns Hopkins University, C. W. Thornthwaite Associates, Laboratory of Climatology for figures 6.7, 6.13; the Leeds Local Executive Committee of the British Association and M. W. Beresford for figures 1.17, 1.18; the Institute of British Geographers, Longman, and P. R. Crowe for figures 3.3, 8.12; the McGraw Hill Book Company, New York, for figures 2.16, 3.4, 3.5A, 8.13, 8.14; Macmillan, London and Basingstoke, for figures 2.13, 8.16; the Malaysian Ministry of Agriculture and Fisheries, Drainage and Irrigation Division, Kuala Lumpur for figure 7.11; the Massachusetts Institute of Technology Press for figure 2.7; Associated Book Publishers Ltd (Methuen) for figures 1.11, 2.15, 4.2, 9.3; Mezhdunarodhaja Kniga, Moscow for figures 10.4, 10.8; Microforms International Marketing Corporation for figure 8.11; the Munitalp Foundation, Nairobi, for figure 2.10; the North-Holland Publishing Company for figure 5.7; the New Zealand Meteorological Service, Wellington, for figures 6.6, 6.8, 6.9; Pergamon Press Ltd, G. W. Hurst and L. P. Smith for figure 9.12; the Royal Meteorological Society for figures 1.6, 2.9, 2.12, 2.19, 2.20, 3.5B, 4.7, 4.8, 6.1, 6.2, 6.3, 6.5, 6.10, 6.11, 7.4, 7.10, 8.1, 8.3, 8.6, 8.7, 8.8, 8.9, 8.10, 8.15, 9.1, 9.5, 9.7; Springer-Verlag for figure 9.8; *Tellus*, Stockholm, for figures

[1] Numbers refer to figures in this book.

*Acknowledgements*

2.14, 8.4, 8.5; the *Journal of Tropical Geography* (University of Singapore) for figures 5.8, 7.6, 7.7, 7.8, 7.9, 7.12A; UNESCO, Paris, for figures 2.24, 4.11, 4.12, 4.13, 4.14, 9.4, 10.3; Weidenfeld & Nicolson Ltd and the McGraw-Hill Book Company, New York, for figure 2.1; John Wiley & Sons Inc. for figures 3.6, 3.7, 4.3; John Wiley & Sons Inc. and the Smithsonian Institute Press for figure 10.1; the World Meteorological Organization, Geneva, for figure 4.9.

The author and publishers further acknowledge the modification and reproduction of non-copyright material from the following:

The United States Department of Commerce, National Oceanic and Atmospheric Administration, Environmental Data Services for figures 1.14, 1.21, 1.22A, 1.23, 2.2, 2.4, 2.5, 2.6, 2.21, 2.22, 2.23, 4.6, 10.5; the United States Department of Commerce, National Oceanic and Atmospheric Administration, Environmental Data Service and Jacob Bierknes for figure 4.4; and William M. Gray for figures 5.1, 5.2, 5.3, 5.4, 5.5, 5.6, 7.1, 7.3.

Full citations to books, journals, authors and publishers may be found by consulting the acknowledgements at the foot of the figures and the sections of references at the end of each chapter. Full acknowledgements for photographs used are given in the plate captions.

# Part I
# Introduction

*Plate 1  Above:* Cumulus convection is an important mechanism for mixing water vapour upwards from the surface and dispersing it throughout the lower troposphere. The photograph shows small cumulus clouds over Iran. *Reproduced by permission of Aerofilms Ltd.  Below:* Cloud cover over the Indian Ocean on 8 November 1970, as seen from weather satellite ITOS1. The equatorial zone is relatively cloud-free, most cloud being over the Bay of Bengal. *Reproduced by permission of the United States Department of Commerce, National Oceanic and Atmospheric Administration.*

I

# 1
# The climatic environment

Traditionally climatology has been concerned with the collection of statistics which express the average state of the atmosphere. Unfortunately some of these statistics have little relevance to the natural environment as experienced by man. In this book attention has therefore been concentrated upon those aspects of climatology that are environmentally important, and these include temperature, precipitation, evapotranspiration, solar radiation, and wind. Climate is examined as part of the natural environment and the climatic environments of the world are studied by describing their nature and also by considering their causes.

To understand the climatic environment it is necessary to study climate on a variety of scales varying from the local to the global. In this chapter the factors which determine local and microclimates are examined and in chapter 2 the nature of the major climatic regions of the earth is considered, which basically means a study of the general circulation of the atmosphere. The remainder of the book is concerned with detailed studies of the regional climatic environments of the world.

## The boundary layer

From observations of fluid flow through pipes, Reynolds (1883) distinguished between two types of motion, laminar motion and turbulent motion. He introduced some dye into the middle of a liquid current flowing through a thin glass tube and observed its behaviour. When the flow velocity was low, a well-defined dye filament was visible, the fluid layers appearing to follow parallel courses; but when the flow velocity was increased, the filament was torn apart and the dye distributed throughout the whole fluid, whose motion had now become very irregular. The first type of flow is called laminar, and the second turbulent.

Atmospheric motions near the earth's surface are practically always turbulent, even when the air is very stable. But this turbulent motion cannot extend to the ground because a layer of air, called the laminar boundary layer, adheres with great tenacity to the surface. It is at most a few millimetres thick, and usually much less than that. Above the laminar boundary layer the average wind speed slowly increases, until at some height above the ground it becomes almost constant. This height at which surface friction becomes unimportant, depends on the roughness of the earth's surface, but is normally 1,000–2,000 m.

## Introduction

The wind velocity varies with height in the atmosphere well above the ground, but these variations, which are usually small, are caused by temperature or pressure gradients and are not the result of surface friction. In this discussion of surface friction, it is assumed that horizontal temperature and pressure gradients are constant with height above the layer which is under the influence of surface friction.

A comparatively small amount of solar radiation is absorbed in the atmosphere, and estimates show that 50 per cent reaches the surface of the earth as illustrated in figure 2.1. Radiation absorbed by the atmosphere is dispersed in the lower 10–20 km layer, whereas absorption in the soil takes place in the thin upper layer. This means that the troposphere receives practically no direct heat from solar radiation whereas the temperature of the soil surface may vary greatly due to fluctuations in the solar radiation. These temperature variations spread onwards and downwards from the surface of the soil and form, in the lower part of the atmosphere and the upper layer of the soil, two thermal boundary layers. The thickness of these layers depends on the thermal diffusivity of the media and on the nature of the solar radiation variations. For instance, 12-hour temperature variations (midnight to noon and noon to midnight) averaged irrespective of sign, usually diminish with height and reflect the influence of the daily oscillations of the solar radiation on the temperature of the lower layer of the atmosphere. An example is shown in figure 1.1 and indicates a thickness of the atmospheric thermal boundary layer of about 2 km.

In addition to the thermal effect on the atmosphere, the surface of the earth also has an influence of a dynamical nature which is caused by the lower layers of the air current adhering to the ground surface. Under the influence of this adherence not only the speed but also the direction of the air-stream vary (see figure 1.2 where an example is given of the variation with height of wind speed and direction). From figure 1.2 it appears that above 2 km, if the horizontal pressure gradient is independent of the height, the variations of the wind speed and direction are inconsiderable, and so the dynamical influence of the earth's surface is apparent up to a height of 1·5 to 2 km.

The comparatively large influx of radiative energy and the adherence of the air to the surface thus produce characteristic features in the temperature and wind fields in the lowest layers of the atmosphere which are not found at greater altitudes. This layer in

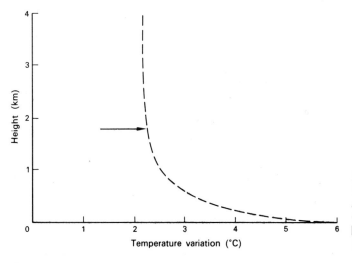

**Figure 1.1** Vertical profile of 12-hour temperature variations averaged irrespective of sign. Observations made at the Pavlovsk Aerological Observatory, USSR. (*After Laikhtman, 1964.*)

4

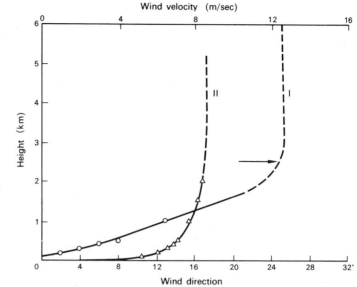

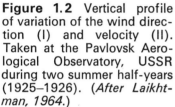

**Figure 1.2** Vertical profile of variation of the wind direction (I) and velocity (II). Taken at the Pavlovsk Aerological Observatory, USSR during two summer half-years (1925–1926). (*After Laikhtman, 1964.*)

which the influence of the earth's surface is of paramount importance is known as the planetary boundary layer of the atmosphere, and though typically about 2 km in depth, it can vary from place to place and with the season. The structure of the planetary boundary layer is complex; it consists of several sublayers with distinctive properties which will be discussed in more detail later.

It is the various interactions through the planetary boundary layer as summarized in

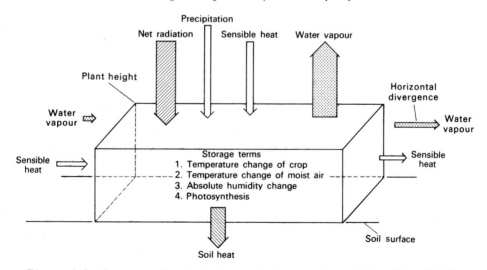

**Figure 1.3** Energy and water balance of a land surface. (*After King, 1961.*)

figure 1.3 that form our climatic environment, for this is in fact determined by the energy and the water balances in the planetary boundary layer.

The effective heat sources are short-wave solar radiation, and the heat advected by the winds. The incoming short-wave radiation heats the surface of the ground but subsequently part of this heat is re-radiated as long-wave radiation and part flows in the form of sensible heat downwards into the soil and upwards into the atmosphere. Some of the heat at the soil surface is used in evaporating water. Since energy is required to carry out this change of state, this energy is known as latent heat. When the water vapour condenses in the atmosphere, the latent heat is given up and used to warm the atmosphere. The interplay of the various components just described constitutes the heat balance of the ground surface.

Evapotranspiration is a slow continuous process, but precipitation occurs in short bursts and is generated by disturbances in the free atmosphere above the planetary boundary layer. The interplay of evapotranspiration and precipitation forms the water balance of the earth's surface.

Clearly, sensible heat and water vapour have to be transported upwards through the planetary boundary layer into the free atmosphere. In the laminar boundary layer of about 1 mm depth, which covers most surfaces, this transfer will be by the process of molecular diffusion, because in this layer there is no mechanical turbulence. Above the laminar boundary layer there is normally marked turbulence, and turbulent eddies are enormously more effective in transporting sensible heat and water vapour than the process of molecular diffusion. The eddies can take the form of general turbulent mixing or of penetrative convection. General turbulent mixing which is evident in the fluctuations of wind speed and direction is greatest within the planetary boundary layer. Penetrative convection which consists of convection cells or bubbles of rising air is not restricted to air near the surface but can cause vertical mixing throughout the depth of the troposphere.

The various components of the climatic environment will now be discussed in more detail.

## Radiation

### 1  Radiation laws

Any object not at a temperature of absolute zero ($-273°$C) transmits energy to its surroundings by radiation, that is, by energy in the form of electromagnetic waves travelling with the speed of light and requiring no intervening medium. Electromagnetic radiation is characterized by its wavelength, of which there is a wide range or spectrum extending from the very short X-rays through the ultra-violet and the visible to infra-red, microwaves and radio waves. The wavelengths of visible light are in the range $0.4$ $\mu$m to $0.74$ $\mu$m ($1$ $\mu$m $= 10^{-4}$ cm).

A perfect black body is one which absorbs all the radiation falling on it and which emits, at any temperature, the maximum amount of radiant energy. This definition does not imply that the object must be black, because snow which reflects most of the visible light falling on it is an excellent black body in the infra-red part of the spectrum. For a perfect all-wave black body the intensity of radiation emitted and the wavelength distribution depend only on the absolute temperature, which is the temperature measured from absolute zero, and in this case the Stefan-Boltzmann law applies. This law states that the flux of radiation from a black body is directly proportional to the fourth power of its absolute temperature, that is,

$$F = \sigma T^4$$

where F is the flux of radiation, T is the absolute temperature, and $\sigma$ is a constant

$(5.670 \times 10^{-8} \text{ mW/cm}^2(\deg \text{ K})^4)$. It can also be shown by the Wien displacement law, that the wavelength of maximum energy $\lambda_m$ (figure 1.4) is inversely proportional to the absolute temperature,

$$\lambda_m = \frac{a}{T}$$

where $a$ is a constant $(2.898 \times 10^{-3} \deg \text{ K})$.

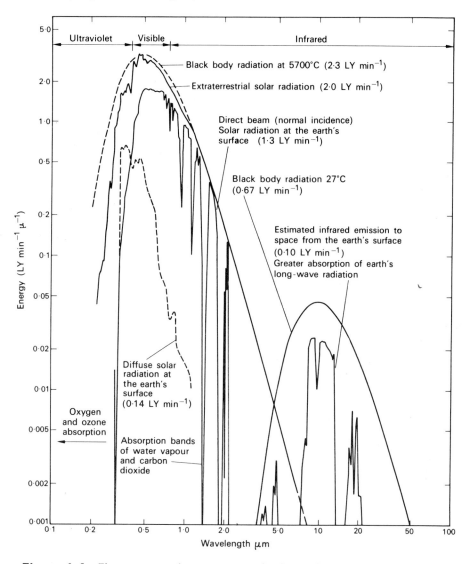

**Figure 1.4** Electromagnetic spectrums of solar and terrestrial radiation. The black body radiation at 5700°C is reduced by the square of the ratio of the sun's radius to the average distance between the sun and the earth in order to give the flux that would be incident on the top of the atmosphere. (*After Sellers, 1965.*)

## Introduction

### 2 Short-wave radiation

If the sun is assumed to be a black body an estimate of its effective radiating temperature may be obtained from the Stefan-Boltzmann law, and from this law the sun appears to have an effective surface temperature of 5750°K. For the sun the wavelength of maximum emission is near 0·5 μm, which is in the visible portion of the electromagnetic spectrum, and almost 99 per cent of the sun's radiation is contained in the so-called short wavelengths from 0·15 to 4·0 μm; of these short wavelengths, 9 per cent are in the ultra-violet (less than 0·4 μm), 45 per cent are in the visible (0·4 to 0·74 μm), and 46 per cent are in the infra-red (greater than 0·74 μm).

### 3 Long-wave radiation

The surface of the earth, when heated by the absorption of solar radiation, becomes a source of long-wave radiation; this is because the average temperature of the earth's surface is about 285°K, and therefore most of the radiation is emitted in the infra-red spectral range from 4 to 50 μm, with a peak near 10 μm, as indicated by the Wien displacement law. This radiation is sometimes called terrestrial radiation, because it is emitted by the earth and atmosphere, and sometimes thermal radiation, because it is a form of heat energy.

### 4 Albedo

It is necessary to distinguish clearly between reflected and re-radiated radiation. If the radiation is directly reflected there is no change in the wavelength, and so short-wave solar radiation is reflected as short-wave radiation. If the radiation is absorbed by the surface

**Table 1.1**  Values of albedo for various surfaces

| Surface | Albedo (percentage of incoming short-wave radiation) |
| --- | --- |
| Black soil, dry | 14 |
| Black soil, moist | 8 |
| Ploughed field, moist | 14 |
| Sand, bright, fine | 37 |
| Dense, dry and clean snow | 86–95 |
| Sea ice, slightly porous milky bluish | 36 |
| Ice sheet, covered by a water layer of 15–20 cm | 26 |
| Woody farm covered with snow | 33–40 |
| Deciduous forest | 17 |
| Tops of oak | 18 |
| Pine forest | 14 |
| Desert shrubland | 20–29 |
| Swamp | 10–14 |
| Prairie | 12–13 |
| Winter wheat | 16–23 |
| Heather | 10 |
| Yuma, Arizona | 20 |
| Washington, D.C. (September) | 12–13 |
| Winnipeg, Manitoba (July) | 13–16 |
| Great Salt Lake, Utah | 3 |

*After Kung, Bryson, and Lenschow (1964); and Kondratyev (1954).*

*Plate 2* Aerial photographs may be used to estimate the albedo of the ground: those areas reflecting a high percentage of the incident radiation (high albedo) appear light, while areas of low albedo appear dark. If the exact conditions of the film, exposure, development and printing are known, it is possible to relate photographic shades to albedo. The photograph shows part of the Yorkshire Pennines; the depths of shading suggest variations in the albedo of the differing surfaces which include bare rock, pasture, woods and a small lake. Bendelow (1969) used aerial photographs to determine the albedo of Morecombe Bay at low tide. *Reproduced by permission of Meridian Airmaps Limited and the West Riding County Council.*

and then re-radiated, the wavelength of the re-radiated radiation will vary according to the Stefan-Boltzmann and Wien laws. Some of the short-wave radiation falling on the surface of the earth is reflected, and the ratio of the incoming to the reflected surface short-wave radiation is known as the albedo. Some typical albedo values are given in table 1.1; the albedo of a black body is zero.

## 5 Emissivity

The earth's surface is commonly assumed to emit and absorb in the infra-red region as a grey body, that is, as a body for which the Stefan-Boltzmann law takes the form

$$F = \epsilon \sigma T^4$$

where the constant of proportionality $\epsilon$ is defined as the infra-red emissivity or, equivalently, the infra-red absorptivity (one minus the infra-red albedo). Some typical infra-red emissivities for various surfaces are given in table 1.2, where the values are expressed as

**Table 1.2**  Infra-red emissivities

| Surface | Per cent |
|---|---|
| Water | 92–96 |
| Snow, fresh fallen | 82–99·5 |
| Ice | 96 |
| Concrete, dry | 71–88 |
| Desert | 90–91 |
| Grass, high dry | 90 |
| Fields and shrubs | 90 |
| Oak woodland | 90 |
| Pine forest | 90 |
| Leaves and plants | |
| 0·8 $\mu$m | 5–53 |
| 1·0 $\mu$m | 5–60 |
| 2·4 $\mu$m | 70–97 |
| 10·0 $\mu$m | 97–98 |

*After Sellers (1965).*

percentages since the emissivity of a black body is one. From table 1.2 it appears that practically all surfaces, including snow, have emissivities of between 90 and 95 per cent, corresponding to infra-red albedos of 10 to 5 per cent.

## 6 Greenhouse effect

The atmosphere is nearly transparent to short-wave radiation from the sun, of which large amounts reach the earth's surface, but it readily absorbs infra-red radiation emitted by the earth's surface, the principal absorbers being water vapour (5·3 to 7·7 $\mu$m and beyond 20 $\mu$m), ozone (9·4 to 9·8 $\mu$m), carbon dioxide (13·1 to 16·9 $\mu$m), and clouds (all wavelengths). Only about 9 per cent of the infra-red radiation from the ground surface escapes directly to space, mainly in the 'atmospheric window' (8·5 to 11·0 $\mu$m); the rest is absorbed by the atmosphere, which in turn re-radiates the absorbed infra-red radiation, partly to space and partly back to the surface. The main receiver of heat from the sun is thus the ground surface, while most of the heat loss to space is from the lower troposphere. This means that there must be a general temperature gradient from the ground surface to the lower troposphere, so as to allow the flow of heat from the surface to the atmosphere and then to space. The earth's surface is thus at a higher average temperature than that which would be observed if the atmosphere were perfectly transparent or completely absent. It has been estimated by Fleagle and Businger (1963) that if the atmosphere were absent, the earth's surface would be 30°–40°c cooler than it is at present. The term 'greenhouse effect' is often applied to this particular phenomenon, because it is analogous to that effect which is supposed to operate in a greenhouse where the glass is transparent to short-wave radiation but nearly opaque to infra-red radiation.

## 7  Radiation balance

Of particular interest to the meteorologist is the so-called net radiation, which is the difference between the total incoming and the total outgoing radiation. The net radiation indicates whether net heating or cooling is taking place; the net radiation will normally be negative at night indicating cooling, but during the day it may be negative or positive depending on the balance of the incoming and the outgoing radiation.

In figure 1.5 which illustrates the components of the net radiation balance, it is assumed that the atmosphere and soil are dry, that is, there is no evaporation nor condensation. The simplest case occurs at night, because there is no incoming short-wave radiation, but instead there is a continuous long-wave radiation loss. At night the soil surface emits long-wave radiation ($R_{L\uparrow}$) to the atmosphere, where water vapour and carbon dioxide absorb large amounts of it, which in turn is partly re-radiated downwards ($R_{L\downarrow}$) to be re-absorbed by the soil surface. The difference between $R_{L\uparrow}$ and $R_{L\downarrow}$ is the net radiation loss ($R_N$), which in this example represents a cooling, and consequently there is a flow of sensible heat from the atmosphere and the lower layers of the soil towards the soil surface.

During the day the situation is more complex because of the incoming short-wave radiation, which can take two forms: direct radiation from the sun and diffuse sky radiation. The term global radiation is used for the sum of the direct and the diffuse short-wave radiation received by a unit horizontal surface. Short-wave radiation in the atmosphere is scattered and reflected by molecules, dust particles, clouds, etc., so that a large percentage of the short-wave radiation reaching the surface will not be coming from the

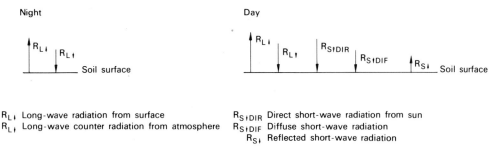

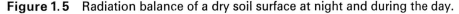

$R_{L\downarrow}$  Long-wave radiation from surface
$R_{L\uparrow}$  Long-wave counter radiation from atmosphere

$R_{S\downarrow DIR}$  Direct short-wave radiation from sun
$R_{S\downarrow DIF}$  Diffuse short-wave radiation
$R_{S\downarrow}$  Reflected short-wave radiation

**Figure 1.5**  Radiation balance of a dry soil surface at night and during the day.

direction of the sun. Indeed in some parts of the world, the Arctic for instance, the diffuse sky radiation can form a major part of the incoming short-wave radiation. Short-wave radiation on reaching the surface can be either absorbed or reflected. The reflected radiation represents a complete energy loss, because it does not heat the surface, and it depends on the albedo of the surface. Of the short-wave energy absorbed, some will be re-radiated as long-wave radiation and some will be transferred as sensible heat into the soil and the atmosphere.

The long-wave interactions will be similar to those at night, so the net radiation is given by

$$R_N = (R_{L\downarrow} + R_{S\downarrow DIF} + R_{S\downarrow DIR}) - (R_{L\uparrow} + R_{S\uparrow})$$

where the notation is the same as in figure 1.5.

Figure 1.6 from Monteith and Szeicz (1961) illustrates, for temperate latitudes, the relative importance of the various radiative transfers throughout the day in clear weather. The surface consists of grass with an albedo of 26 per cent, a value which has been found to be characteristic of many green crops giving a nearly complete coverage of the soil; therefore approximately 74 per cent of the incident solar radiation is absorbed. The long-wave radiation loss from the surface is greatest around mid-day when the grass is hottest, and least around the time of minimum temperature, which is commonly just before dawn in clear weather. The diurnal variation of the grass temperature is about 20°C, which is about 7 per cent of the absolute temperature. From the Stefan-Boltzmann law a variation of about 28 per cent is to be expected in the long-wave radiation loss from the surface, and this is close to the observed value. The variation in the long-wave radiation from the atmosphere is much less because it depends largely on the temperature of the water vapour and carbon dioxide in the lowest 200 m or so of the atmosphere and this varies but little in comparison with ground surface temperature. The lowest 100 m of the atmosphere contributes strongly to the long-wave radiation reaching the surface and each

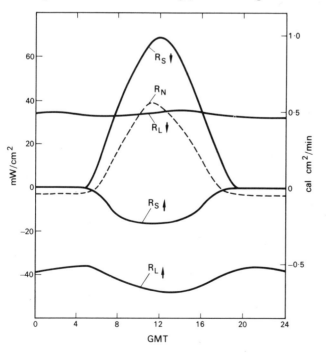

**Figure 1.6** Diurnal variation of the components of the radiation balance for a clear August day at Rothamsted (52°N) over a thick stand of grass 40 cm high. (*After Monteith and Szeicz, 1961.*)

100 m above has rapidly diminishing influence so that conditions above 1,000 m are usually of little significance with a clear sky. This is well illustrated by table 1.3, from Lettau and Davidson (1957), which gives values of atmospheric downward long-wave radiation from a sounding at O'Neill, Nebraska.

The net outgoing long-wave radiation from the ground surface has two basic components: the total long-wave emission from the surface, which is a function of the surface emissivity and temperature, and the counter long-wave radiation from the atmosphere, which is a function of air temperature, water vapour content, and the cloud cover. The downwards atmosphere long-wave radiation under cloudless conditions has been found to be closely related to the temperature and the vapour pressure measured in a Stevenson Screen, and this would be expected from the values given in table 1.3. Various empirical

**Table 1.3** Total counter radiation (O'Neill, Nebraska)

| Per cent | Originating below |
|---|---|
| 9·3 | 0·1 m |
| 15·9 | 0·4 |
| 20·3 | 0·8 |
| 25·8 | 2·0 |
| 35·0 | 6·0 |
| 44·6 | 20·0 |
| 58·9 | 100·0 |
| 74·6 | 400·0 |
| 84·8 | 1,000·0 |
| 98·5 | 4,000·0 |

*After Lettau and Davidson (1957).*

formulas for estimating the counter radiation have been suggested and among these are the following:

$$R_{L\downarrow} = \sigma T^4 (0.805 - 0.235) (10^{-0.052e}) \quad \text{(Ångström, 1916)}.$$

$$R_{L\downarrow} = \sigma T^4 (0.52 + 0.065\sqrt{e}) \quad \text{(Brunt, 1941)}.$$

$$R_{L\downarrow} = 5.31 (10^{-14}T^6) \quad \text{(Swinbank, 1963)}.$$

where $R_{L\downarrow}$ is the atmospheric counter long-wave radiation, T is the screen temperature, e is the screen vapour pressure, and $\sigma$ is a constant equal to $5.67 \times 10^{-8}$ mW/cm$^2$(deg K)$^4$.

When considering the effect of cloud on atmospheric counter radiation there are two main components, which are cloud amount and the temperature of the cloud base; since high-level clouds have much lower base temperatures than low clouds, it therefore follows that cloud height is also important. Cloud layers are normally considered to radiate as black bodies, and therefore the effect of cloudiness on the net long-wave radiation ($R_{L\uparrow}$ − $R_{L\downarrow}$) is often taken into account by multiplying the estimate for a clear sky by a factor $(1 - \lambda c)$ in which c is the fractional cloudiness and $\lambda$ depends on cloud height. For low clouds $\lambda = 0.8 - 0.9$, for medium cloud $0.6 - 0.7$ and for cirrus cloud about $0.2$. Normally dense low clouds are most effective in retarding temperature falls at night, while cirrus clouds have little effect.

It has been found from experiments by Monteith and Szeicz (1961) that changes in the net radiation ($R_N$) received by bare soil or vegetation during cloudless summer days are related to global radiation ($R_s$) by the equation

13

$$R_N = \frac{(1 - a)}{(1 + \beta)} \cdot R_s + L_o$$

where $a$ is the short-wave albedo; $\beta$ is a heating coefficient defined as the increase in long-wave radiation loss per unit increase of $R_N$; and $L_o$ is the net radiation as $R_s \to O$. Values of $\beta$ vary with climate and depend as much upon atmospheric radiative properties as surface conditions, but for Britain, Monteith and Szeicz (1962) suggest that $\beta$ is probably close to 0·1 for agricultural crops and natural vegetation which completely cover the ground and are never short of water. If transpiration is physiologically restricted or if ground cover is incomplete, $\beta$ may lie between 0·1 and 0·2 and for very dry soil between 0·3 and 0·4. Negative values of $\beta$ have, however, been reported by some authors such as Stanhill, Hofstede, and Kalma (1966). For Britain, Gadd and Keers (1970) suggests values of $L_o$ equal to $-$0·05$R_s$. The calculation of net radiation from sunshine duration is considered later in the section on precipitation and evaporation.

## 8  Relief
So far, only the net radiation from a plane surface with an unobstructed horizon has been discussed. Because the effective radiating temperature of the atmosphere is usually lowest directly overhead, the net outgoing long-wave radiation is greatest in this direction and decreases with increasing zenith distance to zero on the horizon. According to Dubois (1929) and Linke (1931) the ratio of the net outgoing radiation from a basin ($R_{NLB}$) to that from a flat plain ($R_{NLP}$) is given by

$$\frac{R_{NLB}}{R_{NLP}} = (1 - \cos^{(\gamma+2)}Z_1)$$

where $\gamma$ is a constant which depends on the surface vapour pressure, and $Z_1$ is the zenith distance to the rim of the basin. Values of this ratio for different values of $Z_1$ and for $\gamma$ equal to 0·3 and 0·8 are listed in table 1.4. Table 1.4 helps to explain the unusual warmth

**Table 1.4**  Values of the ratio of the effective out going radiation from a basin to that from a flat plain

| $Z_1$ (degrees) | $\gamma = 0.3$ | $\gamma = 0.8$ |
|---|---|---|
| 0 | 0·000 | 0·000 |
| 10 | 0·035 | 0·042 |
| 20 | 0·133 | 0·160 |
| 30 | 0·282 | 0·332 |
| 40 | 0·458 | 0·526 |
| 50 | 0·638 | 0·710 |
| 60 | 0·797 | 0·856 |
| 65 | 0·862 | 0·910 |
| 70 | 0·915 | 0·950 |
| 75 | 0·955 | 0·977 |
| 80 | 0·982 | 0·993 |
| 85 | 0·996 | 0·999 |
| 90 | 1·000 | 1·000 |

where $\gamma = 0.11 + 0.0255e$
and $e$ is the surface vapour pressure in millibars

*After Dubois (1929) and Linke (1931).*

experienced in deep valleys on hot summer days. For example, the net outgoing long-wave radiation from a basin whose rim extends $40°$ above the horizon ($Z_1 = 50$) will be only 64–71 per cent of that from a flat plain, the deficit in outgoing radiation being made up by sensible heat transfer to the atmosphere. The values in table 1.4 also apply to a tree obscuring the sky from the zenith to $90° - Z_1$. A tree canopy provides effective frost protection for the underlying surface by shielding it from the coldest portion of the sky toward which radiative losses are greatest.

Not only does relief affect the loss of long-wave radiation but it can also influence the amount of short-wave radiation received; Geiger (1961, 1969) has made a study of this particular topic. The influence of slope upon the amount of short-wave radiation received may be illustrated by figure 1.7 which is intended to show diagramatically, for latitude $50°N$, the direct solar radiation incident on slopes up to an angle of just over $30°$ as a function of the time of the year. Because of the symmetry of easterly and westerly slopes, only easterly slopes need be shown. The greatest differences between slopes coincide with the start of the growing season in spring; in winter and summer differences are comparatively small. The screening of the horizon by hills or mountains can also reduce the daily isolation totals. Where mountains tower above the horizon sunrise occurs later and sunset earlier, but conversely the length of day may be extended at high latitudes owing to the depression of the horizon.

So far only the interaction of direct solar radiation and slope has been considered, but diffuse sky radiation represents an additional source of radiation for all slopes and comes from all directions. Kondratyev and Manolova (1955, 1960) have shown that for slopes up to an angle of $30°$ the sky radiation can be regarded as independent of azimuth. It thus

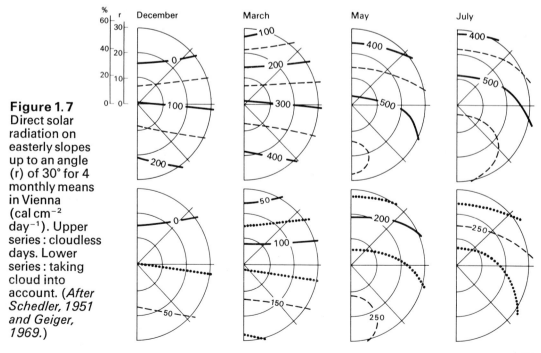

**Figure 1.7** Direct solar radiation on easterly slopes up to an angle (r) of $30°$ for 4 monthly means in Vienna (cal cm$^{-2}$ day$^{-1}$). Upper series: cloudless days. Lower series: taking cloud into account. (*After Schedler, 1951 and Geiger, 1969.*)

decreases the relative differences between slopes due to direct radiation and this compensating effect will be more effective the greater the proportion of the sky radiation in the global radiation. In high latitudes the diffuse sky radiation can form a major part of the global radiation and therefore differences in slope are unimportant.

## Convection and mechanical turbulence

### *1  Laminar boundary layer*

Because the laminar boundary layer of air adheres strongly to the earth's surface, it prevents the turbulent motions of the atmosphere from extending to the ground. Within the laminar boundary layer, heat, water vapour, and momentum are transferred vertically by molecular processes only, and the rates of transfer are proportional to the vertical gradients of temperature, specific humidity and horizontal air velocity. For example, the vertical transfer of sensible heat (H) is given by

$$H = -\rho C_p K_h \frac{\partial \theta}{\partial z}$$

where $\rho$ is the air density; $C_p$ is the specific heat of air at constant pressure; $K_h$ is the thermal diffusivity of air ($0 \cdot 16$ to $0 \cdot 24$ cm$^2$ sec$^{-1}$); $\theta$ is the potential temperature; and z is the height.

If the sun's heat absorbed at the ground surface were propagated through the air purely by molecular conduction then the diurnal temperature wave would vanish at only 4 m above the surface and at a height of 2 m the maximum temperature would occur at midnight. But the diurnal variation of air temperature extends well above 1,000 m, so clearly a much more effective process than molecular diffusion must operate in the free atmosphere. This more effective process of heat transfer is in fact the general mechanical turbulence of the atmosphere. Because of the low thermal diffusivity of air, temperature gradients within the laminar boundary layer must be high, and probably reach values of $30°$c mm$^{-1}$ in hot deserts.

### *2  Atmospheric turbulence*

Atmospheric turbulence near the ground is manifested by rapid, and sometimes violent, variations of wind speed and direction. Since turbulence is a very disorganized motion of the air, it is not possible to predict the individual fluctuations from past history. In the wind tunnel, a mean wind can be defined and measured at any point in the flow without too much difficulty, but there have been many arguments concerning what constitutes mean flow and what constitutes turbulence in the atmosphere. A mean wind cannot be defined without specifying the time over which the average flow is defined; therefore the mean wind may be the average for a minute, a day, or a year. Fluctuations about the mean are called turbulence.

Normally, the rates at which momentum, heat, and water vapour are being transported vertically through a level near the surface (at say 1 m) by turbulence are equal to the rates at which the same three quantities are being transferred to or from the earth's surface by molecular processes. Molecular diffusion is most important very close to the ground, where the vertical gradients of temperature, specific humidity, and wind speed are greatest, but its importance, decreases rapidly with height and it is insignificant above 1 or 2 mm. Turbulence cannot extend to the ground, because the vertical velocity vanishes

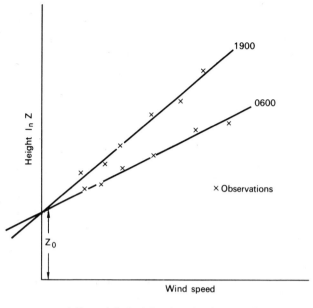

**Figure 1.8** Schematic illustration of the roughness length ($Z_0$). Average profiles of wind speed with height for a number of observations at 0600 and 1900 local time when near neutral stability conditions prevail. It is assumed that the observations are made in the first 20 m over short grass.

at the surface, but its importance increases rapidly with height in the lowest few centimetres. Clearly, there must be a thin layer of air close to the ground where both turbulence and molecular diffusion are of equal importance as transporting processes.

### 3   Surface layer

The characteristics of the wind profile in the first 20 m above the ground can best be explained by reference to an example. Figure 1.8 shows the average variations of wind speed with height over short grass. At 0600 and 1900, neutral stability conditions prevail and the temperature is almost constant with height; therefore the wind speed increases almost linearly with the logarithm of the height, at least below 20 m. That is

$$\frac{u_2 - u_1}{\ln z_2 - \ln z_1} = \text{Constant}$$

where $u_1$ and $u_2$ are the wind speeds at heights $z_1$ and $z_2$ respectively.

If it is assumed that the wind speed becomes zero at some height $z_0$ above the surface then it follows that

$$u_z = (\text{Constant}) \ln \left( \frac{z}{z_0} \right)$$

when $u_z$ is the wind speed at the height z. The quantity $z_0$ is called the roughness length and in figure 1.8 it is given by the value of the intercept of the lines with the ln z axis. The intercept has the same value at both 0600 and 1900, since it is mainly a function of the type of surface. Some typical values of roughness lengths, for a selection of surfaces, are shown in figure 1.9. For tall vegetation, the above equation becomes

$$u_z = (\text{Constant}) \ln \left( \frac{z - d}{z_0} \right)$$

where d is called the zero-plane displacement, since the wind speed goes to zero at the height $z_o + d$, rather than at $z_o$. The concept of the zero-plane displacement can be useful for measuring the roughness of an irregular surface at some level above the ground, for example, in the case of an irregular forest canopy where some trees project above the general level of the canopy. The relationship between the roughness length and the vegetation height h can, according to Tanner and Pelton (1960), be expressed by

$$\log z_o = a + b \log h$$

where common (base 10) logarithms are used. A variety of values have been obtained for the constants a and b; according to Tanner and Pelton, $a = -0.883$ and $b = 0.997$, but Kung (1961, 1963) suggests that $a = -1.24$ and $b = 1.19$, and Sellers (1965) that $a = -1.385$ and $b = 1.417$. However the variation in these values is probably not significant. In practice for natural vegetation that yields to the wind, the roughness length tends to decrease with increasing wind speed, presumably because at high speeds the vegetation bends and presents a relatively flat and smooth surface to the air flow.

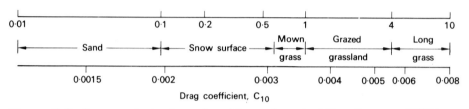

**Figure 1.9** Some typical values of roughness length. (*After Deacon, 1969.*)

## 4 *Friction layer*

Starting from the ground surface, the increase of wind speed with height is first of all in accordance with the logarithmic law in the first 20 m and the direction is at an angle to the upper wind direction. With a further increase in height, the wind increases more gradually and turns into the upper wind direction. Providing the pressure pattern is not changing rapidly, the wind at the top of the planetary boundary layer approximates to the geostrophic wind ($V_g$), that is to say, the wind blowing parallel to the isobars with the requisite speed for the force due to the horizontal pressure gradient (grad p) to be balanced by the deflecting force caused by the earth's rotation (Coriolis force). The geostrophic wind relationship for latitude $\phi$ is:

$$V_g = \text{grad } p/2\rho\Omega \sin \phi$$

where $\Omega$ is the angular velocity of the earth. The layer below the level at which the observed wind is in general or close approximation to the geostrophic wind is often known as the friction layer, and the atmosphere above the friction layer is usually called the free air. As suggested above, the friction layer which corresponds to the dynamic part of the planetary boundary layer, can be divided into a number of sub-layers (see figure 1.10), and these are the spiral layer, the surface layer, and the laminar boundary layer. The laminar boundary layer is the layer where flow is laminar and the viscous stresses dominate the eddy stresses. The surface layer is the layer in which the wind speed varies with height according to the logarithmic law and the eddy stress is an order of magnitude larger than

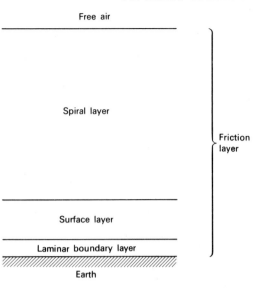

Free air

Spiral layer

Friction layer

Surface layer

**Figure 1.10** Schematic illustration of the sublayers of the friction layer. The friction layer corresponds to the dynamic part of the planetary boundary layer.

Laminar boundary layer

Earth

the horizontal pressure force. The spiral layer, which has not yet been considered, is a layer in which the eddy stress has approximately the same order of magnitude as the pressure gradient and Coriolis forces.

*5 Spiral layer*

Below the top of the friction layer the wind blows obliquely across the isobars with a component towards lower pressure at a speed below that of the geostrophic wind. The angle of obliqueness increases and the speed decreases with increasing depth into the friction layer. This wind spiral with height is analogous to the turning of ocean currents as the effect of wind stress diminishes with increasing depth, and is known as the 'Ekman spiral', an example of which is shown in figure 1.11. The variation of wind speed and

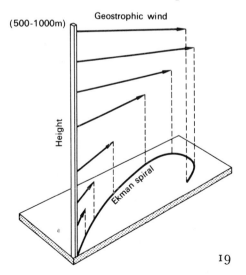

(500-1000m)

Geostrophic wind

Height

Ekman spiral

**Figure 1.11** Schematic illustration of the Ekman spiral of wind with height. (*After Barry and Chorley, 1968.*)

19

direction in the spiral layer can be explained by the assumption that the eddy viscosity $K_M$ is constant with height. This also indicates that the depth of the friction layer is proportional to $(K_M/\Omega \sin \phi)^{\frac{1}{2}}$, where the eddy viscosity $K_M$ is defined by

$$K_M = \frac{k^2 u_z z}{\ln(z/z_o)}$$

where k is Karman's constant, and the other symbols are as previously defined. The above equation indicates that large values of $K_M$ are to be expected with rough surfaces, strong winds, and strong instability and that these will create deeper friction layers than smooth surfaces, light winds, and stable conditions. The Coriolis parameter is weaker in lower latitudes and this suggests that deeper friction layers may exist in the tropics as compared with higher latitudes.

The influence of the surface on wind speed is shown by the following calculations by Lettau (1959) and Deacon (1969). Assuming a geostrophic wind speed of 25 m/sec at latitude 43°, Deacon showed that over scrub with a roughness length of 100 cm, the wind at 10 m above ground level would be 7 m/sec; but if the scrub were cleared and replaced by bare soil with a roughness length of 0·1 cm, the 10 m wind speed would increase to 13 m/sec. He comments that in a semi-arid climate a wind speed of 13 m/sec is sufficient to cause soil to start blowing away, and so the vital role that wind breaks and shelter belts play in such situations becomes clear.

## 6   Convection

Ground surface is not horizontally uniform, even within a short distance, because of irregularities such as changes in slope, soil type, or land use. As a result of these small irregularities, temperature differences of perhaps 0·5°c develop over short distances during sunshine. The warmer areas on the ground surface transfer their heat to the atmosphere, which in turn becomes warmer, less dense, and starts to rise as buoyant parcels of air which are known as thermals. When they are only a few metres above the ground, their heat supply is cut off, and thus during the few minutes now to be considered, the air parcels in a thermal conserve their heat, assuming that there is no mixing with the surrounding air.

As they ascend (figure 1.12), they are subject to a gradually decreasing atmospheric

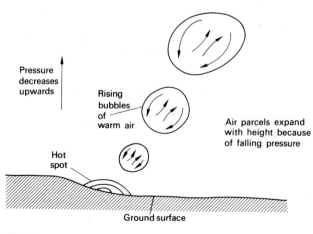

Figure decorations:
Pressure decreases upwards

Rising bubbles of warm air

Air parcels expand with height because of falling pressure

Hot spot

Ground surface

**Figure 1.12** Schematic illustration of bubbles of relatively warm air rising through the atmosphere.

pressure and therefore expand. The rate of decrease of pressure p with height z is given by

$$dp = -g\rho dz$$

where g is the acceleration due to gravity. Assuming that the perfect gas law applies to dry air, pressure p can be expressed as

$$p = \rho RT$$

where R is the gas constant for dry air, and T is the temperature (in °K). Combining the two above equations, they become

$$\frac{dp}{p} = \frac{-g}{R\overline{T}} dz$$

where $\overline{T}$ is the temperature of the surrounding air.

If T is the temperature of the ascending parcel of air, and it is assumed that the ascent takes place without an exchange of heat between the parcel and its environment (an adiabatic process), then the First Law of Thermodynamics gives

$$\frac{dT}{T} = \frac{AR}{C_p} \frac{dp}{p}$$

where A is the reciprocal of the mechanical equivalent of heat, and $C_p$ is the specific heat of air at constant pressure.

Combining the last two equations gives an expression for the rate of change of temperature with height:

$$\frac{-dT}{dz} = \frac{Ag}{C_p} \frac{T}{\overline{T}}$$

Under normal circumstances T is very nearly equal to $\overline{T}$ and the rate of decrease of temperature with height for an ascending parcel of dry air can be approximated by $Ag/C_p$. This has a value of about 9·8°c/km and is known as the dry adiabatic lapse rate.

If the air parcel is saturated or becomes saturated during ascent, the dry adiabatic lapse rate no longer applies, since heat will be gained by the parcel through the latent heat released as water vapour condenses into droplets. Since the amount of water vapour condensed depends on the temperature and pressure of the air, the saturated adiabatic lapse rate will not be constant. At low temperatures and pressures, the saturated adiabatic lapse rate is almost equal to the dry adiabatic lapse rate but at high temperatures and pressures it is considerably less. A typical value of the saturated adiabatic lapse rate is for 20°c and 1000 mb about 4·4°c/km.

As shown in figure 1.13, when a large bubble of air, initially unsaturated, is forced aloft, it cools at the dry adiabatic lapse rate until condensation occurs, and then it cools at the saturated adiabatic lapse rate, the condensed water vapour becoming visible as cloud, normally of the type known as cumulus.

If the surrounding atmosphere has an environmental lapse rate which is greater than the dry adiabatic lapse rate, then the ascending parcel of air will always be warmer than its environment. It will consequently be lighter and buoyant and will continue to ascend, once an initial upward motion takes place. Under these conditions the surrounding atmosphere is said to be unstable. On the other hand, if the environmental lapse rate is

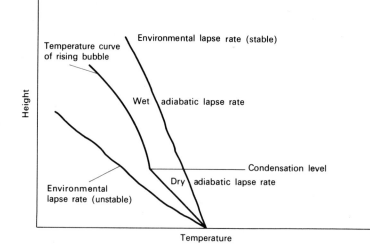

**Figure 1.13** Schematic illustration of lapse rates and atmospheric stability. Environmental lapse rate is not normally constant with height and is largely dependent on individual air-masses. Temperature curve of rising air bubble is controlled by the basic laws of physics and is fixed for a given set of initial conditions. It is assumed that all changes in the rising bubble are adiabatic.

smaller than the dry adiabatic lapse rate, the surrounding atmosphere would be warmer and lighter than an ascending parcel. Therefore any initial upward motion of the parcel would tend to be damped out and the parcel would return to its original position. Such an atmosphere is said to be stable. If the environmental lapse rate and the dry adiabatic lapse rate are equal, the atmosphere is said to have neutral stability; most of the atmospheres previously considered have had neutral stability.

In the explanation of stability it was assumed that no mixing took place between the rising air parcel and the atmosphere, but this is not true in the real atmosphere. The rising parcel will mix with the environment and cease to ascend when it becomes indistinguishable from the surrounding atmosphere, thereby losing its buoyancy. In this way convection forms a very effective method of transfering heat and water vapour from the surface to the lower atmosphere.

Turbulence in general in the atmosphere is partly controlled by the stability, for when the atmosphere is stable, turbulence is on a small scale with no thermals and the transfer of heat and water vapour upwards from the surface is slow. When an inversion is present, that is, when temperature increases with altitude, then the atmosphere is very stable and the transfer of heat and water vapour upwards through the inversion is very slow. In contrast, if the atmosphere is unstable there is a large amount of convective activity, turbulence is marked, and under these conditions heat and water vapour are transferred rapidly upwards. In a city under inversion conditions, it is often very hazy and smoky because there is little vertical exchange of air. If the atmosphere becomes slightly unstable, it often becomes reasonably clear because of the vertical turbulent mixing throughout the lower troposphere. In summary it can be said that atmospheric as well as surface conditions control the rate of supply of energy and water vapour to the lower troposphere.

## Precipitation and evaporation

Expressed in the most basic terms, the water balance at any point on the earth's surface represents the interaction of precipitation, evapotranspiration, and the stored moisture below the earth's surface. Since evapotranspiration is an energy-consuming process, it has a strong influence on the energy budget. The relation of evapotranspiration to precipitation is also a major factor governing climate and the characteristics of soils and plant communities occuring in any region. This point was made by Thornthwaite (1948) when he wrote, 'We cannot tell whether a climate is moist or dry by knowing the precipitation alone. We must know whether precipitation is greater or less than the water needed for evaporation and transpiration. Precipitation and evapotranspiration are equally important climatic factors.'

### 1 Evapotranspiration

In meteorology, evaporation is the change of liquid water or ice to water vapour, and it proceeds continuously from the earth's free water surfaces, soil, snow, and ice-fields. Transpiration is the process by which the liquid contained in the soil is extracted by plant roots, passed upwards through the plant, and discharged as water vapour to the atmosphere; the rate of transpiration is highest during the day and falls almost to zero during night hours. Evapotranspiration is the combined process of evaporation from the earth's surface and transpiration from vegetation. Potential evapotranspiration is the maximum amount of water vapour that can be added to the atmosphere under the given meteorological conditions from a surface covered by green vegetation with no lack of available water.

The evapotranspiration from a land surface depends primarily on the radiant energy supply to the surface, but may be limited by the rate of movement of water to the evaporating surface. Evapotranspiration from natural surfaces is a physical process in which liquid water is vaporized and carried into the atmosphere. According to King (1961), three simultaneous dynamic processes are involved, which are

1   a flow of water vapour by turbulent and molecular diffusion from the evaporating surface to the atmosphere;
2   a flow of heat by radiation, convection, and conduction to the evaporating surface and the removal therefrom as latent heat of vaporization; and
3   a flow of water through the soil and plants to the evaporating surface.

The water vapour from an evaporating surface first passes through a thin laminar layer next to the surface where only molecular diffusion takes place, and is then transported upwards by turbulent motions. The basic equation for the mean vertical transfer of water vapour (E) is

$$E = -\rho K_v \frac{\partial q}{\partial z}$$

where $K_v$ is the eddy diffusivity for water vapour, and q is the specific humidity (mass of moisture per unit mass of moist air). According to Thornthwaite and Holzmann (1939), evapotranspiration can be calculated using the following propositions:

1   the transfer factor for momentum is identical with that for water vapour:

$$K_M = K_v$$

2   the shearing stress is constant with height;

3    the wind velocity $u_z$ at the height z may be expressed by a logarithmic wind function of the form

$$u_z = \frac{1}{k} \left(\frac{\tau_0}{\rho}\right)^{\frac{1}{2}} \ln \left(\frac{z + z_0 - d}{z_0}\right)$$

where $\tau_0$ is equal to $\rho K_M \frac{\partial u}{\partial z}$. These propositions lead, in a neutral atmosphere, to the following equation for evapotranspiration $(E_t)$:

$$E_t = \frac{\rho k^2 (u_2 - u_1)(q_1 - q_2)}{\left[\ln \dfrac{(z_2 + z_0 - d)}{(z_1 + z_0 - d)}\right]^2}$$

Assuming an air pressure of 76 cm Hg and an air temperature of 20°c, and subject to the boundary conditions,

$$z_2 = 200, \; u_2 = u, \; e_2 = e_a, \; z_1 = d, \; u_1 = 0, \; e_1 = e_s,$$

then

$$E_t = \frac{13 \cdot 65}{\left[\ln \dfrac{200 + z_0 - d}{z_0}\right]^2} u(e_s - e_a)$$

$$E_t = f(z_0, d)u(e_s - e_a)$$

where u is the wind velocity measured at 2 m height in m sec$^{-1}$, $e_s$ and $e_a$ the vapour pressure in mm Hg respectively at the apparent evaporating surface and at 2 m height.

The last equation is very similar to the empirical Dalton equation, which relates evapotranspiration to the prevailing wind velocity and the vapour pressure gradient or degree of atmospheric saturation. The Dalton equation may be stated as follows:

$$E_t = f(\bar{u})(e_s - e_a)$$

where $f(\bar{u})$ is an empirically derived function of the wind velocity, usually given in the form

$$f(\bar{u}) = a(1 + b\bar{u})$$

where $\bar{u}$ is the time averaged wind velocity measured at standard height and a and b are constants.

Brutsaert (1965) has suggested that in a neutral atmosphere evapotranspiration can be expressed as

$$E_t = f(z_0, d)u^{0 \cdot 75}(e_s - e_a)$$

There are difficulties in making practical use of equations of this type, because some of the terms are unknown or difficult to measure.

Evapotranspiration requires a supply of energy to be used as latent heat of vaporization and this is the second of King's three dynamic processes. Neglecting the amount of energy used for photosynthesis and not taking into account the supply of advective energy, the energy balance can be written as follows:

$$R_N = (1 - r)R_S - R_{NL} = LE_t + K + S + G$$

where $R_N$ is the net radiation, r is the albedo of the surface, $R_S$ is the global short-wave radiation, $R_{NL}$ is the net long-wave radiation, L is the latent heat of vaporization, $E_t$ is the evapotranspiration, K is the sensible heat transfer to the atmosphere, S is the sensible heat transfer to the soil, and G is the storage of heat in the crop.

Crops generally have albedos varying between 0·20 and 0·26 according to Monteith (1959) and Rijtema (1965), but by contrast the albedo of a free water surface is about 0·05. If long time periods of several days are being considered, the heat transfer to and from the soil and the heat storage of the crop can both be neglected.

The main difficulty in using the energy balance approach arises from the distribution of energy between latent heat and sensible heat transfer to the atmosphere.

The ratio of the sensible heat flow to the latent heat flow is known as the Bowen ratio, and can be written as

$$\beta = \frac{K}{LE_t} = \gamma \left(\frac{K_H}{K_V}\right)\left(\frac{T_s - T_a}{e_s - e_a}\right)$$

where $\gamma$ is the psychrometer constant (about 0·65), $K_H$ is the eddy conductivity, $T_s$ and $T_a$ are respectively the surface and air temperature, and $e_s$ and $e_a$ are the vapour pressure respectively at the surface and in the air.

With the Bowen ratio, the expression for the energy used for evapotranspiration becomes:

$$LE_t = \frac{R_N - S}{1 + \beta}$$

Both the aerodynamic and energy balance approaches require detailed and difficult measurements, but Penman (1948) has succeeded in combining both methods in a formula for the calculation of evapotranspiration from standard meteorological data. He neglected the storage of heat in the soil, assumed a saturated vapour pressure at the surface, and derived the following equation:

$$E_t = \frac{\Delta R_N/L + \gamma E_a}{\Delta + \gamma}$$

where $\Delta$ is the slope of the temperature-vapour pressure curve at air temperature and $E_a$ (aerodynamic evaporation) is calculated from an aerodynamic relationship in the form

$$E_a = (e_s - e_a)f(u)$$

Aerodynamic relationships of this type have been discussed earlier, when a variety of forms were suggested. The values of the constants will clearly depend on the nature of the surface, and one particular formulation for $E_a$ adopted by Penman (1956) is

$$E_a = 0·165(V_s - V_a)(0·8 + u_2/100)$$

in which $E_a$ is in mm/day, $V_s$ and $V_a$ are respectively the saturated and actual vapour pressures at screen level in mbar and $u_2$ is the wind speed at a height of 2 m in km/day.

Normally values of the net radiation balance are not available for use in Penman's equation, and therefore under these conditions, the net radiation has to be calculated indirectly. Penman suggested the following empirical formula for calculating the net radiation from climatological observations including the observed hours of bright sunshine:

$$R_N = R_{ss}(1 - r)(a + b\,n/N) - \sigma T^4(0·56 - 0·08\sqrt{V_a})(0·1 + 0·9\,n/N)$$

where $R_{ss}$ is the solar radiation incident on an horizontal surface in the absence of the earth's atmosphere, n/N is the ratio of actual to possible hours of bright sunshine, a and b are constants, and $\sigma T^4$ is the full black body radiation at mean air temperature.

Because of changes in cloud structure, the constants a and b are not universal and vary from region to region; suitable regional values are suggested in table 1.5, and Glover and McCulloch (1958) consider that while b is effectively constant, a varies with the latitude $\phi$ up to 60°:

$$a = 0.29 \cos \phi$$

If the albedo (r) is taken as 0.25, according to Monteith (1959) the calculated net radiation applies to a short green crop.

**Table 1.5**  Regional regression coefficients for use in Penman's equation

| Locality | Latitude | Regression Constants a | b |
|---|---|---|---|
| Rothamsted (England) | 51·8°N | 0·18 | 0·55 |
| Gembloux (Belgium) | 50·6°N | 0·15 | 0·54 |
| Versailles (France) | 48·8°N | 0·23 | 0·50 |
| Mt Stromlo (A.C.T. Australia) | 35·3°S | 0·25 | 0·54 |
| Dry Creek (South Australia) | 34·8°S | 0·30 | 0·50 |
| Singapore | 1·0°N | 0·25 | 0·476 |

*After Black, Bonython, and Prescott (1954) and Chia (1969).*

The third of the dynamic processes involved in evapotranspiration, suggested by King, is the flow of water through the soil and plants to the evaporating surface, and this flow will be controlled by the apparent diffusion resistance of the soil and crops. According to Rijtema (1968) the apparent diffusion resistance depends on the following factors:

1  soil area covered by crops and leaf area;
2  light intensity in relation to stomatal opening;
3  availability of soil moisture and transport resistances for liquid flow in the plant.

Few data are available concerning the effect of partial plant cover of the soil on transpiration, but Marlatt (1961) considers that row crops covering more than 50 per cent of the soil act as a full cover crop in relation to evapotranspiration. The effect of light intensity on stomatal resistance is well understood in the laboratory, but is difficult to study in the field. Evidence exists (Burgy and Pomeroy, 1958 and Van Bavel *et al.*, 1963) that under conditions of high light intensity, stomatal resistance has no influence and evapotranspiration is an externally controlled process.

The soil water storage capacity is primarily a function of the rooting depth of the vegetation and the water retention characteristics of the soils on which it grows. The amount of water in the soil which is available for plant growth is generally considered to be that retained at soil water potentials below 15 atmospheres. Often, the available water content of the soil is defined as the volumetric difference in water content at soil water potentials of 15 atmospheres and 0·3 atmospheres. For most plants, growth ceases at soil water potentials above 15 atmospheres, but evapotranspiration will continue at a greatly reduced rate. Clearly the evapotranspiration rate will partly depend on the available water, which in a soil is normally finite and limited. If evapotranspiration is calculated using Penman's equations, a value will be obtained which assumes that there is always a continuous and unrestricted supply of water. In practice the water supply may be restricted and the actual evapotranspiration rates will be considerably less than the values calculated using Penman's equations. The evapotranspiration resulting under conditions of unlimited and unrestricted water supply is known as the potential evapotranspiration. Because soil water conditions are normally unknown, it is usual in climatological discussions to quote the estimated potential evapotranspiration rather than the actual evapotranspiration.

Little has been said so far about the size of the evaporating surface, and this particular problem is best illustrated by considering the extreme case of an oasis in a perfectly dry desert. Figure 1.14 illustrates the energy balance of the oasis and of the desert. The net radiation over the desert is lower than that over the oasis, because the desert sands have a higher albedo than the vegetation of the oasis, and therefore it would appear that the oasis should have a higher temperature than the desert. In the desert region all the available net radiation is transferred as sensible heat to the atmosphere and ground, there being

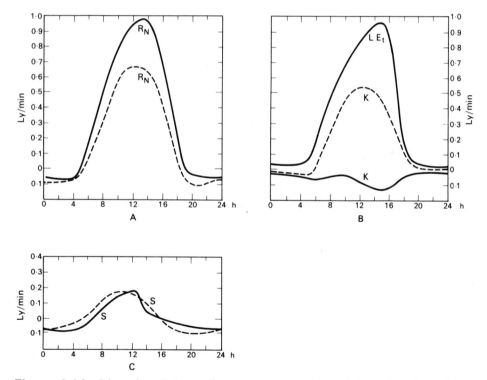

**Figure 1.14** Diurnal variations of energy balance in an irrigated oasis and in a semi-desert. Observations from the Turkestan semi-desert and an adjacent oasis. Solid line : oasis (irrigated) ; dashed line : semi-desert. A : diurnal variation of net radiation. B : diurnal variation of sensible heat transfer to the atmosphere (K) and latent heat transfer ($LE_t$). C : diurnal variation of sensible heat transfer to the soil. (*After Budyko, 1956.*)

no water available for evaporation; but in the oasis large amounts of energy are used in evaporating water and therefore the sensible heat fluxes are small. The air arriving over the oasis is dry, and so evaporation rates are high; often they are so high that sensible heat will be drawn from the atmosphere to supply energy for evapotranspiration. Under these conditions the oasis will be cooler than the desert and evapotranspiration rates will be above those calculated from simple radiation balance considerations. If a large oasis is considered, the very high evapotranspiration rates will only be found along the windward edge, because as the air moves deeper into the oasis it will become more moist and slightly cooler and the sensible heat transfer from the atmosphere to the ground will cease. The example quoted is extreme, but clearly the evapotranspiration from a surface will be partly controlled by the size of the surface and partly by the nature of the surrounding conditions. Unusually high evapotranspiration rates from a small surface are often said to arise because of an oasis effect.

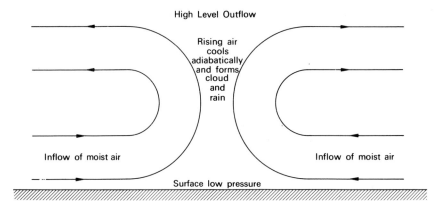

**Figure 1.15** Schematic illustration of a precipitation-forming disturbance.

*2 Precipitation*

Evapotranspiration is a slow continuous process and represents a water loss from the soil and the atmosphere boundary layer. In contrast, precipitation is intermittent and represents a water gain to the soil. Most precipitation is due to disturbances in the middle or upper troposphere which cause it to form in the middle troposphere and to fall through the planetary boundary layer to the ground surface. A very simplified picture of a precipitation-forming disturbance is shown in figure 1.15. Moist air converges into the system near the surface and then rises in the central areas, where cloud and precipitation are formed, since rising air cools adiabatically. Therefore the formation of significant precipitation requires a source of moist air and a cause of convergence and ascent within the moist air. The amount of water held in the atmosphere as water vapour is small, yet continuous high rates of precipitation occur which would normally exhaust the water vapour content of air within the storm very quickly; this implies that there must be a continuous flow of moist air into a precipitating storm. When it is considered that even the most efficient storm can only convert a small amount of the available atmospheric water vapour into precipitation, the problem of moisture flow into the precipitating system is seen to be of some importance. Convergence and ascent of the moist air are most often associated with low-pressure

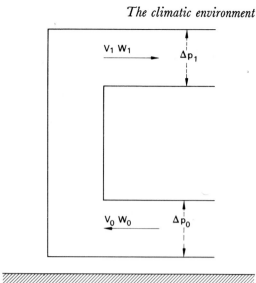

**Figure 1.16** Schematic illustration of a simple storm model. See text for explanation.

systems in the atmosphere; these can assume a range of scales varying from thunderstorms through depressions to tropical storms.

A very simple storm model, illustrated in figure 1.16, consists of a convergence cell with inflow in the bottom layer, outflow in the top layer, and a zone of non-divergence in the middle. If inflow of velocity $V_0$ and precipitable water content $W_0$ occurs in the bottom layer of depth $\Delta p_0$ and a corresponding outflow of velocity $V_1$, precipitable water content $W_1$, and depth $\Delta p_1$ in the top layer, then the released precipitable water $W_e$ is given by

$$W_e = W_o - \frac{\Delta p_o W_1}{\Delta p_1}$$

and the rainfall intensity i in this cell of radius r is

$$i = \frac{2V_o}{r} W_e$$

Precipitable water content is discussed in detail in chapter 2, but it can be estimated from

$$W = \frac{\bar{q}\Delta p}{g}$$

where $\bar{q}$ is the mean specific humidity of a layer of depth $\Delta p$. This is an extremely simple model, but when an allowance is made for scale, it does describe a large number of rain-forming systems. In particular, models of this type describe tropical hurricanes extremely well.

Individual storms contribute widely differing amounts to the annual precipitation in all parts of the world, but according to Riehl (1965) the percentage supplied by large and small storms appears to be fairly constant irrespective of location. Thus figure 1.17, representing two locations in England has some general significance. The daily precipitation values at Bradford and Wakefield for the years 1916–1950 were arranged in decreasing order of size. Both the number of days on which precipitation fell and the corresponding precipitation values were turned into cumulative totals starting with the highest, and each total was expressed as a percentage of the grand total. The Bradford curve indicates that 50 per cent of the total precipitation fell on 12 per cent of the days with precipitation; the corresponding value for Wakefield fell on 14 per cent of the days. This suggests that at least half of the annual precipitation results from a few large rainfalls. Riehl (1965) has constructed similar diagrams for Colorado and Kenya with similar results: 25 per cent of the storms over the upper Colorado Basin contribute half of the annual precipitation. These results indicate why the reliability of precipitation is often so low, for if only two or three large storms fail to materialize in a year, drought may follow. They also show why a study of the environment near the ground must include some synoptic meteorology.

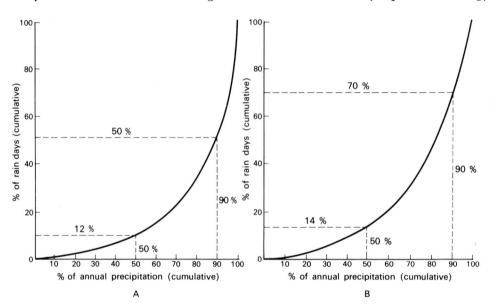

**Figure 1.17** Percentage contributions of daily precipitation totals of various sizes to annual precipitation. For construction of diagrams all daily precipitation totals, 1916–1950, were ordered according to size, from highest to lowest rank. A: Bradford, Yorkshire. B: Wakefield, Yorkshire. (*After Lockwood, 1967.*)

## 3   Water balance

The interplay of precipitation and evapotranspiration forms the water balance, and a typical water-balance diagram is shown in figure 1.18. The station is Huddersfield Oakes in northern England, where the average precipitation values are for the period 1916–1950 and the average potential evapotranspiration values are calculated by the Penman method for the period 1956–1962. At Huddersfield Oakes, mean precipitation exceeds mean potential evapotranspiration for the winter part of the year, and there is excess moisture for surface runoff. From mid-April until the beginning of July, mean potential evapotranspiration exceeds mean precipitation and during this period plants use water stored in the soil and rivers flow using water stored in underground rocks. It appears that in the average year the available water content of the soil is never completely exhausted during the summer dry period at Huddersfield Oakes, and therefore the actual evapotranspiration will never fall far below the potential evapotranspiration. When precipitation first exceeds evapotranspiration in July, the rain water is used mainly to recharge the soil water to field capacity, and only when the soil water is fully replaced in mid-September does significant surface runoff result from precipitation. Significant runoff will occur, regardless of the soil moisture conditions, if the intensity of rainfall exceeds the rate at which infiltration takes place into the soil, but this condition is unusual in most temperate latitude climates. These changes in the water balance can be observed in the landscape. For example in the average year significant surface runoff will only take place at Huddersfield Oakes between mid-September and mid-April; only during these months will ditches flow and standing water appear on the surface. The annual variation in water balance at Huddersfield Oakes is very similar to that at other middle latitude locations. Further examples of these will be discussed in later chapters.

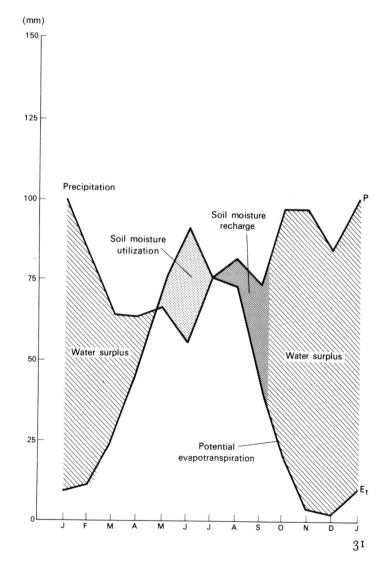

**Figure 1.18** Water-balance diagram for Huddersfield Oakes, Yorkshire. *(After Lockwood, 1967.)*

*Introduction*

## Local microclimates

*1   Surface conditions*

The climate of a particular region depends upon a number of factors, including its latitude, its altitude, and its proximity to oceans, lakes, and mountains. Microclimates within the particular region will depend on such factors as slope aspect, vegetation, and the general nature of the ground surface. Flohn (1969) considers that local climate is governed by the various factors of radiation, heat budget, and water budget. He considers that global radiation, the albedo, and the radiation balance on one hand, and on the other, the duration of flow of sensible heat and latent heat from the earth's surface to the air and vice versa are particularly important. Temperatures are largely controlled by the amount of available sensible heat. For example, in a dry climate with a large net radiation gain, practically all the available energy is converted into sensible heat which warms the atmosphere. This is why day temperatures are so high in subtropical deserts. In some parts of the world the water balance is markedly positive at some times of the year and negative at others. This means that the year is divided into rainfall seasons which can influence the life of the region; good examples of this are observed in India.

The advection of heat by air-streams can change the local energy balance. At certain times of the year there is a continuous net radiation loss in middle and high latitudes, but energy is supplied, and consequently temperatures maintained, by the advection of heat by the atmosphere from the tropics. Air-streams can also change the heat balance because their cloud amounts vary and because of this also their transparency to radiation. According to Flohn, advection accounts for 37–39 per cent of the temperature variations near Berlin in summer and 74–85 per cent in winter; but in Africa during the dry season the values fall to 5–10 per cent.

*2   Urban climates*

Changes in surface conditions will clearly modify the local radiation and water balances and therefore change the local climate slightly. A large city causes a direct modification of surface conditions and can therefore be expected to change the local climate. Four main factors are responsible:

1   the change in the surface configuration and roughness between the open countryside and the city;
2   the modification of the heat balance by changes in the surface albedo and net heat loss between the open countryside and the city;
3   the modification of the normal rural water balance by the city;
4   the release of heat by the burning of fossil fuels.

Urban structures have a considerable effect on the movement of air both by producing turbulence as a result of their roughening of the surface and by the channelling effects of streets. Yamamota and Shimanuki (1964) have estimated that the average roughness length of Tokyo is about 165 cm, which is considerably higher than over the open countryside, but not as high as in a dense forest. On average strong winds decrease towards city centres. Chandler (1965) states that for London, with winds of more than 1·5 m/sec, the mean annual difference between speeds outside and at the centre is about 13 per cent.

The heat balance of a city differs from that of the open countryside because of a variety of factors, including differences in the albedo and in the heat storage capacity of the two areas. Sunlight penetrates through windows into buildings where, since the surfaces are dry, all the radiation is converted into sensible heat which is stored in the structure of the building. This stored heat, together with a variable amount of heat from central heating, is released slowly throughout the 24-hour period and warms the atmosphere. Many city streets act as small valleys and it has already been seen that the outgoing long-wave radiation from a valley is greatly reduced. The increased warmth of the buildings is therefore often trapped at street level by horizontal long-wave radiation between buildings. Pollution, which is often excessive over cities, cuts down the incoming short-wave radiation but it also decreases the long-wave radiation loss. The result of these changes in the heat balance is that cities tend to be warmer than the surrounding rural areas, and this is shown by figure 1.19 in which examples of day and night temperature maps of London are reproduced.

Within urban areas the rapid removal of surface runoff through drains modifies the water balance. Soil moisture storage is decreased and large bodies of standing water are usually absent. Local evaporation will be low and more energy will be available for sensible heat transfer. The air within a city will therefore have a slightly lower relative humidity than air in rural areas; this is illustrated for London in figure 1.20.

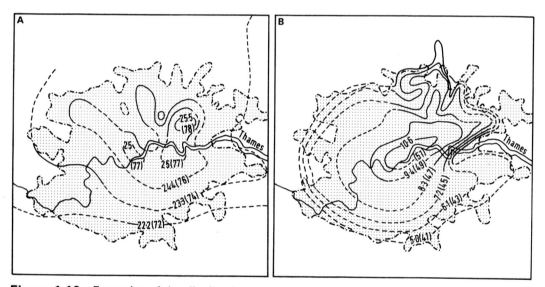

**Figure 1.19** Examples of the distribution of maximum and minimum temperature in London (°C with °F equivalents in brackets). A: distribution of maximum temperature in London, 3 June 1959. B: distribution of minimum temperature in London, 14 May 1959. (*After Chandler, 1965.*)

## 3 Energy balance

The climatic environment at any location is largely governed by the energy and water balances. This statement is best illustrated by a study of these balances in sample climatic regions. Figure 1.21 illustrates (after Budyko, 1956) the annual heat balances at Manãos, South America (3° 08′s, 60° 01′w) and over the Pacific Ocean near New Guinea (0°, 150°e). These are typical continental and oceanic equatorial sites. At Manãos the net

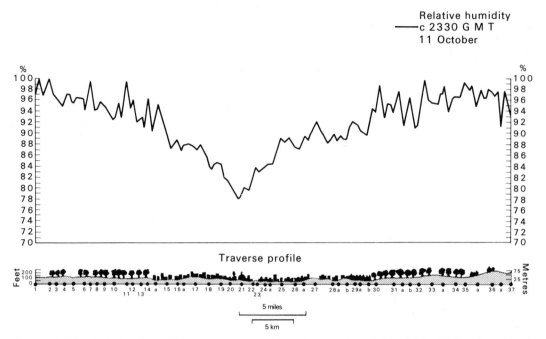

Relative humidity
——c 2330 G M T
11 October

Traverse profile

5 miles

5 km

**Figure 1.20** Relative humidity traverse across London on the night of 11–12 October 1961. (*After Chandler, 1965.*)

radiation changes only slightly during the year, but there are two maximums, a small one in spring and a more significant one in the autumn. These are connected with the increase of total radiation during the equinox period, when in equatorial latitudes the average altitudes of the sun are greatest. Similar comments can be made about the net radiation at the Pacific site. Because of the abundant precipitation at Manãos, the soil is normally

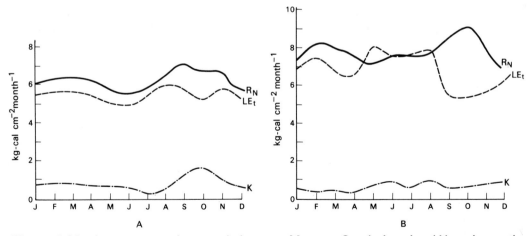

**Figure 1.21** Average annual energy balance at Manaos, South America (A) and over the Pacific Ocean near New Guinea (B). R: net radiation. $LE_t$: latent heat transfer to the atmosphere. K: sensible heat transfer to the atmosphere. (*After Budyko, 1956.*)

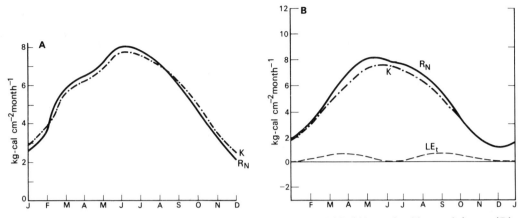

**Figure 1.22** Average annual energy balance at Aswan, UAR (A), and at Yuma, Arizona (B). Key as for figure 1.21. (*After Budyko, 1956 and Sellers, 1965.*)

always moist and the major portion of the net radiation is used for evaporation. The annual march of latent heat transfer is almost parallel to the annual march of the net radiation; therefore the sensible heat transfer to the atmosphere and soil is small in all months of the year. Again, similar comments can be made about the oceanic site in the Pacific where obviously there will be no shortage of water for evaporation. Humid equatorial climates normally show little temperature variation throughout the year and often have no marked seasons. Temperatures are relatively high but often considerably lower than those found in the subtropics, and this is a direct result of the energy balance.

The tropical desert climate forms an interesting contrast to the humid equatorial climate. Figure 1.22 illustrates the annual heat balances at Aswan, Africa (24° 02′N, 32° 53′E) and Yuma, USA (32° 40′N, 114° 39′W). Under conditions of consistently small cloud amounts in tropical deserts, the net radiation is effected mainly by the altitude of the sun in its annual march, and so purely astronomical factors determine the march of net radiation. The greatest values of the net radiation at Aswan and Yuma are smaller than its higher values in other parts of the subtropics, at, for instance, sites like Saigon (figure 1.23).

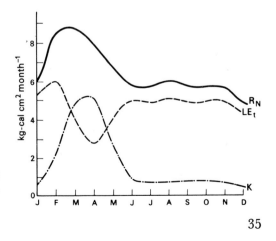

**Figure 1.23** Average annual energy balance at Saigon. Key as for figure 1.21. (*After Budyko, 1956.*)

## Introduction

The highest values of incoming short-wave radiation are observed in the subtropical deserts but the lower net radiation values arise because of the somewhat higher surface albedos in deserts and also because of the very large values of outgoing long-wave radiation from strongly heated desert surfaces. Similarly, the net radiation maximums at Aswan and Yuma barely exceed the maximums observed at some middle latitude sites, of which a good example is Copenhagen (figure 1.24). Precipitation values at Aswan and Yuma are very low and therefore the heat expenditures for evaporation during the year are close to zero. As a result of this, the sensible heat transfer is very large and approaches, in its value, the net radiation. Because of the high sensible heat transfer in deserts, day-time temperatures are high and exceed those found nearer to the equator.

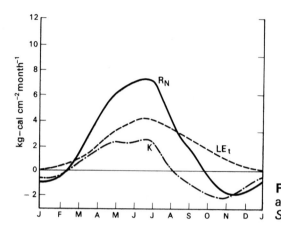

**Figure 1.24** Average annual energy balance at Copenhagen. Key as for figure 1.21. (*After Sellers, 1965.*)

The diagram for Copenhagen (55° 43′N, 12° 34′E) is interesting because the net radiation becomes negative in the winter, as does the sensible heat transfer from the atmosphere. The marked changes in the net radiation and the sensible heat transfers suggest large seasonal changes in temperature and these are normally observed in middle latitudes. The negative radiation balance in winter is compensated by atmospheric heat advection, which is strong at Copenhagen and so the seasonal temperature changes are not particularly large.

### 4    Plant leaves

Although only the large-scale aspects of local climate have been considered, it is possible to apply the principles discussed in this chapter to very small-scale features such as plant leaves. The energy flow between a plant leaf and its environment has been described by Gates (1968) in the following mathematical form:

$$R_a = \epsilon \sigma T_e^4 \pm C - LE_t \pm M$$

where $R_a$ is the radiation absorbed by surface of leaf; $T_e$ is the surface temperature of leaf in °A; $\sigma$ is the Stefan-Boltzmann radiation constant; $\epsilon$ is the surface emissivity of leaf; C is the energy gained or lost by convection; L is the latent heat of water; $E_t$ is the transpiration rate of the leaf surface; and M is the basic metabolism, which can be neglected for many plants.

36

Gates suggests that the convection term can be written as follows for forced convection in a wind:

$$C = 5{\cdot}93 \times 10^{-3} \frac{(V)^{\frac{1}{4}}}{D} (T_e - T_a)$$

where V is the wind speed in cm/sec; D is a characteristic dimension of the leaf, usually the mean width measured in cm; and $T_a$ is the characteristic air temperature.

From equations of this type it is possible, knowing the appropriate constants, to calculate leaf temperatures under various conditions. The importance of the convection term depends on the wind speed and the shape and size of the leaf. Obviously the temperature of a flat leaf which is normal to the incident will be strongly influenced by that radiation, while a cylinder, such as a bank, twig, or pine needle, will be much less strongly coupled to the incident radiation. Usually, large convection coefficients are associated with small organisms and hence their temperature is very close to that of the air.

Linacre (1967) has found that there is a tendency for the difference between leaf and air temperatures to be positive in cool weather, and negative in hot. The cross-over temperature for well-watered, thin-leaved plants exposed to noonday sunshine is about 30°C, a value which may not apply to the leaves of all plants in all climates. He considers that the optimum temperature for the growth of many plants is near 31°C, and so leaf anatomy and morphology would evolve to bring leaf temperatures towards their optimum at the time of day when most sunlight is available for photosynthesis.

Priestly (1966) has demonstrated empirically that the upper limit for air temperatures at screen height over well-watered terrain is about 33°C. According to Linacre this results from the tendency of plant leaves to reach an equilibrium temperature. In a large well-watered region it is likely that the ground will be covered by thin-leaved vegetation. The extent of the region will be such that the advection of energy can be neglected. In such a case, the air temperature can rise only by the flow of heat from the leaves, and this will occur while the leaves are warmer. In such a region, therefore, the air temperature must be close to the upper limit for leaf temperatures.

# References

ÅNGSTRÖM, A. 1916: Über die Gegenstrahlung der Atmosphäre. *Meteorologische Zeitschrift* **33**, 529.

BARRY, R. G. and CHORLEY, R. J. 1968: *Atmosphere, weather and climate*. London: Methuen.

BENDELOW, V. C. 1969: A determination of the albedo of Morecambe Bay north of a line from Aldingham to Base. *Meteorological Magazine* **98**, 305.

BLACK, J. N., BONYTHON, C. W., and PRESCOTT, J. A. 1954: Solar radiation and the duration of sunshine. *Quarterly Journal Royal Meteorological Society*, **80**, 231.

BRUNT, D. 1941: *Physical and dynamical meteorology*. London: Cambridge University Press.

BRUTSAERT, W. 1965: A model for evaporation as a molecular diffusion process into a turbulent atmosphere. *Journal Geophysical Research* **70**, 5,017.

BUDYKO, M. I. 1956: *The heat balance of the earth's surface*. Translated by N. I. Stepanova. Washington: US Weather Bureau.

BURGY, R. H. and POMEROY, C. R. 1958: Interception losses in grassy vegetation. *Transactions American Geophysical Union* **39**, 1,095.

CHANDLER, T. J. 1965: *The climate of London*. London: Hutchinson.

*Introduction*

CHIA LIN SIEN 1969: Sunshine and solar radiation in Singapore. *Meteorological Magazine* **98**, 265.

DEACON, E. L. 1969: Physical processes near the surface of the earth. In Flohn, H. (editor), *General climatology* **2**. Amsterdam: Elsevier Publishing Co., p. 39.

DUBOIS, P. 1929: Nächtliche effektive Ausstrahlung. *Beiträge Zur Geophysik* **22**, 41.

FLEAGLE, R. G. and BUSINGER, J. A. 1963: *An introduction to atmospheric physics*, New York: Academic Press.

FLOHN, H. 1969: *Climate and weather.* London: Weidenfeld and Nicolson.

GADD, A. J. and KEERS, J. F. 1970: Surface exchanges of sensible and latent heat in a 10-level model atmosphere. *Quarterly Journal Royal Meteorological Society* **96**, 297.

GATES, D. M. 1968: Energy exchange in the biosphere. In Eckardt, F. E. (editor), *Functioning of terrestrial ecosystems at the primary production level. Proceedings of Copenhagen Symposium.* Paris: UNESCO, p. 33.

GEIGER, R. 1961: *Das Klima der bodennahen Luftschicht.* Braunschweig: Friedr. Vieweg und Sohn. English translation: Scripta Technica, Inc. 1965. *The climate near the ground.* Cambridge, Mass.: Harvard University Press.

GEIGER, R. 1969: Topoclimates. In Flohn, H. (editor), *General climatology*, **2**. Amsterdam: Elsevier Publishing Co.

GLOVER, J. and MCCULLOCH, J. S. G. 1958: The empirical relation between solar radiation and hours of sunshine. *Quarterly Journal Royal Meteorological Society* **84**, 172.

KING, K. M. 1961: Evaporation from land surfaces. In *Proceedings of Hydrolology Symposium No. 2: Evaporation.* Ottawa: National Research Council of Canada, p. 55.

KONDRATYEV, K. YA. 1954: *Radiant solar energy.* Leningrad: Gidrometeoizdat.

KONDRATYEV, K. YA. and MANOLOVA, M. P. 1955: K voprosu o prikhode rasseiannoǐ i summarnoǐ radiatsii ra poverkhnost' sklona, Leningrad, *Meteorologiia i Gidrologiia* **6**, 31.

KONDRATYEV, K. YA. and MANOLOVA, M. P. 1960: The radiation balance of slopes. Phoenix: *Solar Energy* **4**, 14.

KUNG, E. C. 1961: Derivation of roughness parameters from wind profile data above tall vegetation. In *Studies of the three-dimensional structure of the planetary boundary layer*, Annual Report, 1961. Madison: Department of Meteorology, University of Wisconsin, p. 27.

KUNG, E. C. 1963: Climatology of aerodynamic roughness parameter and energy dissipation in the planetary boundary layer of the northern hemisphere. In *Studies of the effects of variations in boundary conditions on the atmospheric boundary layer*, Annual Report 1963. Madison: Department of Meteorology, University of Wisconsin.

KUNG, E. C., BRYSON, R. A., and LENSCHOW, D. H. 1964: Study of a continental surface albedo on the basis of flight measurements and structure of the earth's surface cover over North America. *Monthly Weather Review* **92**, 543.

LAIKHTMAN, D. L. 1964: *Physics of the boundary layer of the atmosphere.* Jerusalem: Israel Program for Scientific Translations.

LETTAU, H. H. 1959: Wind profile, surface stress and geotrophic drag coefficients in the atmospheric surface layer. *Advances Geophysics* **6**, 241.

LETTAU, H. H. and DAVIDSON, B. 1957: Exploring the atmosphere's first mile. Vol. **2**: *Site description and data tabulation.* New York: Pergamon Press.

LINACRE, E. T. 1967: Further notes on a feature of leaf and air temperatures. *Archiv Meteorologie Geophysik und Bioklimatologie, Ser. B,* **15**, 422.

LINKE, F. 1931: Die nächtliche effektive Ausstrahlung unter verschiedenen Zenitdistanzen. *Meteorologische Zeitschrift* **48**, 25.

LOCKWOOD, J. G. 1967: Climate. In Beresford, M. W. and Jones, G. R. J. (editors), *Leeds and its region.* Leeds: British Association for the Advancement of Science.

MARLATT, W. E. 1961: The interactions of microclimate, plant cover and soil moisture content affecting evapotranspiration rates. Department of Atmospheric Science, Colorado State University Fort Collins, *Technical Paper* **23**.

MONTEITH, J. L. 1959: The reflection of shortwave radiation by vegetation. *Quarterly Journal Royal Meteorological Society* **85**, 386.

MONTEITH, J. L. and SZEICZ, G. 1961: The radiation balance of bare soil and vegetation. *Quarterly Journal Royal Meteorological Society* **87**, 159.

MONTEITH, J. L. and SZEICZ, G. 1962: Radiation temperature in the heat balance of natural surfaces. *Quarterly Journal Royal Meteorological Society* **88**, 508.

PENMAN, H. L. 1948: Natural evaporation from open water, bare soil, and grass. *Proceedings of Royal Society*, London, **A193**, 120.

PENMAN, H. L. 1956: Evaporation. An introductory survey. *Netherlands Journal Agricultural Science* **4**, 9.

PRIESTLEY, C. H. B. 1966: The limitation of temperature by evaporation in hot climates. *Journal Agricultural Meteorology* **3**, 241.

REYNOLDS, O. 1883: An experimental investigation of the circumstances which determine whether the motion of water shall be direct or sinuous, and of the law of resistance in parallel channels. *Philosophical Transactions Royal Society*, London, **174**, 935.

RIEHL, H. 1965: *Introduction to the atmosphere.* New York: McGraw-Hill.

RIJTEMA, P. E. 1965: *An analysis of actual evapotranspiration.* Wageningen: Agricultural Research Report **689**.

RIJTEMA, P. E. 1968: Derived meteorological data: transpiration. In *Agroclimalogical methods. Proceedings of Reading Symposium.* Paris: UNESCO, p. 55.

SCHEDLER, A. 1951: Die Bestrahlung geneigter Flächen durch die Sonne. *Zentralanstalt für Meteorologie und Geodynamik, Jahrbuch* **87D**, 52.

SELLERS, W. D. 1965: *Physical climatology.* Chicago: University of Chicago Press.

STANHILL, G., HOFSTEDE, G. F., and KALMA, J. D. 1966: Radiation balance of natural and agricultural vegetation. *Quarterly Journal Royal Meteorological Society* **92**, 128.

SWINBANK, W. C. 1963: Long-range radiation from clear skies. *Quarterly Journal Royal Meteorological Society* **89**, 339.

TANNER, C. B., and PELTON, W. L. 1960: Potential evapotranspiration. Estimates by the approximate energy balance method of Penman. *Journal Geophysical Research* **65**, 3391.

THORNTHWAITE, C. W. 1948: An approach toward a rational classification of climates. *Geographical Review* **38**, 55.

THORNTHWAITE, C. W. and HOLZMANN, B. 1939: The determination of evaporation from land and water surfaces. *Monthly Weather Review* **67**, 4.

VAN BAVEL, C. H. M., FRITSCHEN, L. J. and REEVES, W. E. 1963: Transpiration by Sudan grass as an externally controlled process. *Science* **141**, 269.

YAMAMOTA, G. and SHIMANUKI, A. 1964: Profiles of wind and temperature in the lowest 250 metres in Tokyo. *Tôhoku University Science Report, Series* **5**, *Geophysics* **15**, 111.

# 2
# General circulation of the atmosphere

The planet earth receives heat from the sun in the form of short-wave radiation, but it also radiates an equal amount of heat to space in the form of long-wave radiation. This balance of heat gained equalling heat lost only applies to the planet as a whole over several annual periods; it does not apply to any specific area for a short period of time. The equatorial region absorbs more heat than it loses, while the polar regions radiate more heat than they receive. Nevertheless, the equatorial belt does not become warmer during the year, nor do the poles become colder, because heat flows from the warm to the cold region, thus maintaining the observed temperatures. An exchange of heat is brought about by the motion of the atmosphere and the upper layers of the oceans, thus forming the general circulation of the atmosphere and oceans.

Interest in the general circulation of the atmosphere arises because it is one of the major factors controlling the distribution of world climatic zones. In the tropics vast areas have easterly winds with a large equatorial component, which are the so-called trade winds noted for their generally fine, sunny weather. Poleward of the trades there are regions of subsiding air with generally anticyclonic conditions, which form the great deserts of the planet. In middle latitudes the average atmospheric motion is towards the east, but the flow is highly disturbed. Lastly, the polar regions with their high radiation loss have generally subsiding air-masses.

## Controlling factors of the general circulation

The general circulation of the atmosphere is essentially determined by the distribution of radiation over the earth's surface and the rotation of the planet about its axis.

### *1  Distribution of radiation*

Figure 2.1 shows a generalized model of the annual radiation balance of the atmosphere. The assumption is made that 100 units of radiant energy are received at the top of the atmosphere, and the model then traces the various interactions that the radiation undergoes in the atmosphere. It is interesting to note that nearly 50 per cent of the incoming radiation penetrates through the atmosphere and is absorbed by the earth's surface. The atmosphere is therefore semi-transparent to the incoming short-wave radiation, but in contrast it is almost opaque to infra-red radiation which is absorbed strongly by water

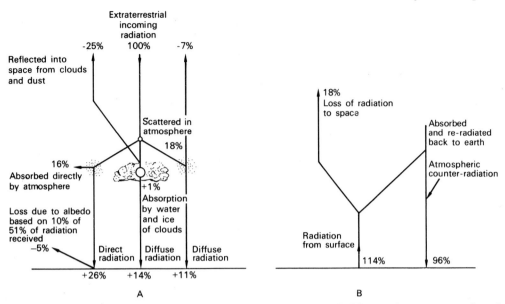

**Figure 2.1** The annual radiation budget of the planet earth. The exchanges are referred to 100 units of incoming solar radiation at the top of the atmospheres. A: Short-wave solar radiation. B: Long-wave terrestrial radiation. (*After Flohn, 1969, and others.*)

vapour and carbon dioxide. The earth's surface, which is warmed by short-wave radiation, subsequently warms the atmosphere, which continually loses infra-red radiation to space.

*a Global radiation* Global radiation is the sum of all the short-wave radiation received both directly from the sun and indirectly from the sky, on a horizontal surface. Generalized isolines of average annual global radiation are shown in figure 2.2 which conveys a very

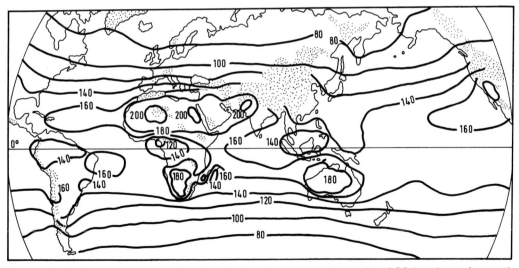

**Figure 2.2** Average annual global radiation. Isolines at intervals of 20 k cal cm$^{-2}$ year$^{-1}$. (*After Budyko, 1956.*)

general picture of the distribution of global radiation. It looks deceptively simple because the original map was based on few observations which, particularly over the oceans, were restricted to isolated island stations. On land in latitudes higher than 35°N and s there is a gradual decrease of global radiation towards the polar regions, where little short-wave radiation can be expected in winter because it falls mostly in the time interval between spring and autumn. Similarly, in higher latitudes over the oceans there is also much cloud and so little short-wave radiation reaches the surface. The equatorial belt over the continents also shows a relative minimum of radiation because of the cloudiness and frequent rainfall in the convergence zone between the trade winds. The areas receiving most global radiation are found in the subtropics where there are unusually clear skies because of the prevailing anticyclones; these are also zones with desert conditions.

*b Albedo* Radiation reflected directly back to space from the earth constitutes a loss of available energy to the earth–atmosphere system. The distribution of albedo values over the earth's surface must therefore be considered together with the global radiation. Mean annual values of planetary albedo are shown in figure 2.3 as the ratio of the incoming short-wave radiation to the reflected short-wave radiation. On this particular map planetary albedo values show a relative maximum at the equator, minimums in the two subtropical belts, and an increase with latitude from the subtropical minimums up to values of about 60 per cent at 70° latitude. These high planetary albedo values in polar latitudes are probably due to the large amount of cloud and also to the extensive ice and snow fields.

The map shown in figure 2.3 is based on the work of Kondratyev (1962), but more recent estimates by Vonder Haar and Suomi (1971) using meteorological satellite data suggest that the entire earth and both hemispheres are darker and warmer than earlier estimates had shown, this being particularly so in the tropics. The mean planetary albedo

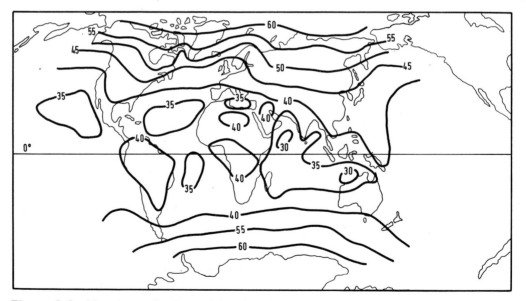

**Figure 2.3**   Mean annual values of the albedo of the earth's surface as seen from space (*After Robinson, 1966 and Kondratyev, 1962.*)

of the earth is now considered to be about 30 per cent and the mean equivalent black body temperature to be 254°κ. Each hemisphere has nearly the same planetary albedo and infra-red loss to space on the mean annual time scale, and this points to the dominant influence of clouds on the energy exchange with space, since the surface features of the two hemispheres are quite different. The departure of the estimates by Vonder Haar and Suomi from the earlier values for the deep tropics is of special significance because it implies that there is a greater energy input at low latitudes than was previously thought.

*c   Radiation balance*   The most important parameter in the study of the energy balance of the atmosphere is not the incoming radiation but the net radiation, which is made up of the difference between the incoming solar radiation mostly in or near the visible, and the outgoing terrestrial radiation in the infra-red. Figure 2.4 shows the geographical distribution

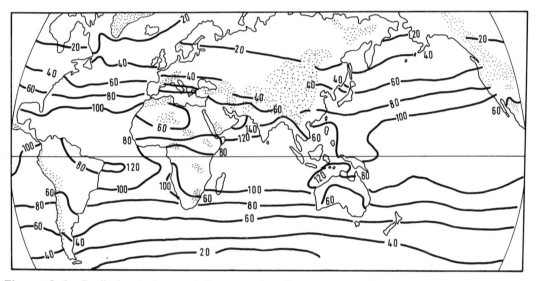

**Figure 2.4**   Radiation balance of the ground surface, annual values at intervals of 20 k cal cm$^{-2}$ year$^{-1}$. (*After Budyko, 1956.*)

of annual net radiation at the earth's surface only, for the atmosphere is excluded. This figure reveals that the annual means of the net radiation balance over the greater part of the earth's surface are positive, and thus signifies that the absorbed short-wave radiation is greater than the long-wave outgoing radiation. This pattern is the result of the greater transparency of the atmosphere for short-wave radiation in comparison with long-wave radiation, and the excess of energy at the earth's surface is transferred to the atmosphere by turbulent heat exchange and by evaporation. Figure 2.5 shows the annual expenditure of heat at the surface because of evaporation; the annual turbulent heat exchange between surface and atmosphere is shown in figure 2.6. These charts suggest that the main method of dissipation of the excess heat at the earth's surface is by heat absorption in evaporation. The chart of turbulent heat exchange shows that the sensible heat fluxes directed from the earth's surface towards the atmosphere are usually much bigger than the fluxes directed

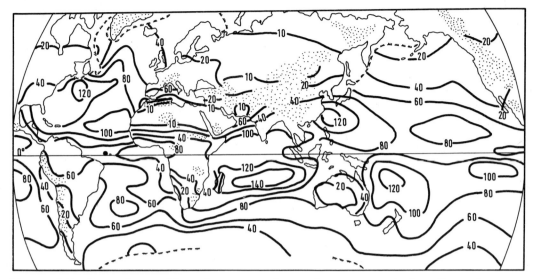

**Figure 2.5** Expenditure of heat for evaporation and evapotranspiration (latent heat exchange), annual values at intervals of 20 k cal cm$^{-2}$ year$^{-1}$. (*After Budyko, 1956.*)

from the atmosphere to the earth. For a considerable part of the earth's surface the turbulent loss of heat is comparatively small, but it reaches large values on land in desert zones and over the oceans in the region of warm currents.

The loss of heat by the surface-atmosphere system is determined principally by the longwave radiation from the troposphere, which makes the main contribution to the outgoing radiation. The radiation of the earth's surface to space forms less than 10 per cent of the

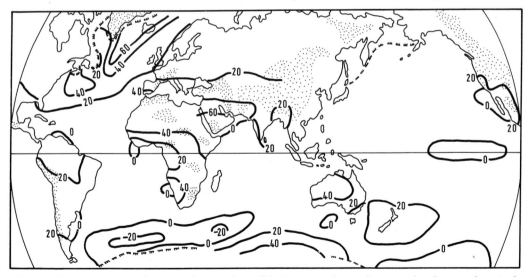

**Figure 2.6** Turbulent heat exchange (sensible heat exchange), annual values at intervals of 20 k cal cm$^{-2}$ year$^{-1}$. (*After Budyko, 1956.*)

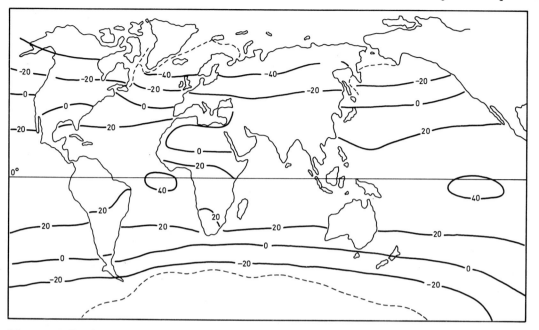

**Figure 2.7** Geographical distribution of annual net radiation balance of the surface-atmosphere system. Isolines at intervals of 20 k cal cm$^{-2}$ year$^{-1}$. (*After Budyko and Kondratyev, 1964.*)

total outgoing radiation, and the infra-red radiation from the stratosphere makes a still smaller contribution. Figure 2.7 shows the geographical distribution of the annual net radiation balance of the surface–atmosphere system. The most characteristic feature of this net radiation balance is its proximity to a zonal distribution, reaching maximum positive values of 40k cal/cm$^2$/yr at the equator and falling to minimum negative values of 60k cal/cm$^2$/yr at the poles. Positive values, mostly found between 40°N and 40°s, imply a heat gain by the surface–atmosphere system, while negative values imply a general heat loss to space. The radiation balance of the surface–atmosphere system is positive during the whole year only in the narrow equatorial zone between the latitudes 10°N to 10°s, for elsewhere the sign of the radiation balance changes twice a year (illustrated in table 2.1). For about three summer months in a year the radiation balance of the whole of each hemisphere is positive, but in late summer zones of negative balance arise near the poles and then gradually spread toward the equator, reaching latitude thirty after five months; a similar process of retreat begins in the spring.

Recent work by Vonder Haar and Suomi (1971) using satellite measurements shows that significant changes in the earth's radiation budget occur within latitude zones, especially in the tropics. In the tropics some of the greatest minimums of net radiation are found over the oceanic deserts west of South America and Africa. Here the low, bright, warm clouds reflect the solar energy well and also radiate strongly in the infra-red. Maximums of net radiation occur over the clear oceanic regions. The exact significance of these marked zonal variations in net radiation balance is not yet clear, but they must cause interesting differential heating effects within the tropics.

45

*Introduction*

**Table 2.1**  Net radiation balance of the earth–atmosphere system at various latitudes

| | | (cal cm$^{-2}$ day$^{-1}$) | | | |
|---|---|---|---|---|---|
| | | Dec–Feb | Mar–May | Jun–Aug | Sept–Nov |
| °N | 90 | −295 | −175 | −15 | −325 |
| | 80 | −300 | −135 | −10 | −320 |
| | 70 | −320 | −75 | 20 | −280 |
| | 60 | −300 | −25 | 65 | −230 |
| | 50 | −250 | 25 | 100 | −160 |
| | 40 | −190 | 75 | 140 | −95 |
| | 30 | −108 | 115 | 155 | −30 |
| | 20 | −45 | 140 | 140 | 25 |
| | 10 | 25 | 145 | 135 | 70 |
| | 0 | 85 | 110 | 85 | 100 |
| °S | 10 | 140 | 90 | 15 | 120 |
| | 20 | 180 | 20 | −50 | 115 |
| | 30 | 200 | −45 | −125 | 100 |
| | 40 | 175 | −105 | −200 | 70 |
| | 50 | 120 | −185 | −250 | 20 |
| | 60 | 40 | −240 | −295 | −55 |
| | 70 | −45 | −270 | −310 | −130 |
| | 80 | −75 | −275 | −300 | −190 |
| | 90 | −60 | −275 | −295 | −190 |

*After Newell et al. (1969).*

Although there is a long-term global balance between incoming and outgoing radiation, considerable imbalances exist both locally and seasonally. It has already been shown that there is substantial excess of net radiation in low latitudes and a deficit towards the poles. Alternatively, it can be considered that the mean atmospheric temperature in low latitudes is lower, and in high latitudes higher, than those appropriate to the local radiative balance, and that this situation is possible because of the global-scale mixing performed by the general atmospheric circulation. The latter is maintained against frictional forces by the heat energy from the sun, and would probably assume a very simple form but for the modifying influence of the earth's rotation. This simple circulation, in the absence of the earth's rotation, would probably consist of a single direct cell with rising air over the equator and sinking air over the poles. This direct cell arises because differential heating creates variations in atmospheric density and causes pressure gradients which in turn generate atmospheric motions. It is therefore necessary to look at the influence of the earth's rotation on atmospheric motions.

*2  Rotation of the earth*

*a  Coriolis acceleration*  The Coriolis acceleration is an apparent acceleration which air possesses by virtue of the earth's rotation, with respect to axes fixed in the earth. It is everywhere at right angles to the earth's axis and to the air velocity, and in the northern hemisphere acts to the right when viewed along the motion. In a Cartesian co-ordinate system in which x and y are in the horizontal plane to east and north respectively, and z is vertically upwards, the Coriolis acceleration has the following components:

$$CA_x = 2 \, \Omega \, (v \sin \phi - w \cos \phi)$$
$$CA_y = -2 \, \Omega \, u \sin \phi$$
$$CA_z = 2 \, \Omega \, u \cos \phi$$

where $\Omega$ is the earth's angular velocity; u, v, w are the component wind velocities in the x, y, z directions respectively, and $\phi$ is the latitude.

Only the component of the Coriolis acceleration which acts in the horizontal (xy) plane is normally of significance in meteorological dynamics, and the term $2\,\Omega \sin \phi$ is known as the Coriolis parameter (f). This component is such that the earth simulates a flat disc rotating anticlockwise in the northern hemisphere and clockwise in the southern hemisphere. Air moving horizontally outwards from the centre of the disc appears to an observer stationed there to be deflected to the right in the northern hemisphere and to the left in the southern hemisphere. The geostrophic wind ($V_G$) which is the horizontal equilibrium wind representing an exact balance between the horizontal pressure gradient and the horizontal components of the Coriolis force ($fV_G$) was mentioned in chapter 1.

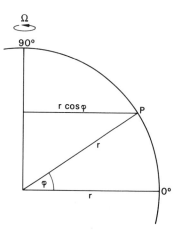

**Figure 2.8** Angular momentum.

*b Angular momentum* The rotation of the earth can affect the atmosphere in a second way. A point P (figure 2.8) of the earth in latitude $\phi$ has a west–east (zonal) velocity of $\Omega\,r \cos \phi$ where r is the distance from the earth's centre. Thus, at this latitude the absolute angular momentum of air of relative zonal velocity u is given by

$$M = \Omega r^2 \cos^2 \phi + rucos\phi$$

where M is the absolute angular momentum per unit mass about the earth's axis. The first term on the right-hand side of the above equation represents the $\Omega$-momentum, i.e., the absolute angular momentum which would be present if the atmosphere were in solid rotation with the earth. The second term is the relative angular momentum associated with the motion relative to the earth in, for example, depressions and anticyclones.

The atmosphere tends to conserve angular momentum, and therefore the wind speed u relative to the earth's surface tends to increase as zonal rings of air move poleward in a simple circulation cell, and to decrease as zonal rings flow towards the equator. Zonal rings flowing poleward in the high atmosphere, where frictional effects are small, will soon acquire large eastward components and this places limits on the poleward extension of the upper meridional flow.

47

*Introduction*

So far a frictionless earth has been assumed, whereas in real life this is not true; nevertheless the general principles just described hold true. A complicating factor is that in the latitude of the trade winds the atmosphere exerts a westward drag on the earth, while in the middle latitudes it exerts an eastward drag. Since the rate of rotation of the earth is almost constant, over an annual period the eastward and westward drags must be equal, and this particular balance consideration implies that there must be alternating zones of easterly and westerly winds together with some atmospheric transport of angular momentum. This is reasonable to suppose, since if winds were all uniformly in the same direction, westerly, for example, the earth's surface would be continually accelerated towards the east.

## 3   Laboratory models

Further information on the basic physical controls of the general circulation can be gained by a study of models of the atmosphere, and these may be either laboratory models or mathematical models. At present, these models are in no way exact facsimiles of the atmosphere, but rather analogues which provide crude parallels to the atmosphere and therefore enable certain simple conclusions to be reached. The models constructed may, of course, in no way resemble the atmosphere, but exist purely to explore theoretical problems. The laboratory experiments which most nearly duplicate the basic motions observed in the atmosphere are those involving a rotating annulus of fluid subjected to axisymmetric heating and cooling. The earliest experiments of this type appear to have been those of Vettin (1885), whose apparatus was a rotating cylinder about 30 cm in diameter and 5 cm deep, containing air. In one experiment ice was placed at the centre and the resulting motion of the air was, in its essential features, a simple direct cell like that envisaged for the atmospheric circulation by Hadley (1735). The actual apparatus can assume a variety of forms, but basically it consists of a convection chamber formed by the annular space between two concentric cylinders of which the vertical walls can be either heated or cooled. Sometimes the walls act as electrodes and the fluid is heated by an electric current.

The whole apparatus is rotated on a turntable, and the resulting motions in the fluid, which can be a mixture of water and glycerol, are made visible by the use of tiny neutrally-buoyant reflecting solid particles such as polystyrene beads. The particle motions are recorded by a camera which rotates with the convection chamber, and so a time exposure will record particle motions relative to the rotating apparatus as streaks. Typically the radius of the outer surface of the inner cylinder varies from 0–4 cm, while the inner surface of the outer cylinder may have a radius of about 9 cm and the fluid depth is about 16 cm, while rotation rates vary from 0 to about 7 rad/sec. Further details of these experiments and the apparatus used may be found in Hide (1969), Hide and Mason (1970), and Douglas, Hide, and Mason (1972).

Though various experimental arrangements are possible, the one of greatest interest in the context of general circulation is that in which the central cylinder acts as a cold source (pole) while the outer cylinder is a warm source (equator). If the temperature gradient is kept constant and the rates of rotation varied, a variety of interesting flow patterns become apparent. At low rates of rotation the flow is completely symmetrical about the axis of rotation, but as the rotation rate increases, the flow becomes non-axisymmetric and waves appear which are arranged around the central axis. This change is illustrated in the accompanying photographs. The fluid flows through the waves in the direction of rotation, with the flow concentrated into narrow bands or streaks, since the velocity is not uniform over the whole wave. It is also observed that the waves progress around the axis in the direction of rotation, and that the flow exhibits regular periodic time-variations, including a change in the number of waves, and shows the regular progression around the wave pattern of sizeable distortions and a wavering in the shape of the flow pattern. As the rate

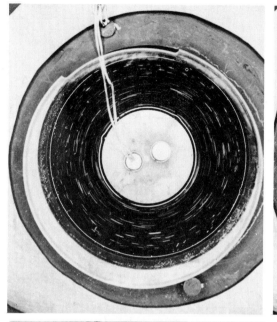

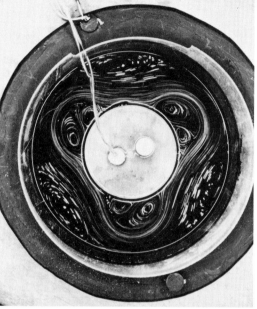

*Plate 3* Streak photographs illustrating three typical top-surface flow patterns of free thermal convection in a wall-heated rotating fluid annulus. The camera was mounted above the convection chamber and rotated with it in an anti-clockwise direction, and therefore the streaks record motions relative to the rotating chamber. In these three photographs the outer rim was heated while the central cylinder was cooled, the temperature difference being kept constant. For further details of the apparatus, see Hide (1969), Hide and Mason (1970) and Douglas, Hide and Mason (1972). *Above left:* Rate of rotation about 0·5 rad/sec; symmetric flow. *Above right:* Rate of rotation about 2 rad/sec; non-axisymmetric flow. *Opposite:* Rate of rotation about 4 rad/sec; non-axisymmetric flow. *Reproduced by permission of the Geophysical Fluid Dynamics Laboratory, Meteorological Office, Bracknell.*

of rotation increases so does the number of waves until eventually the flow becomes ir-regular.

The waves observed in the non-axisymmetric case are fully developed baroclinic waves, which are responsible for the main heat transfer between the warm rim and the cool centre. A baroclinic wave is one which forms in a strongly baroclinic region of the fluid, and the growth of the disturbance is characterized by the ascent of the warmer, and descent of the colder fluid masses, representing a decrease of potential energy and an associated release of kinetic energy. The fluid motions in the baroclinic waves are described by the term 'sloping or slantwise convection', because typical trajectories of individual fluid elements are tilted at an angle to the horizontal that is comparable with but less than the slope of the isotherms.

Flow patterns observed in the annular container appear to be controlled by both the thermal contrast and the rate of rotation, that is to say, by the so-called thermal Rossby number. The thermal Rossby number (R) depends upon the temperature contrast ($\Delta T$) in the annulus and the inverse of the rate of rotation ($\Omega$), resulting in the following:

$$R \propto \frac{\Delta T}{\Omega^2}$$

From the annulus experiments it appears that non-axisymmetric flow occurs with low thermal Rossby numbers, that is with relatively high rates of rotation, while axisymmetric flow occurs with high thermal Rossby numbers. The vertical component ($\Omega \sin \phi$) of the earth's rotation about its axis is dependent upon latitude, being zero at the equator and a maximum at the pole. Thus the middle latitudes could be considered as corresponding to the experiments with small thermal Rossby numbers (large rates of rotation) and the equatorial latitudes to those with large thermal Rossby numbers (small rates of rotation). Different types of atmospheric circulations are observed in tropical and temperate lati-tudes, the latter exhibiting waves and jet streams in the upper atmosphere which are very similar to those observed in the annulus experiments. Thus part of the difference between the equatorial and middle latitude atmospheres is due to the variation of the Coriolis para-meter with latitude, and the annulus experiments suggest that the equatorial atmosphere should be studied separately from the middle and high latitude atmosphere, since in many ways it is fundamentally different.

### 4  *Mathematical models*

The general circulation may also be studied by the use of mathematical models which are integrated on a high-speed computer. In these models the fundamental equations of fluid motion as modified by the assumption of hydrostatic equilibrium are used, together with suitable assumptions about thermodynamics and radiation. The first objective of these mathematical models of the general circulation is to simulate climate on a continental, hemispheric, or global scale. Once this initial goal has been achieved to a reasonable degree of faithfulness to reality, use can be made of mathematical models to study the physical mechanisms that create and maintain climate. This study is accomplished by controlled experiments in which some of the defining parameters of the model are changed while others are kept constant: thus for example, ocean currents could be eliminated or mountain ranges removed. The evolution of these mathematical models of the general circulation can be traced by referring to the following papers: Smagorinsky (1963),

Manabe and Strickler (1964), Smagorinsky *et al.* (1965), Manabe *et al.* (1965), Manabe and Wetherald (1967), Kurihara and Holloway (1967), Manabe (1969), and Holloway and Manabe (1971). The last paper describes a model which has nine levels from 80 m to 28 km above the ground and is integrated with respect to time on a global grid with a mean horizontal resolution of 500 km. The model has realistic continents with smoothed topography and an ocean surface with February water temperatures, while the insolation is for a northern hemisphere winter. In addition to wind, temperature, pressure, and water vapour, the model simulates precipitation, evaporation, soil moisture, snow depth, and runoff. The thermal structure of the atmosphere computed by the model is very similar to that of the actual atmosphere except in the northern hemisphere stratosphere, since the distributions of the major arid regions over the continents, and of the rain belts in both the tropics and the middle latitudes, are successfully simulated.

## 5   Transport processes

A major function of the atmospheric circulation is the transport of heat, water vapour, angular momentum, etc., making it necessary to consider transport processes. If S is the amount of the quantity being studied in a unit volume of air, then the flux of this quantity across a surface of unit area is given by the product VS, where V is the velocity of the air normal to the surface. Now let an over-bar represent the mean value in time or in space, or in both, then at any instance $V = \bar{V} + V'$ and $S = \bar{S} + S'$ where the prime denotes the instantaneous deviation from the mean. Therefore

$$VS = (\bar{V} + V')(\bar{S} + S') = \bar{V}\bar{S} + \bar{V}S' + V'\bar{S} + V'S',$$

and the mean flux is given by

$$\overline{VS} = \bar{V}\,\bar{S} + \overline{\bar{V}S'} + \overline{V'\bar{S}} + \overline{V'S'}$$

The mean value is defined such that

$$\bar{S}' = 0 = \overline{V'}$$

and hence

$$\overline{\bar{V}S'} = \bar{V}\,\bar{S}' = 0 \text{ and } \overline{V'\bar{S}} = \bar{V}\,'\bar{S} = 0$$

The mean flux is therefore given by

$$\overline{VS} = \overline{VS} + \overline{V'S'}$$

This equation can be expressed in the form

$$\text{total flux} = \text{advective flux} + \text{eddy flux}$$

The advective flux is due to the mean motion of the atmosphere while the eddy flux depends on the correlation between V and S as they fluctuate about their mean values. This means that both the mean circulations of the atmosphere and the large-scale eddies have to be studied when considering the transport of heat, water vapour, angular momentum, etc.

*Introduction*

## General meridional circulation

The sun's radiation, through its absorption by the ground, causes a temperature gradient (and therefore a density gradient) between equator and pole, and thereby drives the general circulation. Under these conditions, if the earth did not rotate, a poleward flow of air would exist aloft while a compensating return flow towards the equator would exist at low levels. Such a simple circulation cell is often known as a Hadley cell, but because of the rotation of the earth it does not extend from equator to pole in the atmosphere, but instead it is confined to low latitudes.

*1   Hadley cell*

A schematic representation of the mean meridional circulation in the northern hemisphere during winter, based on work by Palmen (1951), is shown in figure 2.9. The simple Hadley cell circulation is clearly seen south of 30°N. Warm air rises near the equator and flows poleward at high levels in the troposphere, and because of the lack of surface friction angular momentum tends to be conserved. The conservation of angular momentum was discussed in the previous section. The horizontal shear across a zonal current that would be established if air were exchanged meridionally with the conservation of absolute angular momentum is given by

$$\frac{\partial u}{\partial y} = f + \frac{u}{r} \tan \phi$$

where $\frac{\partial u}{\partial y}$ is the horizontal shear across a zonal current; f is the Coriolis parameter; u is the zonal wind speed; r is the distance from the earth's centre; and $\phi$ is the latitude.

Normally the second term, produced by the latitudinal convergence of meridians, amounts to less than 10 per cent of the Coriolis parameter. As air flows poleward in the Hadley cell the zonal wind speed increases in accordance with the above equation, and in the subtropics reaches high values.

Jet streams are defined by the World Meteorological Organization as air currents with quasi-horizontal axes which are thousands of kilometres long, hundreds of kilometres wide, and several kilometres deep. Wind speeds in the core of a jet stream should exceed 30 m/sec, with vertical shears of about 5 m/sec/km and horizontal shears of about 5 m/sec/100 km. The poleward flowing limb of the Hadley cell produces powerful jet streams in the sub-tropics at about the 200 mb level. If the 200 mb wind speed at 30°N is calculated, by

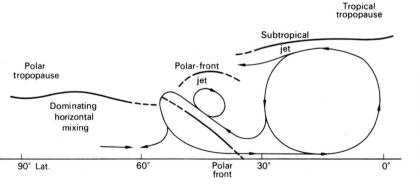

**Figure 2.9** Schematic representation of the meridional circulation and associated jet stream cores in winter. (*After Palmen, 1951.*)

assuming that the angular momentum is conserved in air moving from near the equator, unrealistically high wind speeds are predicted indicating that other factors are probably also important. Nevertheless, in rotating fluid annulus experiments and on the equatorial side of well-developed subtropical jet streams, shears are found which conform to the last equation for $\dfrac{\partial u}{\partial y}$ indicating that conservation of angular momentum is approached in some regions of the atmosphere. The near approach of calculated to observed shears is illustrated in figure 2.10, where the winds were measured by an aircraft flying over East Africa.

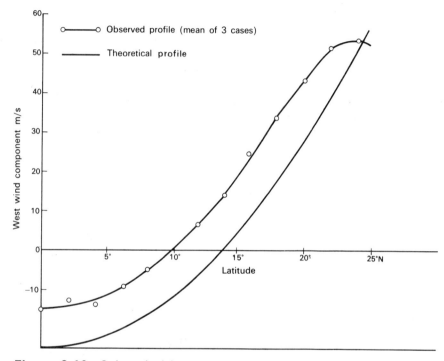

**Figure 2.10** Subtropical jet stream; westerly wind component profiles (*After Johnson and Morth, 1960.*)

Wind speeds in a jet stream can only build up to a certain limiting value, beyond which the resulting turbulence tends to dissipate excessive kinetic energy in the jet core. In practice jet-stream speeds above 100 m/sec are unusual and speeds above 150 m/sec are extremely rare. There is therefore a limit to the latitude range over which air flows meridionally while conserving absolute angular momentum. In the equatorial atmosphere this distance would appear to be between 25° and 35°, since jet streams are found near latitudes 30°N and 30°s.

The position of the subtropical jet stream is clearly marked in figure 2.9 and it extends almost continuously around the globe, except over Asia during summer. Since the mean jet stream can be rationalized as a phenomenon arising within the Hadley cell from the conservation of angular momentum, its position should only vary slightly with the seasons. This can be observed over certain parts of the earth where the jet stream is only slightly

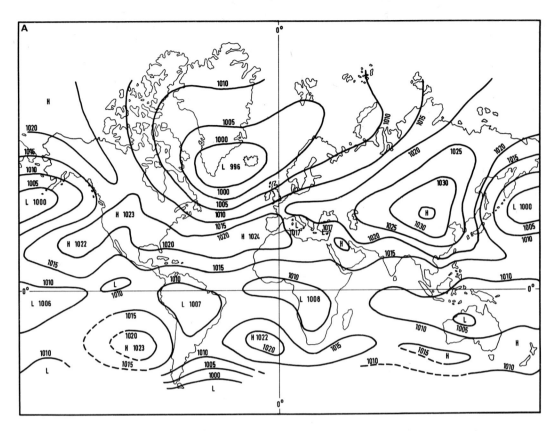

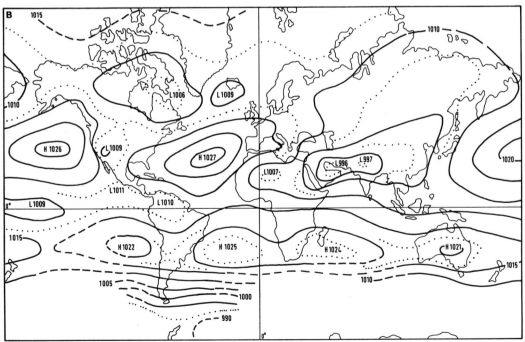

nearer the equator in winter than in summer. High-level convergence takes place into the jet stream, partly because of interial friction limiting wind speeds, and as a result massive subsidence exists below it. This subsidence in turn leads to the generation of large surface high-pressure systems within the subtropics; these are clearly shown on the average pressure charts in figure 2.11. Divergence takes place out of the subtropical highs, the equatorial flow forming both the trade winds and the return flow of the Hadley cell.

Recent investigations have shown that the above picture of the meridional circulation may be over-simplified. Figure 2.12 which is based on research carried out by Newell, Vincent, Dopplick, Feruzza, and Kidson (1969), shows estimates of the mean meridional circulation for the four seasons. The meridional patterns for the mid-seasons clearly correspond to the classical picture, but during the extreme seasons the Hadley cell of the winter hemisphere dominates the circulation pattern of the tropics and the summer Hadley cell of the northern hemisphere disappears altogether. The mean circulation for the extreme seasons is largely the result of the monsoon circulation over Asia and the Indian Ocean. During the northern summer the normal tropical circulation pattern is reversed over Asia, with rising air in the subtropics and a high-level southward flow that destroys the subtropical westerly jet stream and creates an easterly jet stream near the equator. Two weak Hadley cells exist over the Pacific and Atlantic Oceans during this season but the mean meridional circulation is completely dominated by the flow over Asia. Similarly, in winter the mean pattern is determined by the circulation over Asia. Lorenz (1969) considers that a Hadley cell exists in each hemisphere, even in the extreme seasons, at all longitudes except those occupied by the Asiatic monsoon, and that the results shown in figure 2.12 arise from averaging a rather weak Hadley-cell motion extending over four-fifths of the tropics with a strong monsoonal flow in the remaining fifth. Even so, the explanations of the tropical circulation cells and the subtropical jet streams may have been over-simplified. Furthermore, Lamb (1972) considers that the Hadley cell may not be as stable as the classical picture suggests, because the axes of the subtropical high-pressure belts are known to have varied by 1–4° of latitude within the last 100 years, and perhaps even more over longer time-periods.

Though the classical explanation of the subtropical jet stream is true for certain longitudes, it would seem likely that what is being observed at times is a middle latitude jet stream which has been displaced to unusually low latitudes. Evidence exists that the equatorward sections of the large-scale baroclinic waves of the middle latitudes penetrate into the tropics and their associated jet streams may be mistaken for the subtropical jet stream. Since the jet streams in these waves arise from the baroclinicity of the middle-latitude atmosphere, they will be completely independent of the Hadley cell circulation.

The Hadley cell is thus driven by solar radiation but on the broad scale its limits are determined by the speed of rotation of the earth. Energy exchanges within the Hadley cell are discussed in a later section. It imposes a broad-scale climatic zonation on the tropical world. The rising limb of the cell near the equator corresponds on the surface to a belt of relatively low pressure into which the trade winds converge. This low-pressure belt normally contains clouds which can give rise to heavy rainfall, and consequently there is a wet cloudy zone near the equator. The continual subsidence in the subtropical anticyclones

**Figure 2.11**   A: Average atmospheric pressure (mb) at sea level in January (1900–1939). B : Average atmospheric pressure (mb) at sea level in July (1900–1939). (*After Lamb, 1969.*)

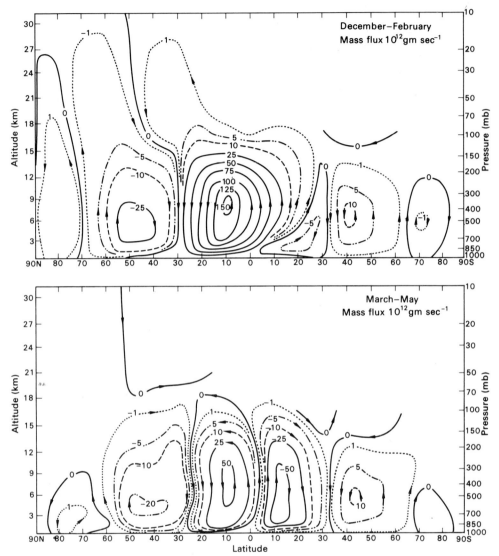

**Figure 2.12** Mean meridional circulation for four

leads to clear skies and therefore a lack of rainfall, and the great deserts of the earth are found below the subtropical jet stream.

### 2 *Middle latitudes*

An indirect circulation cell, that is sinking warm air and rising cool air, is observed in middle latitudes. Temperature falls poleward in middle latitudes, but not in a uniform manner, because horizontal temperature gradients are concentrated in narrow 'frontal zones'. These frontal zones are also the main regions of ascent in the middle latitude cell, and air ascending poleward over fronts should attain large westward components if its

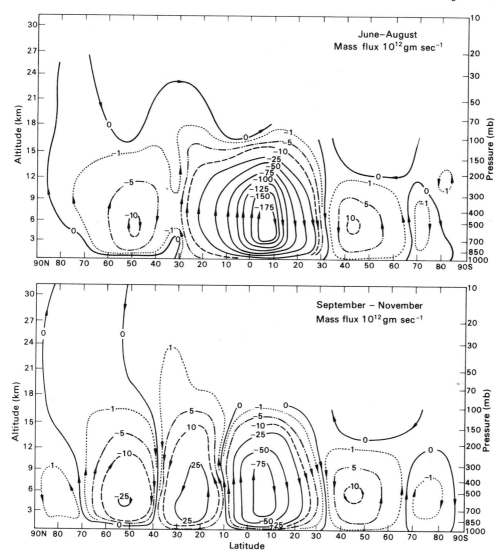

three-month periods. (*After Newell* et al., *1969.*)

absolute angular momentum is conserved. Hence in middle latitudes the general flow is towards the west and jet streams in the upper troposphere are closely associated with frontal zones near the surface. The average meridional circulation in middle latitudes does not appear to be capable of transporting all the heat and momentum required by balance considerations, and so instead, part of this transport is carried out by horizontal eddies. These eddies take the form of large meanders in the upper tropospheric westerlies and closed circulations at the surface.

Compared to the Hadley cell the middle latitude atmosphere is highly disturbed and the suggested meridional circulation in figure 2.10 is largely schematic. The average positions

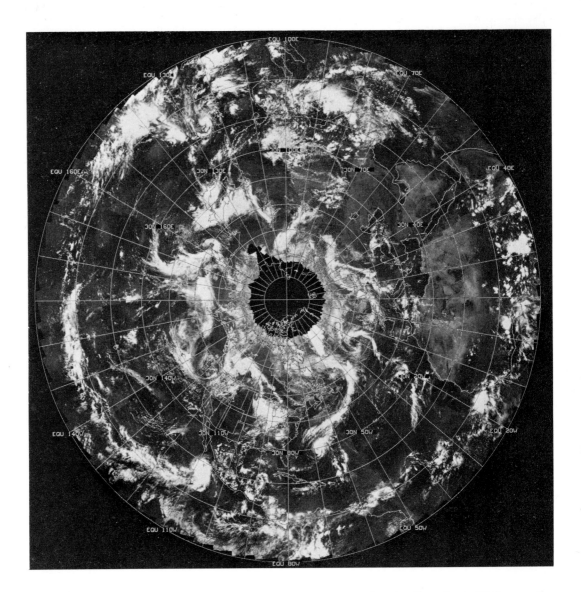

*Plate 4* A mosaic of pictures of the northern hemisphere taken by Essa 9 on 25 September 1971. In general the bright cloudy areas indicate upward motion in the atmosphere, while extensive cloud-free areas are probably the result of widespread subsidence. The Hadley cell implies ascent in a few intense weather systems near the equator and general subsidence in the subtropics, and this is clearly shown over Africa where there are several intense cloud systems south of 10°N but clear skies over the Sahara and the Middle East. A similar cloud distribution is seen over the central Pacific. The largest area of cloud in the subtropics is over India, which is still under the influence of the southwest monsoon. Several middle latitude disturbances with their associated frontal systems are visible. The Greenland ice-sheet can be seen together with snow fields along the northern edges of the continents. *Reproduced by permission of the United States Department of Commerce, National Oceanic and Atmospheric Administration.*

of the very large-scale meanders or waves in the upper troposphere appear to be geographic-ally fixed but the minor waves are more mobile. Near the surface closed circulations exist in the form of depressions and anticyclones which drift slowly towards the east. The depressions normally contain the frontal zones along which much of the middle latitude ascent takes place, so the feature marked as the polar front in figure 2.9 is highly variable in character and position. The picture sketched above for the middle latitudes compares well with that conveyed by rotating fluid annulus experiments with low thermal Rossby numbers. It appears the disturbed weather of the middle latitudes arises partly from the high values of the Coriolis parameter found in these regions, and not completely from the large thermal contrasts.

The Hadley cell tends to be relatively undisturbed and, therefore, the mean pressure charts in figure 2.11 tend to be mirrored by the daily charts. The highly disturbed state of the middle latitude atmosphere implies that average charts are little more than statistical abstractions, and the average features shown in the pressure patterns in figure 2.11 will not necessarily be repeated on daily charts. The travelling disturbances tend to give rise to evenly distributed precipitation, which does not reach such high values as those found in the equatorial regions because of the lower moisture content of the air.

Weak subsidence exists over the polar regions, and the polar cell can be regarded as another direct cell. In these regions surface cooling is normally intense, and a marked in-version normally exists just above the surface and tends to insulate it from changes in the free atmosphere. Because of the intense cold in the polar regions the moisture content of the atmosphere is low and therefore precipitation is slight.

## Circulation of water in the atmosphere

Figure 2.13 outlines the general hydrological cycle which includes overland and under-ground flow as well as water vapour transport in the atmosphere, but in this particular section interest will centre on the latter. In chapter 1 it was suggested that the water balance

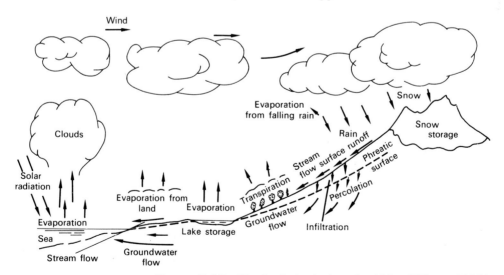

**Figure 2.13** The hydrological cycle. (*After Wilson, 1969.*)

## Introduction

is an important part of the climatic environment, and it is therefore important to look at water vapour transport in the atmosphere.

### 1  Precipitable water content

A very good measure of the amount of moisture in the atmosphere is the precipitable water content, which is the amount of rainfall that would be generated if all the moisture in the atmospheric column were condensed and fell to the ground. The precipitable water content (W) of a unit column of air at a given instant above a point on the earth's surface is expressed by

$$W = \frac{1}{g}\int_{o}^{p_0} q\, dp$$

where g is the acceleration of gravity; q is the specific humidity; and $p_0$ the mean value of the surface pressure.

The total horizontal transport of water vapour (Q) above a point on the earth's surface is given by

$$Q = \frac{1}{g}\int_{o}^{p_0} q\, V dp = VW$$

where V is the vertically averaged wind in the column.

For a unit column of air extending from the earth's surface to the top of the atmosphere, the water vapour balance equation can be written

$$\frac{\partial W}{\partial t} + \nabla \cdot Q = \Sigma$$

where $\Sigma$ represents the net sources of water substance in the atmospheric column. The first term represents the change of precipitable water content with time, while the second term represents the net flux of water vapour into or out of the column. Obviously the primary source of water vapour in the atmosphere is the evaporation ($E_t$) of water on the earth's surface, while the primary sink is precipitation (P). The atmospheric transport of water in the solid or liquid phases is very small compared with the flux of water vapour, thus $\Sigma$ is very nearly given by the excess of evaporation over precipitation.

Averages of atmospheric water balance or atmospheric water vapour flux are not readily available, but a useful study of the northern hemisphere during the IGY year 1958 has been made by Starr, Peixoto, and Crisi (1965), and much of the following discussion is based on their work.

Figure 2.14 shows the mean precipitable water vapour content ($\overline{W}$) for the year 1958. In general, there is a continuous decrease of precipitable water vapour content from equator to the north pole. This is to be expected as the maximum amount of water vapour held by the atmosphere is dependant on air temperature and the mean temperature of the atmosphere decreases from equator to pole. The maritime and continental influences are clearly evident; for instance, the subtropical deserts of North Africa and the Middle East stand out as dry areas. Much of the water vapour content of the atmosphere is held in the lowest layers, and therefore the areas of high terrain also appear as dry areas. The driest area shown on the chart is in the Arctic, where the very low mean air temperatures lead to a mean precipitable water vapour content of less than 0·5 cm north of 80°N. The areas of

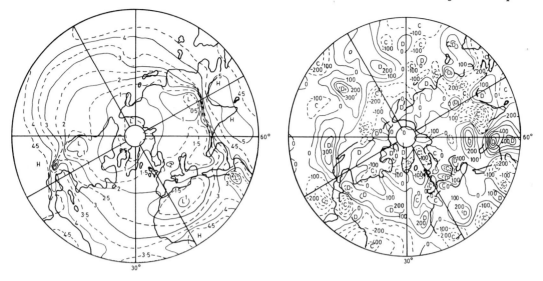

**Figure 2.14** *Left*: Average precipitable water content for the year 1958. Units: gm cm⁻² or equivalent depth of rainfall in cm. *Right*: Average distribution of the horizontal divergence of water vapour flux for the year 1958. The isopleths (full lines for divergence and dashed for convergence) are entered for intervals of 100 cm year⁻¹. Divergence indicates that evaporation is greater than precipitation. (*After Starr* et al., *1965.*)

highest precipitable water vapour content are found in the equatorial region of South America, the Indian Ocean, and equatorial west Africa.

*2   Water vapour divergence*

Starr, Peixoto, and Crisi (1965) have produced a chart showing the horizontal divergence of the vertically integrated total annual flux of water vapour ($\nabla . \overline{Q}$), and a simplified version is produced in figure 2.14. This figure shows a pattern of alternating areas of convergence and divergence, where convergence indicates that precipitation exceeds evaporation while divergence indicates the opposite. The amount of water stored in the atmosphere is small when compared with normal precipitation totals, and so there must be a flow of water vapour into high precipitation areas. This implies that areas of high evaporation must exist and that these will supply water vapour to the high precipitation areas. Figure 2.15 shows mean annual surface evaporation, while mean annual precipitation is illustrated in figure 2.16; and these two diagrams should be studied together with figure 2.14.

Mean annual evaporation is highest in the subtropics and decreases both towards the equator and the poles. The highest evaporation rates are found over the subtropical oceans where the clear skies allow a large influx of radiation and there is an unrestricted supply of surface water. Very high evaporation rates are found in the subtropics when additional energy is available from warm ocean currents, and this particularly applies on the western sides of oceans. Mean annual precipitation is greatest in the equatorial trough and falls to very low values in the subtropical anticyclones, which is the reverse of the distribution of mean annual evaporation in the tropical world. Secondary precipitation maximums exist over the oceans of the middle latitudes, but generally amounts decrease towards the poles and the interiors of the continents.

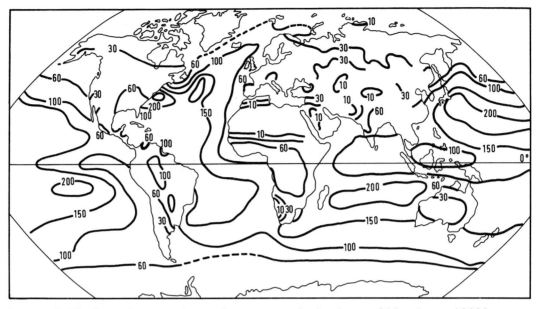

**Figure 2.15**   Annual evaporation and evapotranspiration in cm. (*After Barry, 1969.*)

In figure 2.14 which can be considered as an integration of the mean annual evaporation and precipitation charts, areas of general convergence are shown near the equator indicating an excess of precipitation over evaporation due to the convergence of the trade winds from both hemispheres, while general divergence is found in the subtropical anticyclones. In the middle and high latitudes there are alternating areas of convergence and

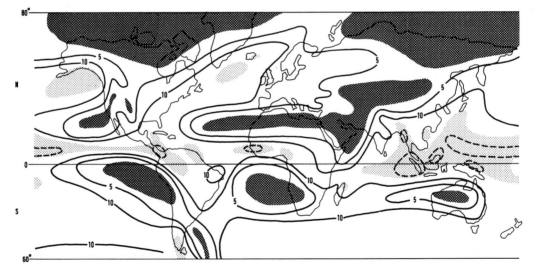

**Figure 2.16**   World distribution of annual precipitation (decimetres), simplified to show only major regimes; oceanic rainfall estimated. Areas with precipitation below 2·5 and above 20 decimetres shaded; dashed lines: 30 decimetres. (*After Riehl, 1965.*)

divergence. Two areas of strong convergence within the tropics are of special interest. The first which extends from southwestern Arabia generally westward and southward across equatorial eastern and central Africa contains the headwaters of the Blue Nile, several tributaries of the White Nile, the upper parts of the Congo, and Ubangi rivers; and there are also several rivers flowing southward from this area through Kenya. The second convergence zone which is over north-central India, extends northward through Kashmir to the Pamirs and Altai Mountains west of Sinkiang. The heavy rains over India are well known and the Indian convergence zone covers the headwaters of such extensive river systems as the Indus, Ganges, Brahmaputra, Salween, Mekong, and Yangtze.

The subtropical anticyclonic regions of the Atlantic and Pacific Oceans show extensive areas of divergence, which is to be expected since they are among the major source regions of moisture for the equatorial trough. The Pacific anticyclone has a marked cellular structure, and thus in the subtropical north Pacific there are several centres of marked divergence between Marcus Island and Mexico. Also associated with the subtropical anticyclones are three areas of divergence of particular interest; these are over the central Mediterranean Sea, over Iran, and over Mauritania in west Africa. The central Mediterranean divergence extends southward over the desert areas of Libya and Algeria and this whole area is known for its dryness. The Mediterranean acts as an important source of atmospheric moisture, indicated by its relatively high salinity. The divergence over west and central Africa coincides with scanty precipitation and with the cold Canary or North African current. Of particular interest are the areas of divergence over the deserts of west and central Africa, Arabia, the Middle East and Iran, because it is difficult to conceive of these areas as being major sources of atmospheric moisture. Starr and Peixoto (1958) in their study of 1950 discovered similar divergence over these same deserts, and Starr, Peixoto, and Crisi (1965) in their study of 1958 found themselves unable to explain the feature. The moisture is perhaps being supplied by the evaporation of water stored underground, or results from computational errors arising from the strongly divergent flow.

The mid-latitude regions of the northern hemisphere show many areas of divergence and convergence, although the over-all picture is one of general convergence. Prominent areas of convergence are associated with the extratropical storm tracks across the North Atlantic and North Pacific Oceans, and in particular the areas of convergence over the eastern United States and the Gulf Stream, and between Iceland and Greenland, are clearly related to polar front depressions. A similar comment can be made about a long and extensive area of convergence that extends from the East China Sea across the entire North Pacific Ocean to the west coast of North America. The arctic regions north of 60°N show a patchwork of small areas of weak convergence and divergence, but the general picture is that there is southward transport of water vapour across both latitudes 80° and 70°N.

As a summary of figure 2.14, the latitudinal distribution of the average annual meridional transfer of water vapour is shown in figure 2.17, where the data presented suggest the existence of a meridional circulation with at least two major cells. The transport of water vapour towards the equator in the Hadley cells is very clearly shown, as is the poleward transport in the middle latitudes.

*3   Water vapour sources*
It might be considered that the transport of water vapour by the atmosphere was unimportant because the bulk of the precipitation resulted from local evaporation. The

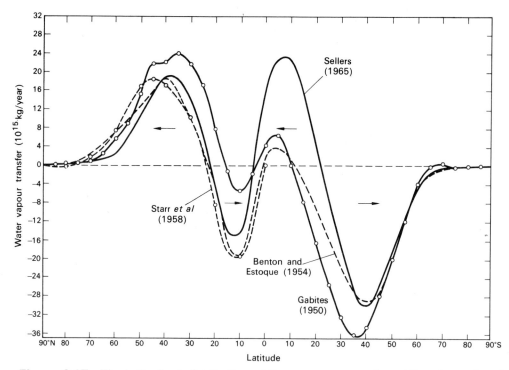

**Figure 2.17** The latitudinal distribution of the average annual meridional transfer of water vapour in the atmosphere. (*After Sellers, 1965.*)

previous discussion should have made it clear that this is unlikely, but it is an idea worth considering because it could have implications concerning the control of the environment. If a large proportion of the precipitation consisted of locally evaporated water vapour, the possibility would exist of controlling the amount of precipitation by altering local evaporation sources. It might be possible to increase precipitation in arid areas by building reservoirs or by planting a large number of water consuming trees; alternatively in humid areas it might be possible to decrease the precipitation by draining ponds and clearing forest. It is claimed that in certain equatorial countries the precipitation has decreased because of the clearing of jungle for agriculture. The evidence available at present indicates that the local water vapour contribution to precipitation is usually very small. Drozdov and Grigor'eva (1963) consider that in European Russia the local moisture accounts for only slightly more than 10 per cent of the total annual precipitation. Benton, Blackburn, and Snead (1950), who have studied the annual water balance of the Mississippi River Basin which covers almost half of the United States, calculated that only 8–9 per cent of the water vapour for precipitation came from the land areas of the Mississippi Basin, 3–6 per cent from land areas outside the basin, the remaining 85–89 per cent being of oceanic origin. From these studies, Benton *et al.* (1950) concluded for temperate regions that 'increasing or decreasing precipitation even slightly by local regulation of land use is therefore out of the question'. However this conclusion may not apply to the tropics where there are moisture source regions with only relatively slight air motion.

## 4  Rivers

Comment which has been made about the apparent relationship between areas of convergence in figure 2.14 and the headwaters of rivers, can now be explored further by examining the water balance of a small area of the earth's surface. The water balance of a unit area of the earth's surface over a unit time period may be expressed by

$$P + SF_{IN} + G_{IN} = SW + GW + E_t + SF_{OUT} + G_{OUT}$$

where SW is the change in soil water storage per unit time; GW is the change in ground water storage per unit time; $SF_{IN}$ and $SF_{OUT}$ are the water flows over the surface into and out of the unit area respectively; $G_{IN}$ and $G_{OUT}$ are the ground water flows into and out of the unit area respectively; and the remaining terms have been previously defined. If this equation is slightly rearranged it becomes,

$$P - E_t = SW + GW + (SF_{OUT} - SF_{IN}) + (G_{OUT} - G_{IN})$$

Since the soil water and ground storage capacities have finite maximum values which are usually relatively small, there will be a flow of surface and ground water out of the unit area if precipitation exceeds evaporation. As the surface flow normally takes the form of rivers, this suggests that rivers will be generated under the areas of convergence (illustrated in figure 2.14) that occur over land.

## Circulation of energy in the atmosphere

### 1  Energy content

The total energy content of one gram of moist air may be expressed by the following equation:

$$Q = \frac{AV^2}{2} + Bgz + C_p T + Lq$$

where Q is the total energy content, V is the wind speed, g is the acceleration due to gravity, z is the height above mean sea level, $C_p$ is the specific heat of air at constant pressure, T is the temperature in degrees Kelvin, L is the latent heat of condensation of water vapour, q is the specific humidity, and A and B are conversion constants appropriate to the units employed.

If the vertical motion of the parcel of air is frictionless and adiabatic and if it is assumed that there is no mixing with the surrounding atmosphere, then there is a direct conversion between the potential energy (gz) and the sensible heat content ($C_p$ T). It is, therefore, convenient to consider these two quantities together under the heading of total potential energy (Margules, 1903). The total potential energy is, typically, 100 to 1000 times greater than the kinetic energy ($V^2/2$) of the atmosphere; for example, in a powerful jet stream the kinetic energy content will be equivalent to 0·6 to 0·9 cal/g, but the total potential energy at jet stream level may be between 75 and 80 cal/g. Normally the kinetic energy content is about equal to the probable error in estimating the total potential energy and may, therefore, be neglected in estimations of the total energy content (Q). The total potential energy content is normally about 10 times greater than the latent heat content (Lq); in a very warm, moist atmosphere the latent heat content may be between 10 and 12 cal/g while the total potential energy will be about 70 cal/g.

## Introduction

If condensation is taking place in a rising air parcel, there is a conversion of latent heat into total potential energy. If it is assumed that there is no mixing with the surrounding atmosphere and the energy losses due to radiation and the fall-out of condensed water are ignored, the energy content of the parcel will remain constant during ascent. The graph of Q for such a parcel will be a vertical straight line as in figure 2.19. The amount of temperature fall attributable mainly to radiation cooling in the atmosphere depends on several factors, but falls of between 1 and 2°C are probably normal, and this is equal to a cooling rate of between 0·25 and 0·5 cal/g per day. Air subsiding very rapidly under conditions of no mixing will maintain an approximately constant Q content, but if the air is subsiding slowly, as in an anticyclone, it may take 15 to 20 days to sink from the 300 mb level to the 600 mb level, and under these conditions the air may have cooled gradually during the descent by 5 to 10 cal/g. Profiles of Q through a slowly sinking air-mass will, therefore, show a decrease in Q values from the top to the bottom. If no energy is supplied to the atmosphere by the earth's surface and if descent takes place almost to the surface, there will be a minimum in the value of Q at the surface. However, in anticyclones there is normally a low-level inversion and below the inversion active convection may take place. Moreover if the anticyclone is over a warm surface, energy can be transported upward by convection and the layer containing the minimum Q values will be lifted from the surface into the lower atmosphere.

## 2  Energy transport

In recent years much attention has been given to the problem of energy transport by the atmosphere and oceans. In the past, oceanic energy transport was assessed as being small, that is not more than 10 per cent of the total energy transfer. Recent computations indicate, however, that the oceans may carry a much larger fraction of the total energy transfer, reaching perhaps 25 per cent in the subtropics and middle latitudes. Comment was made earlier that the total flux of an atmospheric quantity can be divided into two components, which are the advective flux and the eddy flux. Figure 2.18 shows the eddy sensible heat flux as computed by Mintz (1951) for January 1949, and the eddy latent heat flux as computed by Starr, Peixoto, and Livadas (1957) and Van de Boogaard (1964); this last computation applies only to a single day, 12 December 1967, hence the flux may be unrepresentative. This diagram suggests that the eddy fluxes are largest in the middle latitudes and decrease toward the tropics, where they may approach zero. The large eddy fluxes in the middle latitudes reflect the importance of air-mass exchanges, because cold, dry air normally flows from the poles and warm, moist air from the tropics. Within the tropics where the atmosphere tends to be uniform over large areas and the flow is relatively undisturbed, it appears that the mean meridional circulation of the Hadley cell is the prime mechanism of net heat exchange, and this suggests that the mean advective flux is relatively more important within the tropics than in the middle latitudes.

There is within the tropics general low-level flow towards the equator, and since there is a large exchange of energy between the earth's surface and the atmosphere, obviously there must be a general advection of energy towards the equator. This energy is mainly in the form of latent heat (Lq) carried in the lower layers of the trade winds rather than as total potential energy. The Hadley cell circulation implies that energy must be carried vertically aloft in the equatorial zone and then exported at high altitudes to the subtropics.

66

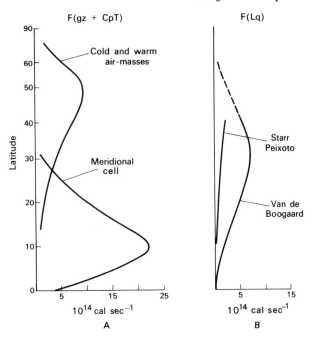

**Figure 2.18** A: poleward flux of sensible heat by eddies in January 1949 (Mintz, 1951) and poleward transport of potential energy plus sensible heat by the mean meridional circulation in winter. B: poleward flux of latent heat by eddies after Starr, Peixoto, and Livadas (1957) and after van de Boogaard (1964). (*After Riehl, 1969.*)

Figure 2.19 illustrates the mean height distribution of total energy content (Q) in both the equatorial trough zone and at 20° latitude away from it. The diagram shows that in both cases the Q values are high near the surface, and then decrease to a minimum near 700 mb, after which they increase again. In the equatorial zone large quantities of energy are exported aloft and the mean meridional circulation demands ascent, but a uniform, gradually rising circulation would produce cooling above the level of minimum Q. Further-

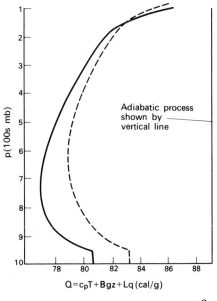

**Figure 2.19** Vertical profiles of total energy content. Solid profile is for 20° latitude from equatorial trough. Dashed profile is for equatorial trough itself. (*After Riehl and Malkus, 1958.*)

67

more, the upper troposphere is continually cooling because of radiation losses and this energy has to be replaced from below to maintain the shape of mean Q curve. It would appear therefore that the picture of a Hadley cell with general ascent in the equatorial trough is incorrect.

Riehl and Malkus (1958) have tried to solve the obvious dilemma concerning the equa-

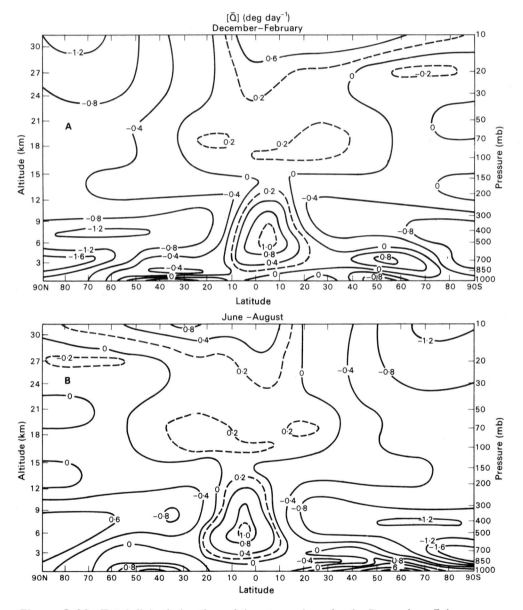

**Figure 2.20** Total diabatic heating of the atmosphere for A : December–February, B : June–August.

torial limb of the Hadley cell by their 'hot tower hypothesis'. According to these authors the rising portion of the meridional cell is concentrated in the restricted regions of the towering cloud banks within tropical storms. In ordinary cumulus clouds there is much mixing with the surrounding atmosphere, but within the general area of equatorial disturbances there are imbedded giant cumulonimbus whose central cores are protected from

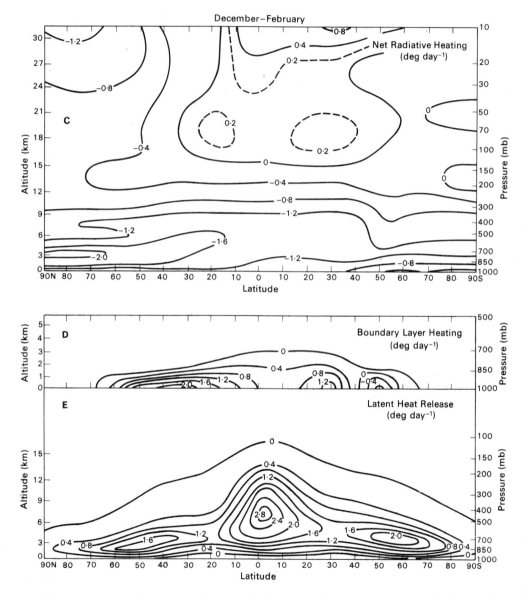

C, D, E: components of the diabatic heating of the atmosphere for December–February. Units: °C day$^{-1}$. (*After Newell* et al., *1969.*)

dilution by the large cross-sections of the towers. Normal cumulus clouds rarely penetrate far into the atmosphere but the giant cumulonimbus provide a means whereby high-energy content air from near the ocean surface can be pumped to great heights to balance the heat losses and provide for the calculated energy exports. In these clouds water vapour is condensed and precipitated as rainfall and the latent heat released is converted into total potential energy which is transported to the high troposphere. According to Malkus (1962) only about 1500–5000 active giant cumulonimbus clouds are needed within the 10° latitude width equator belt to provide for its high-level poleward energy export. It thus appears that the rising limb of the Hadley cell is restricted to a few very active synoptic disturbances within the equatorial trough.

If ascent near the equator is limited to synoptic disturbances, the descent in the subtropics is general and widespread. In the subtropics the sinking air-masses carry total potential energy gained in the equatorial trough and feed it into the middle latitude circulations. The meridional transfer of total potential energy has a double maximum in both hemispheres, one in the subtropics between 15 and 25° and the other in high middle latitudes between 50 and 60°. The first is located on the poleward side of the tropical rain belt and the second on the poleward side of the most intense cyclonic activity along the polar front. There is little direct heating of the atmosphere by the transfer of sensible heat from the surface, for instead, most atmospheric warming between 70°N and 70°S is due to the release of latent heat by condensation and much of this occurs between 10°N and 10°S (figure 2.20). Poleward of 70°N and 70°S most warming occurs by the advection of warmer air from lower latitudes.

The observed radiation distribution in the tropics raises the interesting question as to whether the Hadley cell is strictly a direct cell, that is to say one in which the potential energy represented by the juxtaposition of relatively dense and light air-masses is converted into kinetic energy as the lighter air rises and the denser air sinks. In the normal direct cell the rising air is found over the heat source, but in the tropics the greatest inputs of solar energy are not found near the equator but in the subtropical deserts where there is massive subsidence. It is only when the energy exchanges within the tropics are studied in detail that the nature of the Hadley cell becomes clear. Values of total diabatic (non-adiabatic) heating for the extreme seasons are illustrated in figure 2.20, where it is seen that the largest positive rates are observed in the middle troposphere between 10°N and 10°S. Figure 2.20 further indicates the magnitudes of the components of the diabatic heating during the northern winter season. It is observed that the net radiative heating of the troposphere is negative, indicating that it is losing heat by long-wave radiation. The transfer of sensible heat into the planetary layer is small except over the hot deserts of the northern hemisphere, leaving the release of latent heat as the main form of atmospheric heating. During the northern winter the release of latent heat occurs mainly in three localities, one of which is found over the equator and the other two in middle latitudes. The maximum in diabatic heating observed over equatorial regions is therefore mainly the result of latent heat released by condensation in the equatorial trough; since it occurs in the rising limb this confirms that the Hadley cell is a direct cell. As described earlier, the large influx of solar radiation into the subtropics does not heat the atmosphere directly but instead is used to evaporate water from the tropical oceans; this is advected by the trade winds into the equatorial trough where it is condensed and the latent heat released to the atmosphere. The Hadley cell is, therefore, only driven indirectly by solar radiation.

## General circulation and the climatic environment

The general circulation of the earth's atmosphere can be an object of study in its own right, but here we are more interested in it as a determining factor of the atmospheric environment. The importance of the general circulation in the study of the atmospheric environment arises in two ways: first, it is the chief transporting agent of water vapour and therefore is an important determining factor of local water balances; and second, by its transport of heat, it greatly modifies the local energy balance resulting from purely radiational factors.

It has frequently been illustrated how important the heat and water balances are in determining the atmospheric environment, but it is difficult to quantify these ideas. Budyko (1956) relates the net radiation ($R_N$) available to evaporate water vapour from a

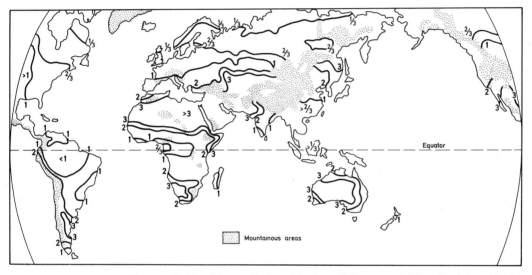

**Figure 2.21**   The radiational index of dryness. (*After Budyko, 1956.*)

wet surface to the heat required to evaporate the mean annual precipitation (P), and he calls this ratio $R_N/PL$ (where L is the latent heat of vaporization) the 'radiational index of dryness'. It has a value of less than unity in humid areas, greater than unity in dry areas and is a concept which is similar to Thornthwaite's (1948) moisture index. Budyko's radiational index of dryness could be a possible starting point for the quantification of ideas about the climatic environment. The parameter $R_N/PL$, which determines the relative values of the heat and the water balances, also appears to partly determine the locations of boundaries of the principal botanical zones. Figure 2.21 shows the distribution of $R_N/PL$. This diagram suggests that the lowest values of the index (up to 0.33) are typical of the tundra zone, the values between 0.33 and 1 of the forest zone, from 1 to 2 of the steppe, and above 2 are typical of the semi-deserts and deserts. Budyko further suggests that more meaningful results can be obtained by considering the annual net radiation balance $R_N$ together with the index $R_N/PL$; the relationship between geobotanical features and the parameter $R_N/PL$ and $R_N$ is illustrated in figure 2.22. In this diagram the solid lines designate the limits of actually observed values of $R_N$ and $R_N/PL$, and within these limits definite values of the parameter $R_N/PL$ separate the principal geobotanical zones—tundra, forest, steppe, semi-desert, and desert.

The zonality of soils is closely associated with that of vegetation, and results obtained for vegetation in figure 2.22 could apply to soil zones as well. According to Budyko, with an increase of the parameter $R_N/PL$, the soil type should change according to the following sequence:

1  tundra soils;
2  podzol soils, brown forest soils, yellow soils, red soils, and laterites (the wide variability of soil types in this group correspond to the wide range of the parameter $R_N$);
3  chernozem and black soils of the savanna;
4  chestnut soils;
5  gray soils.

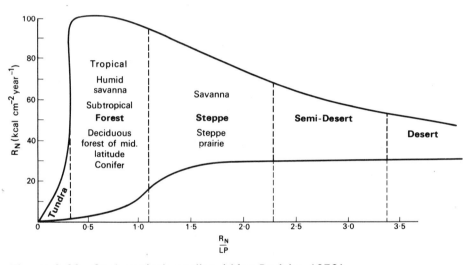

**Figure 2.22** Geobotanical zonality. (*After Budyko, 1956.*)

Similar comments can be made about the zonality of the hydrological regime on land. In this case Budyko suggests that for each set of values of the parameter $R_N/PL$, there corresponds a certain class of values of the runoff coefficient $f_R/P$, where $f_R$ is the annual runoff. The annual runoff approximately equals the difference between precipitation and evaporation, and this is controlled by the divergence of the vertically integrated total annual flux of atmospheric water vapour and also by the net radiation. It appears that for the tundra, where $R_N/PL$ is less than 0·33, the runoff coefficient is greater than 0·7; in the forest zone with $R_N/PL$ lying between values of 0·33 and 1, the runoff coefficient is between 0·3 and 0·7; in the steppe zone with $R_N/PL$ between 1 and 2, it is between 0·1 and 0·3; and in semi-deserts and deserts it is less than 0·1. Absolute values of the annual runoff can be expressed in terms of the parameters $R_N$ and $R_N/PL$, and in figure 2.23 is presented the distribution of annual runoff in a form similar to the graph in figure 2.22.

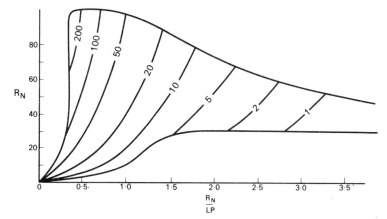

**Figure 2.23** Runoff (cm year⁻¹) in various geographical zones. (*After Budyko, 1956.*)

Budyko has further suggested that there is an optimum value of $R_N/PL$ near 1·0 at which, for a given value of $R_N$, vegetation productivity will be at a maximum. It is assumed that when the parameter $R_N/PL$ is near unity in value, there is just enough energy available to evaporate the mean annual precipitation. As a result of this, transpiration is not curtailed by a lack of moisture, as it would be if $R_N/PL$ were greater than 1·0, nor is the soil water-logged and poorly aerated, as it would be if $R_N/PL$ were much less than 1·0. Sellers (1965) has tested this hypotheses on 151 catchments within the United States, and the results are shown in table 2.2. In this table, values of the parameters $R_N$ and $R_N/PL$ are compared

**Table 2.2** Annual timber growth (m³ per km²) and the radiational index of dryness

| Radiational Index of Dryness | Net Radiation (k cal cm year⁻¹) | | |
|---|---|---|---|
| | ⩽ 40 | 40–60 | ⩾ 60 |
| ⩽ 0·8 | 212 | 207 | 259 |
| 0·8–1·0 | 204 | 216 | 263 |
| 1·0–1·9 | 84 | 88 | 206 |
| ⩾ 1·9 | – | 56 | 138 |

*After Sellers (1965).*

with the annual timber growth; the growth rate increases with increasing values of $R_N$ within each range of values of $R_N/PL$; the growth rate peaks at values of $R_N/PL$ between 0·8 and 1·0 in all but one of the $R_N$ groups.

This problem has been further investigated by Bazilevich, Drozdov, and Rodin (1971), who studied the relationship between vegetation productivity, $R_N$ and $R_N/PL$. They found (figure 2.24) that if the net radiation does not exceed 40 k cal/cm²/year that plant productivity rises quickly following an increase in net radiation. However, with values of net radiation above 40 k cal/cm²/year the vegetation productivity is predominantly influenced by the availability of moisture, and this influence is especially noticeable when values of $R_N/PL$ drop below 1. Similarly, if $R_N/PL$ is equal to 1, then an increase in net radiation from the lowest observed values leads, at first, to a marked rise in the annual productivity, but subsequently there is almost no increment at all. This suggests that values of $R_N/PL$ equal, or close to 1, are not optimal for maximum vegetation productivity in all climatic zones.

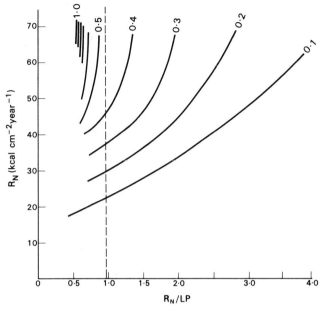
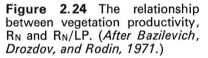

**Figure 2.24** The relationship between vegetation productivity, $R_N$ and $R_N/LP$. (*After Bazilevich, Drozdov, and Rodin, 1971.*)

Evidence therefore exists of a large number of complex interactions between the atmosphere and the earth's surface. These are shown in diagramatic form in figure 2.25, where box 1 indicates the general circulation of the atmosphere which basically is a function of the distribution of net radiation and the earth's rotation. This is of interest principally because it controls the occurrence and distribution of precipitation. Apart from the net radiation, precipitation is the main variable involved in box 2, and this occurs in the radiational index of dryness; therefore the output from box 1 feeds into box 2 via the precipitation factor. It has already been demonstrated that the parameters in box 2 are correlated with the runoff coefficient and the broadscale geobotanical zones, and so the output from box 2 feeds into boxes 4 and 5. An alternative route from box 1 to box 4 is via box 3, which contains the net radiation and the divergence of the atmospheric water vapour flux. These last two parameters determine both evaporation and precipitation, and hence the water balance and the amount of runoff. Since the divergence of the atmospheric water vapour flux is a function of the atmospheric circulation, the connection between box 1 and box 3 is clear. The net radiation and the divergence of the atmosphere water vapour flux control the water balance at any point, and so it would also seem likely that through the water balance they will control the types and rates of weathering and erosion. Thus the output from box 4 feeds into box 6. Both vegetation type and weathering processes partly determine soil type, and therefore the outputs from boxes 5 and 6 feed into box 7.

Besides these, more complex interactions can be added to figure 2.25. The water balance controls the water supply, which is the amount of water available for use by population and industry. The former together with the soil type and the general geobotanical zone control the type and level of farming. It is possible that the availability of food and water partly determines the population level; for example, deserts and tundra areas do not have large populations, whereas the opposite is the case in humid areas. Some of these interactions may seem to bear little relation to the study of the atmosphere but they do illustrate the possible social importance of the atmospheric environment.

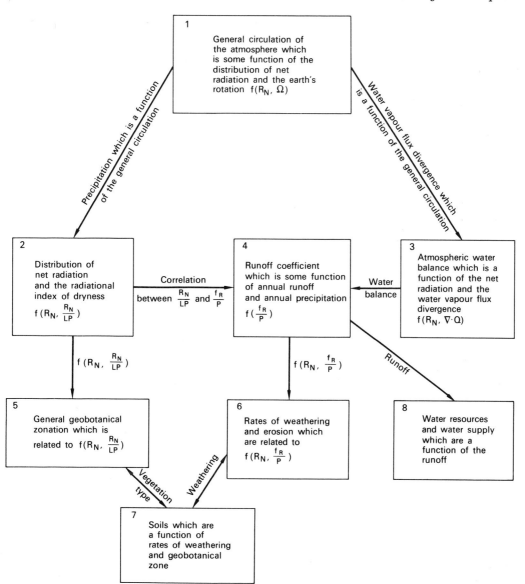

**Figure 2.25** Interactions between the atmosphere and the earth's surface.

## References

BARRY, R. G. 1969: The world hydrological cycle. In Chorley, R. J. (editor), *Water, earth and man.* London: Methuen.

BAZILEVICH, N. I., DROZDOV, A. V. and RODIN, L. E. 1971: World forest productivity, its basic regularities and relationships with climatic factors. In Duvigneaud, P. (editor), *Productivity of forest ecosystems.* Paris: UNESCO.

BENTON, G. S., BLACKBURN, R. T. and SNEAD, V. O. 1950: The role of the atmosphere in the hydro-logic cycle. *Transactions American Geophysical Union* **31**, 61.

BENTON, G. S. and ESTOQUE, M. A. 1954: Water-vapour transfer over the North American continent. *Journal of Meteorology* **11**, 462.

BUDYKO, M. I. 1956: *The heat balance of the earth's surface.* Translated by N. I. Stepanova. Washington: US Weather Bureau.

BUDYKO, M. I. and KONDRATYEV, K. YA. 1964: The heat balance of the earth. *Research in Geophysics. Vol.* **2**: *Solid earth and interface phenomena.* Cambridge, Mass.: MIT, p. 529.

DOUGLAS, H. A., HIDE, R. and MASON, P. J. 1972: An investigation of the structure of baroclinic waves using three-level streak photography. *Quarterly Journal Royal Meteorological Society* **98**, 247.

DROZDOV, O. A. and GRIGOR'EVA, A. S. 1963: *The hydrologic cycle in the atmosphere. (Vlagooborot v atmosfere.)* Leningrad: Gidrometeorologicheskoe Izdatel'stvo.

FLOHN, H. 1969: *Climate and weather.* London: Weidenfeld and Nicolson.

FULTZ, D., LONG, R. R., OWENS, G. V., BOWAN, W., KAYLOR, R. and WEIL, J. 1959: Studies of thermal convection in a rotating cylinder with some implications for large-scale atmospheric motions. *Meteorological Monograph* **4**, *American Meteorological Society*, Boston.

GABITES, J. F. 1950: Seasonal variations in the atmospheric heat balance. Sc.D. dissertation, Massachusetts Institute of Technology, Cambridge, USA.

HADLEY, G., 1735. Concerning the cause of the general trade-winds. *Philosophical Transactions Royal Society* **29**, 58.

HIDE, R. 1969: Some laboratory experiments on free thermal convection in a rotating fluid subject to a horizontal temperature gradient and their relation to the theory of the global atmospheric circulation. In Corby, G. A. (editor), *The global circulation of the atmosphere.* London: Royal Meteorological Society, p. 196.

HIDE, R. and MASON, P. J. 1970: Baroclinic waves in a rotating fluid subject to internal heating. *Philosophical Transactions Royal Society* **268**, 201.

HOLLOWAY, J. L. and MANABE, S. 1971: Simulation of climate by a global circulation model. **1**: Hydrologic cycle and heat balance. *Monthly Weather Review* **99**, 335.

JOHNSON, D. H. and MORTH, H. T. 1960: Forecasting research in East Africa. *Proceedings of Munitalp/WMO Joint Symposium of Tropical Meteorology in Africa*, Nairobi, p. 56.

KONDRATYEV, K. YA. 1962: *Meteorological investigations by means of rockets and satellites.* Leningrad: Gidrometeoizdat.

KURIHARA, Y. and HOLLOWAY, J. L. 1967: Numerical integration of a nine-level global primitive equations model formulated by the box method. *Monthly Weather Review* **95**, 509.

LAMB, H. H. 1969: Climatic fluctuations. In Flohn, H. (editor), *General climatology*, **2**. Amsterdam: Elsevier Publishing Co., p. 173.

LAMB, H. H. 1972: *Climate: present, past and future.* London: Methuen.

LORENZ, E. N. 1967: *The nature and theory of the general circulation of the atmosphere.* Geneva: WMO.

LORENZ, E. N. 1969: The nature of the global circulation of the atmosphere: a present view. In Corby, G. A. (editor), *The global circulation of the atmosphere.* London: Royal Meteorological Society, p. 3.

MALKUS, J. S. 1962: Large-scale interactions. In Hill, M. N. (editor), *The sea.* Vol. **1**. New York: John Wiley, p. 88.

MANABE, S. 1969: Climate and the ocean circulation. **1**: The atmospheric circulation and the hydrology of the earth's surface. *Monthly Weather Review* **97**, 739.

MANABE, S. and STRICKLER, R. F. 1964: Thermal equilibrium of the atmosphere with a convective adjustment. *Journal Atmospheric Science* **21**, 361.

MANABE, S., SMAGORINSKY, J., and STRICKLER, R. F. 1965: Simulated climatology of a general circulation model with a hydrologic cycle. *Monthly Weather Review*, **93**, 769.

MANABE, S. and WETHERALD, R. T. 1967: Thermal equilibrium of the atmosphere with a given distribution of relative humidity. *Journal Atmospheric Science* **24**, 241.

MARGULES, M. 1903: *Über die Energie der Stürme.* Sonderabdruck aus den Jahrbüchern der Zentralanstalt für Meteorologie, Vienna. English translation in C. Abbe (editor) *The mechanics of the earth's atmosphere.* Smithsonian Institute, Miscellaneous Collection **51**, No. 4, 1910.

MINTZ, Y. 1951: The geostrophic poleward flux of angular momentum in the month of January 1949. *Tellus* **3**, 195.

NEWELL, R. E., VINCENT, D. G., DOPPLICK, T. G., FERRUZZA, D. and KIDSON, J. W. 1969: The energy balance of the global atmosphere. In Corby, G. A. (editor), *The global circulation of the atmosphere.* London: Royal Meteorological Society, p. 42.

PALMEN, E. 1951: The role of atmospheric disturbances in the general circulation. *Quarterly Journal Royal Meteorological Society* **77**, 337.

PALMEN, E. 1954: Über die atmosphärischen Strahlströme. Berlin: Institut für Meteorologie und Geophysik, Freie Universität, *Meteorologische Abhandlungen* **2**, 35.

PALMER, C. E. 1951: Tropical meteorology. In Malone, T. F. (editor), *Compendium of meteorology.* Boston: American Meteorological Society, p. 859.

RIEHL, H. 1965: *Introduction to the atmosphere.* New York: McGraw-Hill.

RIEHL, H. 1969: On the role of the tropics in the general circulation of the atmosphere. *Weather* **24**, 288.

RIEHL, H. and FULTZ, D. 1957: Jet streams and long waves in a steady rotating dishpan experiment: structure and circulation. *Quarterly Journal Royal Meteorological Society* **83**, 215.

RIEHL, H. and FULTZ, D. 1958: The general circulation in a steady rotating dishpan experiment. *Quarterly Journal Royal Meteorological Society* **84**, 389.

RIEHL, H. and MALKUS, J. S. 1958: On the heat balance in the equatorial trough zone. *Geophysica* **6**, 503.

ROBINSON, N. 1966: *Solar radiation.* Amsterdam: Elsevier Publishing Co.

SELLERS, W. D. 1965: *Physical climatology.* Chicago: University of Chicago Press.

SMAGORINSKY, J. 1963: General circulation experiments with the primitive equations. **1**: The basic experiment. *Monthly Weather Review* **91**, 99.

SMAGORINSKY, J., MANABE, S. and HOLLOWAY, J. L. 1965: Numerical results from a nine-level general circulation model of the atmosphere. *Monthly Weather Review* **93**, 727.

STARR, V. P., PEIXOTO, J. P. and LIVADAS, G. C. 1957: On the meridional flux of water vapour in the northern hemisphere. *Geofisica Pura e Applicata* **39**, 174.

STARR, V. P. and PEIXOTO, J. P. 1958: On the global balance of water vapor and the hydrology of deserts. *Tellus* **10**, 188.

STARR, V. P., PEIXOTO, J. P. and CRISI, A. R. 1965: Hemispheric water balance for the IGY. *Tellus* **17**, 74.

THORNTHWAITE, C. W. 1948: An approach toward a rational classification of climate. *Geographical Review* **38**, 55.

VAN DE BOOGAARD, H. M. E. 1964: A preliminary investigation of the daily meridional transfer of atmospheric water vapour between the equator and 40°N. *Tellus* **16**, 43.

VETTIN, F., 1885: Experimentelle Darstellung von Luftbewegungen unter dem Einfluss von Temperaturunterschieden und Rotationsimpulsen. *Meteorologische Zeitschrift.* **2**, 172.

VONDER HAAR, T. H. and SUOMI, V. E. 1971: Measurements of the earth's radiation budget from satellites during a five-year period. Part **1**: Extended time and space means. *Journal Atmospheric Science* **28**, 305.

WILSON, E. M. 1969: *Engineering hydrology.* London: Macmillan.

# Part II
# The Low Latitude Atmosphere

Plate 5 *Above left:* An approaching dust cloud in the Sudan. *Reproduced by permission of the Meteorological Office, Bracknell. Above right:* Kilimanjaro seen from the air. Small cumulus clouds can be seen over the plateau of East Africa with light cirrus clouds above. *Reproduced by permission of* John Crofton. *Centre:* The highlands of Tibet act as a high-level heat source, which is a contributory cause of the Indian southwest monsoon. This photograph from Apollo 7 shows the Himalayan Mountains on the southern edge of the highlands. *Reproduced by permission of the United States National Aeronautics and Space Administration, Manned Spacecraft Center. Below left:* Cumulus convection is an important low-level mixing mechanism in the tropics. Small cumulus cloud photographed in West Malaysia. *Below right:* Waterspout seen from RAF Changi, Singapore, 18 December 1965. *Reproduced by permission of J G Gallagher and the Meteorological Office, Bracknell.*

# 3
# The tropical atmosphere

Either the Tropics of Cancer and Capricorn or the 30th parallels, which divide the earth's surface between pole and equator into equal halves, can be taken as the boundaries of the tropical world, though people living in the marginal zones might well consider both definitions somewhat arbitrary because many marginal areas experience tropical weather in summer but near-freezing conditions in winter. In practice it is impossible to draw boundaries which will exactly define the tropical world. In the context of this book it is probably best to define the meteorological tropics as that part of the world where most of the time the weather sequences and climate differ distinctly from those in middle and high latitudes. The subtropical jet streams or the axes of the subtropical anticyclones act as a rough guide to this particular boundary.

The fact that the tropics differ meteorologically from the middle latitudes in a number of important aspects was made clear in the discussion of the general circulation in the previous chapter. Laboratory experiments indicate the importance of the Coriolis parameter in determining the broad nature of regional circulation patterns. Thus, in the atmosphere, the tropics with low values of the Coriolis parameter provide the nearest large-scale approach to a direct circulation, while the middle latitudes with a high value of the Coriolis parameter have an indirect circulation and an unsteady regime dominated by migrating surface highs and lows. In middle latitudes there are marked north–south temperature gradients, and air currents with differing origins will therefore have contrasting temperatures and humidities, but in the tropics temperatures (figure 3.1) tend to be uniform in the horizontal over vast areas, and so contrasting air currents are rare.

A baroclinic atmosphere is one in which surfaces of constant pressure and density intersect, at some level or levels, while the contrasting barotropic atmosphere is one in which the surfaces of constant pressure and density coincide at all levels. A strongly baroclinic atmosphere is one in which there are large horizontal temperature gradients and therefore strong thermal winds. As the middle latitude atmosphere is normally in this state, thermal winds are discussed in detail in chapter 8. The strongly baroclinic nature of the middle latitude atmosphere explains the formation of depressions, jet streams, and fronts, with the latent heat released by the condensation of water vapour only exerting a modifying influence. The tropical atmosphere, with its very small temperature and pressure gradients may, at first sight, seem to be nearly barotropic; certainly the release of latent heat by condensing water vapour is far more important in the formation of disturbances than in middle latitudes.

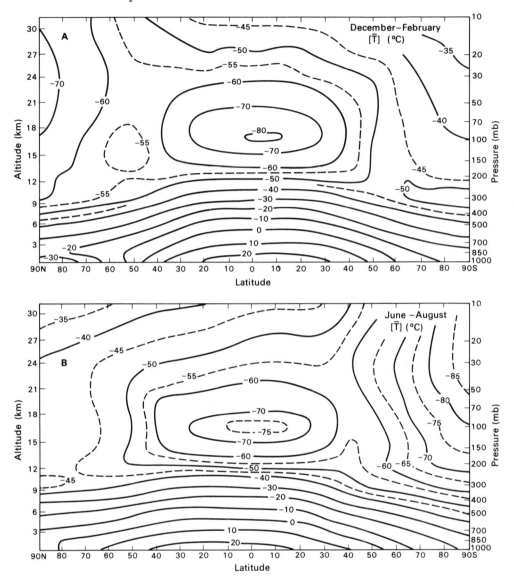

**Figure 3.1** Mean temperature in °C. A: December–February. B: June–August. (*After Newell* et al., *1969.*)

Because the Coriolis parameter is small in the tropics, it follows from the geostrophic wind equation that a small pressure gradient in the tropical atmosphere will give rise to a large geostrophic wind. Weak temperature gradients in the tropics therefore cause small pressure gradients which in turn generate air currents as intense as those found in middle latitudes. Weak temperature gradients may be caused by either local heating arising from the geographical distribution of radiation or by local condensation and the associated release of latent heat. In the tropics, where atmospheric moisture values are often high, condensation rates can be high resulting in the release of large amounts of latent heat, which can be of great importance in the development of circulation systems. In comparing temperature gradients in the tropical and middle latitude atmospheres it is therefore important to remember that the Coriolis parameter is very much smaller in one than in the other, and that this particular parameter has to be used as a scaling factor. The tropical atmosphere is, in many ways, as baroclinic as the middle latitude atmosphere. A final difference between the two atmospheres is that in the tropics the surface winds are normally easterly with varying thermal winds in the lower atmosphere, whereas in the middle latitudes they are generally westerly with a westerly thermal wind.

In figure 2.11, the average isobaric patterns for January and July have been illustrated, and it can be assumed that the mean winds reflect the mean pressure patterns. In the centres of the subtropical anticyclones and in the equatorial trough, winds are normally light and variable; indeed the calms of the subtropics are often known as the 'horse latitudes' and those of the equatorial trough as the 'doldrums'. Between the equatorial trough and the subtropical anticyclones there is a region of easterly winds with a small deflection towards the equator, and these are called the trade winds. Except in the Indian Ocean in summer, the trade winds are very constant in speed and direction reflecting the permanence of the subtropical anticyclones. A seasonal reversal of wind direction takes place over southern Asia and the northern Indian Ocean; these monsoonal regions are considered in detail in a separate chapter.

## Radiation and temperature

Details of the longitudinal variation of net radiation are contained in table 2.1, and the whole problem of the radiation balance of the earth was discussed in chapter 2. Within the zone 10°N–10°S, the net radiation is positive throughout the year, but elsewhere it becomes negative during the winter months. In the zone 25°N–25°S, the winter deficits are small, and negative radiation balances do not normally create problems within the tropics.

Charts of global solar radiation and net radiation for the Indian Ocean and neighbouring continental areas (reproduced in figure 3.2), are mostly based on the results of Mani, Chacko, Krishnamurthy, and Desikan (1965). Unfortunately, charts showing the detailed distribution of radiation over the whole of the tropical world are not available at the time of writing, but the diagrams shown here are considered to be reasonably typical.

Annual values of global solar radiation are given in figure 3.2, where it is seen that the highest values occur over the Tropics of Cancer and Capricorn, in the high-pressure belts of the northern and southern hemispheres. The maximum annual totals appear to be recorded in northeast Africa and Arabia. Because of its general cloudiness, the equatorial zone receives relatively less solar radiation than the subtropics, with minimums over southeast Asia and equatorial Africa. The clear skies of the subtropics not only allow large amounts of radiation to penetrate through the atmosphere to the earth's surface, but also lead to low values of diffuse radiation which can fall to 10–15 per cent of the global radiation in the subtropical summer.

Net radiation is also considered in figure 3.2. The annual distribution of net radiation is less complex than that of global radiation, and over the sea is mainly zonal except in regions affected by warm or cold sea currents. Maximum values are recorded at about 20°N and 20°S over the sea, and unlike the global radiation values over the sea, are always

83

higher than those over land. This is because of differences in albedo and terrestrial radiation; sea surface albedos are normally less than those of land surfaces. Despite the large global radiation values in the deserts the outgoing terrestrial radiation is high, so the net radiation over land remains low. During January the radiation balance is low over the northern hemisphere but increases towards the equator, southwards of which it rises very slowly over the oceans and reaches a maximum near the Tropic of Capricorn. In July the features in the northern and southern hemispheres are reversed, and the maximum values are recorded over the sea in the region of the Tropic of Cancer.

Seasonal changes in net radiation are at a minimum near the equator and amplify slowly towards the subtropics. Similar comments can be made about the global radiation and also, because of the obvious correlation, about temperature. Seasonal variations of temperature are normally extremely small near the equator and only increase slowly with latitude. The highest temperatures are not normally found at the equator itself, but rather in the subtropical deserts where there is a large influx of solar radiation. Details of temperature regimes are discussed in later chapters, but it is important to note here that the large horizontal temperature gradients of the temperate latitudes do not exist in the tropics. This

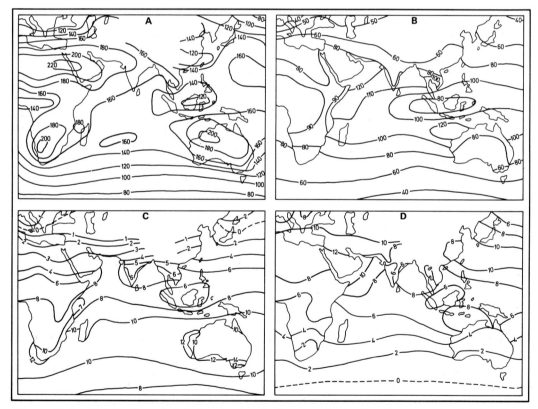

**Figure 3.2** Distribution of global and net radiation over the Indian Ocean and neighbouring land areas. A: annual distribution of global radiation (k cal cm$^{-2}$ year$^{-1}$). B: annual distribution of net radiation (k cal cm$^{-2}$ year$^{-1}$). C: amount of net radiation in January (k cal cm$^{-2}$ month$^{-1}$). D: amount of net radiation in July (k cal cm$^{-2}$ month$^{-1}$). (*After Mani, Chako, Krishnamurthy, and Desikan, 1965.*)

is partly a result of the radiation field but also partly due to the dynamic structure of the tropical atmosphere.

## Air-masses

Since the tropical atmosphere is essentially uniform, a large variety of air-mass types should not be expected; however it is possible to make the normal division on the basis of humidity between maritime and continental air-masses.

Many names have been used to describe air-masses over the tropical oceans—such as, maritime equatorial, maritime tropical, and trade-wind air—and this suggests a wide variety in tropical air-masses. In winter, because modified polar air sometimes reaches the tropical oceans, it might be expected that tropical weather systems would form on account of the clash of contrasting air-masses and that they would contain fronts similar to those found in middle latitudes. The facts are somewhat different for there are few similarities between tropical and high latitude weather systems.

Upper-air observations from tropical islands show changes of temperature and moisture content from day to day. However the amplitude of these changes increases upward, in marked contrast to middle latitudes where they decrease with increasing altitude. Near the surface over the tropical oceans, temperature variations seldom exceed $1°$ or $2°C$; they are entirely within the range of error of the measuring instruments and local modifications caused by cloudiness. Consequently, evidence exists of a nearly uniform air-mass covering vast areas of the tropical oceans, but disturbances containing cloud and rain do form over the tropical oceans. Since most tropical weather systems form within a nearly uniform air-mass, it appears that the air itself, moving with the winds, passes from one side of these disturbances to the other.

The tropical maritime air encountered over the great oceanic regions does not normally possess a deep moist layer, because usually there is an inversion about 2 to 3 km above the surface, below which the moist air is trapped. Above the inversion, the atmosphere is dry and cloudless. Because it is found typically in trade-wind areas, it is known as the trade-wind inversion.

Extreme heating taking place over the subtropical deserts can lead to a second type of tropical air-mass known as tropical continental air of which the main sources are North Africa and the Middle East in summer. Its outstanding characteristic is great heat often reaching $40°$ or $50°C$ at the surface, which usually gives rise to extremely low relative humidities; but the air is often stable and inversions can be found at about 2 km.

In winter the influence of middle latitude air-masses can extend to the subtropics. Since movement of polar air towards the equator leads to divergence, and accordingly a reduction in the depth of the air-stream, the air-stream entering the tropics is normally shallow, covers a wide area, and is easily warmed by the surface below. Thus cold air-masses are soon modified by warming over the tropical oceans and rarely penetrate far into the tropics in a completely unmodified form. For example in winter, air leaves Mongolia as a continental polar air-mass with a temperature between $-10°C$ and $-20°C$, but as it passes over the warm Pacific, it is transformed into a maritime polar air-mass; it finally reaches the Philippines as a maritime tropical air-mass with a surface temperature of about $25°C$.

To summarize, horizontal temperatures appear to be nearly uniform over large regions of the tropical atmosphere, and on that account contrasting air-masses are unusual. A basic characteristic of the tropical atmosphere is its general uniformity in the horizontal, with large areas being covered by similar air-masses. Changes in weather are usually caused by dynamic processes operating within a uniform air-mass rather than by changes in air-mass type.

## Subtropical anticyclones

The two belts of subtropical high pressure at about 30°N and s contain several quasi-permanent anticyclonic cells separated from each other by cols. In the northern hemisphere, the most notable ones are the two oceanic highs, in the Pacific and the Atlantic respectively, and the North African high, which fails to show up at sea level but emerges clearly at the 3 km level. In the southern hemisphere major anticyclones are found over the Pacific, Atlantic, and Indian Oceans.

Subtropical anticyclones show great permanence, for they tend to be located at fixed positions on the globe and undergo only slight seasonal variations, amounting on average to about 5° of latitude. The high-pressure belt is nearest to the equator in winter, but there is a slight asymmetry with regard to the geographical equator, for the southern ridge is situated about 5° latitude closer to it in the mean than the northern one. At the subtropical ridge lines, pressure is practically equal in both hemispheres, varying on average from 1015 mb in summer to 1020 mb in winter.

In chapter 2, the relationship between subsidence out of the subtropical jet streams and the location of the subtropical anticyclones was explored. The continual subsidence in the subtropical anticyclones makes it almost impossible for extensive clouds to form and for precipitation to fall; they are therefore almost rainless. The important arid zones of the world are found along latitudes 30°N and s, where large areas are dominated by anticyclones all the year round. Most of the large deserts of the world are situated in these latitudes; they include the Sahara in North Africa, Arabia and the Syrian desert in west Asia, Death Valley in North America, the Kalahari Desert in southern Africa and the Australian deserts. The great subtropical rainless zones spread over the oceans as well as the land-masses, indicating that the lack of rainfall is due to atmospheric subsidence and not the absence of surface moisture.

Subsidence takes place in a dynamic anticyclone, where the air does not sink right to the surface, but instead spreads out above a layer of relatively cool air in contact with the surface. The boundary between the cool surface air and the sinking warm air appears in upper-air soundings as a marked inversion, which becomes incorporated in the trade winds and is often carried almost to the equator. If an anticyclone exists over a hot desert, such as the Sahara, the lower layers may be distorted by the intense heat and a surface low will appear, but this heat low is a shallow surface feature and the clear subsiding anticyclonic air will be found above it. Subsidence is most pronounced at the eastern ends of the subtropical anticyclonic cells, and it is in these regions that the inversion is nearest the surface.

Climatological data (table 3.1) from Santa Cruz de Tenerife in the Canary Islands and Timimoun in North Africa give some impression of the weather conditions in the subtropical high-pressure belt. At Timimoun the summer temperatures are extremely high, the relative humidity is nearly always low, and precipitation is practically non-existent. Conditions are somewhat less extreme at Santa Cruz because summer temperatures are lower and precipitation generally higher than at Timimoun. The differences between the

**Table 3.1** Climatological data for the subtropical highs

| | Temperature (°C) Average daily | | Relative Humidity Average of observations at 0600 | Precipitation (mm) Average monthly fall |
|---|---|---|---|---|
| | Max | Min | | |
| | *Santa Cruz de Tenerife (maritime)* | | | |
| Dec. | 21 | 16 | 68 | 60 |
| Mar. | 22 | 15 | 65 | 28 |
| June | 26 | 19 | 63 | Trace |
| Sept. | 28 | 21 | 65 | 3 |
| | *Timimoun, Algeria (continental)* | | | |
| Dec. | 20 | 6 | 69 | Trace |
| Mar. | 27 | 11 | 51 | 3 |
| June | 42 | 24 | 28 | Trace |
| Sept. | 39 | 24 | 36 | Trace |

two stations reflect the differences between a tropical-continental and a tropical maritime situation. Oceanic anticyclones act as the major sources of tropical maritime air while the continental ones are the major sources of tropical continental air.

## The trade winds

The trade winds occur on the equatorial sides of the subtropical anticyclones, and are found in the greater part of the tropics (figure 3.3). In general the trade winds blow from ENE in the northern hemisphere and from ESE in the southern, and are noted for extreme constancy

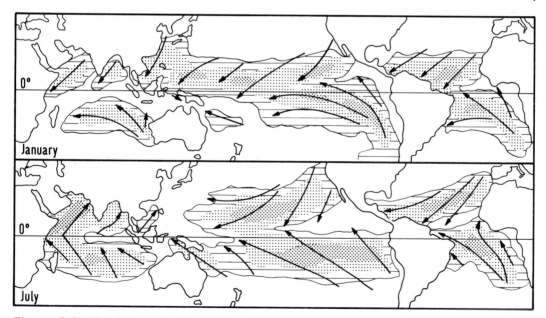

**Figure 3.3** The trade-wind systems of the world in January and July. The isopleths are in terms of relative constancy of wind direction and enclose shaded areas where 50, 70, and 90 per cent of all winds blow from the predominant quadrant with Beaufort force 3 or more (over 3·4 m sec$^{-1}$). (*After Crowe, 1971.*)

in both speed and direction. Indeed, in no other climatic regime on earth do the winds blow so steadily, for this steadiness reflects the permanence of the subtropical anticyclones; normally interruptions in the flow occur only with the formation of a major atmospheric disturbance. As winter is the season when the subtropical highs tend to be most intense, the trade winds are strongest in winter and weakest in summer.

During the summer of 1856 an expedition under the direction of C. Piazzi-Smyth visited the island of Tenerife in the Canary Islands to make astronomical observations from the top of the Peak of Tenerife. On two of the journeys up and down the 3,000 m mountain, Piazzi-Smyth carefully measured the temperature, moisture content, wind direction, and speed of the local trade wind. One set of his observations is illustrated in figure 3.4, which clearly shows an inversion with very dry air above. Piazzi-Smyth found that an inversion was often present and that it was not located at the top of the northeast trade regime, but was situated in the middle of the current; thus it could not be explained as a boundary between two air-streams from different directions. He also noticed that the top of the cloud layer corresponded to the base of the inversion. These observations by Piazzi-Smyth have been confirmed many times and the trade-wind inversion is now known to be of great importance in the meteorology of the tropics.

Broad-scale subsidence in the subtropical anticylones is the main cause of the very dry

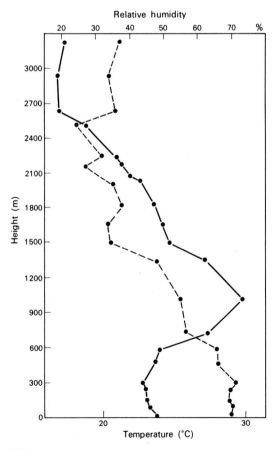

**Figure 3.4** Variation of temperature (solid line, °C) and relative humidity (dashed line, per cent) as observed during descent from Peak of Teneriffe (28°N, 16°W) by Piazzi-Smyth. (*After Riehl, 1954.*)

air above the trade-wind inversion. The subsiding air normally meets a surface stream of relatively cool maritime air flowing towards the equator. The inversion forms at the meeting point of these two air-streams, both of which flow in the same direction, and the height of the inversion base is a measure of the depth to which the upper current has been able to penetrate downward. The main source regions of trade winds and the trade-wind inversion are the eastern ends of the subtropical anticyclonic cells, in particular along the western edges of the Americas and Africa.

Air-streams flowing towards the equator tend to subside and diverge, and this can be considered in terms of the concept of potential vorticity:

$$\text{POTENTIAL VORTICITY} = \frac{\zeta + f}{\Delta p} = \text{CONSTANT}$$

where $\Delta p$ is the pressure difference between two adjacent isentropic surfaces differing in potential temperature by the constant value $\Delta \theta$.

It follows from the above equation that since f decreases towards the equator, the depth $\Delta p$ will also decrease in an air-stream flowing towards the equator, assuming that the relative vorticity $\zeta$ remains constant. From these considerations it might be expected that the trade-wind inversion should decrease in height downwind towards the equator, but over the vast ocean areas trade-cumulus convection 'diffuses' energy and water vapour gained from the oceans to higher levels, and causes the trade-wind inversion to rise, even though the mean vertical motion along trajectories is downward. The trade-wind inversion is not therefore a material surface, and subsiding air above the inversion is slowly incorporated through the inversion into the moist layer below. The increase in height towards the equator of the trade-wind inversion is well illustrated in figure 3.5.

The low level of the trade-wind inversion can have unfortunate effects on cities near the trade-wind source regions. For example in summer, a low-level inversion spreads over Los Angeles and very effectively stops any vertical exchange of air from taking place. The gaseous waste combustion products from industry and domestic households, plus the exhaust fumes from two million motor cars rise into the atmosphere below the inversion and remain trapped within it. Since the surface layer is already naturally hazy, the result is the formulation under strong tropical sunlight of the notorious Los Angeles 'smog' which is derived from a combination of the words 'smoke' and 'fog'. Other tropical cities can suffer from similar smogs due to the trade-wind inversion.

High mountains on tropical islands penetrate through the trade-wind inversion into the dry air above, and as a result desert or semi-desert conditions are found on the mountain tops, with a zone of high rainfall on the lower mountain slopes below the inversion. Hawaii is a good example; in the centre of the island at 3,000 m the annual rainfall is about 250 mm, whereas below the inversion at 1,000 m the annual rainfall is 7,500 mm or above.

## The equatorial trough

The northern and southern trade-winds meet in the equatorial trough, which is a zone of relatively low pressure situated near the equator between the subtropical high-pressure ridges. The trade winds normally become weak near the equatorial trough and they can be replaced by calms or by equatorial westerlies. Indeed, semi-permanent equatorial

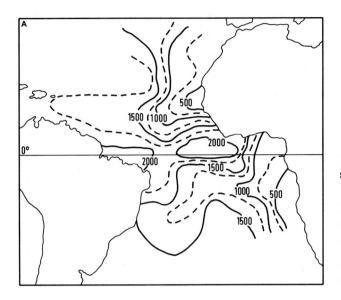

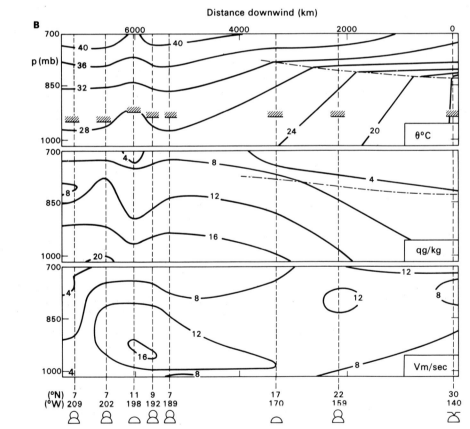

**Figure 3.5** The trade-wind inversion. A: height in metres of the base of the trade-wind inversion over Atlantic. (*After Riehl, 1954*). B: vertical cross-sections of potential temperature, mixing ratio and wind speed from 30°N, 140°W to 70°N, 151°E along approximate air trajectory during steady state situation in May 1958. (*After Sheppard, 1969*).

westerlies are found on the equatorial side of the trough over the Indian Ocean, Indonesia, the eastern Pacific, and Central America. Except over the Indian Ocean, the equatorial trough only changes latitude by small amounts during the year.

Carried along by the flow of the trade winds, the trade-wind inversion enters the equatorial trough, but the height of the inversion increases towards the equator and it often vanishes within the trough. The result is that while convection is restricted within the trade winds, towering cumulus clouds develop in the equatorial trough and give rise to heavy showers, thus often creating a relatively wet cloudy zone as compared with the trade winds.

It should not be thought, however, that the equatorial trough is a region of continuous cloud and rain. Very large areas are dominated by the trade-wind inversion and as a result experience fine dry weather and even a near desert climate. The extremely wet areas in the equatorial trough are restricted to a few favoured localities. Rain storms tend to appear in a rather irregular manner and are often nearly stationary or drift slowly west.

At one time it was thought that a front was formed in the equatorial trough at the boundary between the two converging trade-wind air-streams, but it is now known that fronts in the temperate sense are dynamically impossible within about 5° of the equator. This feature, which was associated with areas of cloud and rain, was known as the 'intertropical front', and early tropical meteorologists had great difficulty in locating it. Properties ascribed by early writers (Palmer, 1951) to the intertropical front include: single, double, or multiple structures; certain fronts bifurcating or having 'tails'; some fronts being virtually continuous and others well broken or even vanishing completely for thousands of kilometres; some moving laterally in a continuous fashion, and others apparently being transferred from one latitude to another without affecting intermediate latitudes. It would appear, therefore, that the concept of the intertropical front is not particularly useful. As convergence zones with cloud and rain do occur within the equatorial trough and are observed by satellites to have the properties described by Palmer, it is necessary to consider the concept of intertropical convergence zones. These zones only occupy a small part of the equatorial trough but can be observed somewhere every day; they are usually associated with large-scale inflow from the tropics or other parts of the equatorial trough. Convergence zones are not necessarily linear in shape but can also be oval, circular, or even irregular. They are revealed by satellite observations as narrow bands of clouds which lie generally between 5° and 15° from the equator and occasionally on both sides of it, the equator itself being cloudless over large parts of the oceans.

Inspection of individual satellite photographs shows that intertropical convergence zones can only rarely be identified as long, unbroken bands of heavy cloudiness. Rather, they are usually made of a number of 'cloud clusters', separated by large expanses of relatively clear skies (Holton, Wallace, and Young, 1971). Chang (1970) and Reed (1970) have presented strong evidence that cloud clusters are, in turn, manifestations of synoptic-scale westward propagating wave disturbances, which are marked by heavy precipitation ($> 2 \text{ cm day}^{-1}$). Fujita *et al.* (1969) and Williams (1970) have shown that cloud clusters represent concentrations of cyclonic vorticity at low levels and anticyclonic vorticity at 200 mb, with strong ascent at middle levels. The clear areas between clusters are marked by weak subsidence together with vertical distributions of vorticity and divergence opposite in sign to those within the clusters. Thus, in a climatological sense, the intertropical convergence zones may be considered to be the loci of cloud clusters associated with westward propagating wave disturbances (Holton, Wallace, and Young).

*The low latitude atmosphere*

Charney (1967) has developed a theory to explain the formation of convergence zones and their relationship to the equator. He believes that organized cumulus convection is controlled through frictionally induced convergence of moisture in the planetary boundary layer, and that the vertical pumping of mass, and therefore of moisture, out of this layer is proportional to the vorticity of the surface geostrophic wind. Since the air at low levels holds the bulk of the moisture and has the greatest potential buoyancy for moist adiabatic ascent in a conditionally unstable tropical atmosphere, cumulus convection and the release of latent heat is largely determined by this vertical pumping and its associated boundary-layer convergence. Thus a local increase of vorticity in a zonally symmetric flow will give rise to the following sequence of events:

1   increased vertical flux of moisture;
2   increased cumulus convection with release of latent heat;
3   a temperature rise;
4   an accelerative pressure–density solenoidal field;
5   increased low-level convergence;
6   a bringing of high angular momentum air from the equatorward side of the disturb-ance into juxtaposition with low angular momentum air from the poleward side;
7   a still greater increase of positive vorticity.

This, suggests Charney, is essentially the instability mechanism which would produce a convergence zone if one did not exist, or maintain one if it already existed.

He considers that the character of equatorial flow depends on a parameter $\eta$ which may be defined as the ratio of the heating of an air column by condensation of the moisture pumped from the frictional boundary layer to the cooling by adiabatic expansion in the rising air:

$$\eta = \frac{Lq}{C_p\, T\, \Delta \ln \theta}$$

when q is the specific humidity of the air at the top of the boundary layer; T is a mean mid-atmosphere temperature; $\Delta \ln \theta$ is a characteristic vertical increment of the logarithm of the mean potential temperature, $\theta$; and the other symbols are as defined previously. When $\eta$ is below a certain critical value, which varies with the other parameters but is roughly between 1 and 4, condensation has only a small effect; but when this threshold value is exceeded, the effect becomes very large and decisively influences the entire circu-lation. The change occurs when the heating due to condensation in the convergence zones overcomes the cooling due to adiabatic expansion. In this sense the instability is analogous to the conditional instability associated with small-scale cumulus convection, but here it pertains to large-scale flow patterns, and for this reason Charney and Eliassen (1964) have called it 'conditional instability of the second kind'.

Charney's theory suggests that convergence zones can exist anywhere between 15°N and 15°S. It is also likely that a small change in sea-surface temperature at a given latitude will influence the position and structure of convergence zones, which is interesting in the light of their unequal distribution within the equatorial world.

Recent work by Pike (1971) suggests that, at least over the sea, the maximum upward motion and precipitation rate associated with the intertropical convergence zone occurs at or very near the latitude of maximum surface temperature. Hubert *et al.* (1969), using a year of satellite data, have concluded that a single pronounced convergence zone is much more common than a pronounced latitudinal double maximum and that it is usually hemispherically very asymmetric when off the equator. This agrees with some results from an interacting atmosphere and ocean model (Pike, 1971) which suggests that a single convergence zone is comparatively stable, on or off the equator.

## Tropical rain-forming disturbances

Even in the equatorial trough, rainfall is not particularly frequent. Because in the trade winds proper the sky will contain only small cumulus clouds, it may be rainless for many days. As dry weather is normal within the tropics, it is necessary to consider under what special circumstances rainfall will occur, which entails a study of tropical synoptic disturbances. Sikdar and Suomi (1971) divide tropical circulations into three broad scales of motion:

1  planetary scale (the equatorial trough and trade-wind regime, subtropical highs and jet streams, monsoons, etc.);
2  synoptic- or large-wave scale (easterly waves, tropical cyclones, waves in the upper troposphere, etc.);
3  meso–convective scale (cumulus clouds which can form cloud clusters).

Much of the rain in the tropical world falls from convective clouds, that is from large cumulus or cumulonimbus. In the most severe storms these clouds are packed side by side to give continuous rain and cloud, but more usually there are wide gaps between the clouds and this leads to a rather random and discontinuous rainfall distribution. Some places will have heavy rainfall, while others nearby will have little or no rain, thus forming a complete contrast to the widespread cloud and rainfall of middle latitudes. Tropical rainfall on many occasions appears to be random, but is nevertheless often caused by some kind of organized disturbance, which may range from a tropical cyclone to a cloud cluster.

Cumulus are the most common tropical clouds and form the outstanding element of the cloud scenery. They are created by convection initiated by solar radiation reacting with the surface and therefore show marked diurnal variations. Their distribution is closely correlated with surface features, though some of the relationships are rather complex. Over most of the tropics the vertical development of cumulus clouds is restricted by the trade-wind inversion, and they do not normally evolve into cumulonimbus. Trade-wind cumulus can give rise to light showers but they do not cause heavy rainfall, which is associated with strong surface convergence and with the absence of the trade-wind inversion, thus allowing a deep unstable moist layer to form.

Related to surface convergences are surface low-pressure systems, ranging from central pressure anomalies of one millibar to those of hurricane strength. Due to mechanisms not yet fully understood, the low-level convergence in these systems becomes concentrated in narrow zones which cover 1–10 per cent of the area of the disturbance, and in which all ascent and precipitation take place in cumulonimbus cloud arrays. The part played by synoptic disturbances in the energy balance of the equatorial trough was discussed in chapter 2.

It has already been stated that tropical synoptic disturbances form within a uniform airmass, and that horizontal temperature gradients within the tropics are generally small.

*Plate 6* Near vertical view of thundercloud over South America from Apollo 9. *Reproduced by permission of the United States National Aeronautics and Space Administration, Manned Spacecraft Center.*

Therefore, differences in air-mass temperature and density cannot be an important source of energy for these disturbances. Condensation of water vapour releasing latent heat appears to be the main source of energy, and this particular energy mechanism was also discussed in chapter 2. To maintain the supply of water vapour it is necessary for a flow of moist air from over a wide area to enter the synoptic system. This inflow can be organized in a variety of forms but once it ceases, the storm decays very rapidly. This suggests that major tropical disturbances such as tropical cyclones will be restricted to the oceans.

Near the equator, in the equatorial trough, intense rainfall is often associated with minor convergence zones in which pressure differences are very small, making them extremely difficult to study. In the subtropics where systems are more clearly marked and consequently more easily observed, there are two particularly well-known pressure systems—the easterly wave and the tropical storm or cyclone.

## 1  Easterly waves

Two types of wave disturbance in the tropical easterlies have been the subject of analysis—the easterly wave of the Caribbean and the equatorial wave of the Pacific. Historically, the Caribbean wave was examined first, and the model was then applied elsewhere. Observations indicate that waves form in the easterly flow over the Caribbean, and that they move west at a speed of 5 to 7 m/sec. The structure of such an easterly wave is shown in figure 3.6. Waves are typically about 15° latitude across, and the easterly winds flow through them from east to west. The wave structure in the easterlies is found to be weakest at sea level, to increase in intensity up to about 4,000 m, and then above this level to again become weaker. About 300 km ahead of the wave trough, the trade-wind inversion reaches its lowest altitude and exceptionally fine weather prevails, indicating intense subsidence. A rapid rise in the altitude of the trade-wind inversion, and therefore an increase in the depth of the surface moist layer, takes place near the trough line; the moist layer attains a maximum depth of well above 6 km in the zone of ascent behind the trough. Here also are found large squall lines, intense rain, and rows of cumulonimbus clouds, but this region is not completely covered by cumulonimbus because being often organized into lines or rows, there are wide zones of subsidence between individual clouds.

The distribution of weather within a wave in the tropical easterlies has been explained by Riehl (1954) in terms of the concept of potential vorticity mentioned earlier. To the east of the wave surface, air is streaming polewards and therefore entering regions with increasing values of the parameter f. As the air enters the trough, the local vorticity ($\zeta$) will become increasingly positive. The potential vorticity equation states that under these conditions the depth of the air-stream must increase to compensate for the change in vorticity. Similar reasoning suggests that ahead of the wave where the surface air is steaming equatorward and away from the trough axis, the vorticity will be decreasing and on that account the vertical depth of the air-stream will decrease. Thus dynamic considerations imply that ascent and convergence should take place behind the trough line and descent and divergence ahead.

Easterly waves can also be explained in terms of the vorticity equation, which while neglecting the less important terms, can be written for pressure co-ordinates as

$$\frac{d}{dt}(\zeta + f) = -(\zeta + f)\,\text{div}_p\,V$$

In the lower layers of the atmosphere air flows westward through the wave, and the above equation states that there will be low-level convergence behind the wave and low-level divergence ahead. At higher levels in the troposphere the winds have less strong easterly components and the wave moves westward relative to the air-stream. Under these conditions the vorticity equation states that there will be high-level convergence ahead of the wave and high-level divergence behind, and in this way the observed vertical circulations with their associated weather patterns are explained.

Easterly waves are well known in the Caribbean and similar systems have been described

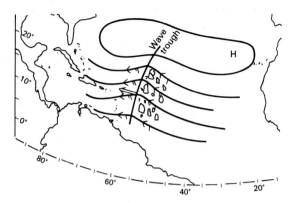

A

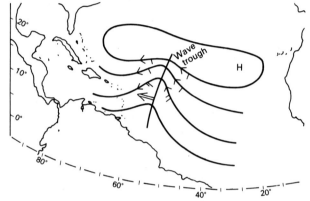

B

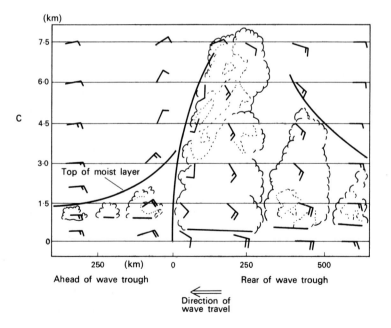

C

**Figure 3.6** *See opposite.*

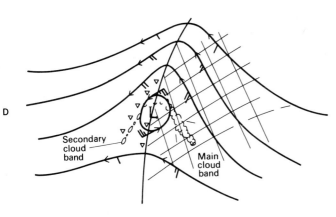

**Figure 3.6** Schematic illustrations of major features of 'easterly wave' type of tropical disturbance. A: Surface streamlines of weak-to-moderate amplitude wave, which typically moves westward in direction of heavy arrow at a speed slightly slower than prevailing trade-wind. The barbs on the wind arrows denote speed, each small one representing 2·5 m sec⁻¹. The area of cloudiness and rain is commonly found to the rear of the trough. B: 500 mb streamline pattern in typical moderate easterly wave. The wave amplitude is greater than at the surface. C: Vertical cross-section from west (*left*) to east (*right*). Cloud forms are shown schematically and not to scale. *Horizontal* winds are denoted by barbed lines, each short barb representing 2·5 m sec⁻¹. Winds blow with barbed lines, east being to the right and west to the left. D: Schematic picture of a very deep easterly wave, showing streamlines and major cloud lands. There is a closed central vortex where the surface pressure may be as low as 1000 mb. The rain area is indicated by hatching and will contain many subsidiary cloud lands. (*After Malkus, 1962.*)

in the west central Pacific. The surges of the trade wind in the northern China Sea are similar to easterly waves, but there is no record of a persistent travelling wave in the China Sea. Though the origins of waves are often obscure, it appears that many over the Caribbean have their sources over Africa.

*2   Tropical storms and cyclones*

Storms with closed circulation systems which can occur anywhere within the tropics outside of the zone 5°N to 5°S, vary from slowly circulating masses of air with scattered cumulonimbus clouds to violent and severe storms. Differences between the various tropical low pressure systems are not easily defined; the World Meteorological Organization has classified low-pressure systems as follows:

1   A tropical depression is a system with low pressure enclosed within a few isobars, and it either lacks a marked circulation or has winds below 17 m/sec.
2   A tropical storm is a system with several closed isobars and a wind circulation from 17 to 32 m/sec.
3   A tropical cyclone is a storm of tropical origin with a small diameter (some hundreds of kilometres), minimum surface pressure less than 900 mb, very violent winds, and torrential rain sometimes accompanied by thunderstorms. It usually contains a central region, known as the 'eye' of the storm, with a diameter of the order of some tens of kilometres, where there are light winds and more or less lightly clouded sky.

Tropical cyclones are given a variety of regional names. In the southwest Pacific and Bay of Bengal the name 'tropical cyclone' is in use; they are known as 'typhoons' in the China Sea, as 'willy-willies' over western Australia, and as 'hurricanes' in the West Indies and in the south Indian Ocean.

Plentiful moisture is a requirement of tropical cyclones, and for this reason their incidence is limited to regions where the highest sea-surface temperatures are found, that is, to the western regions of the tropical oceans in the late summer. Not all incipient tropical low-pressure systems become tropical cyclones; many remain as weak closed circulations with moderate rain and light winds. Indeed, though weak tropical disturbances are relatively common, tropical cyclones are rare, for even in an active year only very seldom will their total number in the northern hemisphere exceed 50, whereas about 20 depressions occur almost daily outside the tropics in winter.

As the structure in figure 3.7 illustrates, ascent within tropical cyclones takes place in cumulonimbus clouds arranged in spirals converging on the central eye. The spirals, which are sometimes hundreds of kilometres long, are at most a few kilometres wide, and the distance between them is about 50 to 80 km near the edge, decreasing towards the central eye. Hence often only a small fraction of the tropical cyclone, not more than 10 per cent, contains the ascent which gives rise to the bulk of the condensation and rainfall.

Close to the centre, where the clouds form a ring around the eye, the strongest winds and heaviest rainfalls combine to produce the storm's full fury, whereas inside the eye, the winds decrease quickly and the heavy rains cease. Indeed modern research has shown that the eye is actually a region of subsiding warm air at the centre of the storm. This warm core seems to be essential to the formation of a tropical cyclone, and its appearance is one of the first signs that a storm is going to turn into a cyclone.

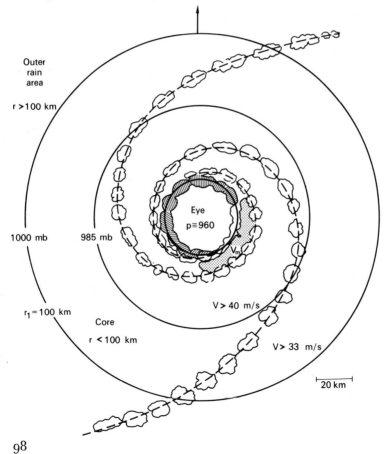

**Figure 3.7** Schematic diagram of core region of typical tropical cyclone of moderate strength. It consists of an 'inner rain area' and an 'eye', with major spiral bands of cumulonimbus clouds along trajectories and forming the eye wall. Stippled region, marked $V_m$, is region of maximum wind speed, and some typical surface pressures and wind speeds are indicated. Direction of storm movement shown by arrow at top. (*After Malkus, 1962.*)

*Plate 7* Hurricane Gladys as seen from Apollo 7 looking towards the southwest with the Island of Cuba in the background. Winds of 40 m/s were reported from the hurricane at this time. The spiral structure of the storm is clearly visible. *Reproduced by permission of the United States National Aeronautics and Space Administration, Manned Spacecraft Center.*

Despite the strong winds, the rate of movement of a tropical cyclone is only about 5 to 8 m/sec. In the northern hemisphere the storm will drift towards the northwest at first, but when the centre passes latitude 20°N, the speed of travel increases, the direction changes to a more northeasterly path, and the system starts to fill. The point at which the direction of motion changes from westerly to easterly is known as the 'point of recurvature'. The

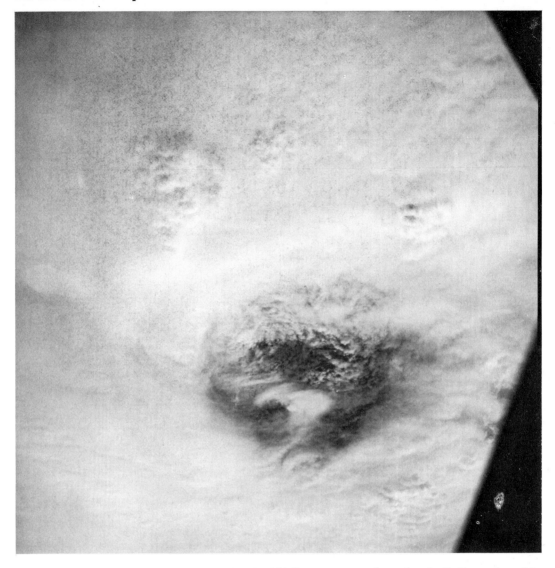

*Plate 8*   The eye of Typhoon Gloria in the Pacific Ocean as seen from Apollo 7. *Reproduced by permission of the United States National Aeronautics and Space Administration, Manned Spacecraft Center.*

recurvature of cyclone tracks in the southern hemisphere is in the opposite sense, storms moving first towards the southwest and later towards the southeast, but they never cross the equator.

As rainfall rates near the centre of a tropical cyclone can exceed 500 mm per day, a continuous inflow of water vapour into the storm is necessary. In the course of its life, a vast quantity of air is drawn into the circulation and funnelled upward. Given a storm with average inflow and a duration of one week, then the air below the trade-wind inversion

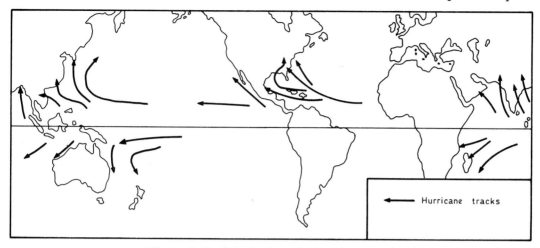

**Figure 3.8** Principal tropical cyclone tracks. (*After Palmen, 1948.*)

contained in a square whose sides are 12–13° latitude long will be raised to 200 mb or higher in the storm. Thus it is clear that the storm will be consuming water vapour previously evaporated over vast areas of the tropical oceans.

Air spiralling toward the storm centre should decrease in temperature if adiabatic expansion occurs during the pressure reduction, which implies that a dense surface fog should be found in the inner regions. However, observations indicate that surface air temperatures within tropical cyclones remain constant or decrease only slightly, and fog is never reported. This suggests that there must be a source of heat and moisture within the storm itself. Greatly agitated by the high winds within the storm, large amounts of water are thrown into the air from the ocean surface in the form of spray. As the air moves towards lower pressure and begins to expand adiabatically, the temperature difference between ocean and air increases; but since the surface of contact is increased to many times the horizontal area of the storm, rapid transfer of sensible and latent heat from ocean to air is made possible. On the outskirts of the storm, the turmoil of the ocean surface is less and this particular process of heat transfer is not operative. It is now obvious that a fair proportion of the water vapour being used by the storm will have originated from the ocean surface within the storm itself.

This discussion should make it apparent that tropical cyclones are strictly maritime systems (figure 3.8); cyclone tracks rarely extend deep into the land-masses. If a tropical cyclone drifts across a large land-mass, the cyclone decays very quickly, normally within one or two days.

## Cloudiness, rainfall, flow patterns, and climatic zones

Figure 3.9 (derived from Sadler, 1968), illustrates monthly mean cloudiness based on two years of satellite data. The cloud patterns may reflect purely surface features or the relative frequencies and intensities of showers and storms taking place throughout the tropics. Sheets of low cloud overlie the cold ocean currents off the west coasts of the African and American continents, but over Africa it has been demonstrated by Sadler that the cloud

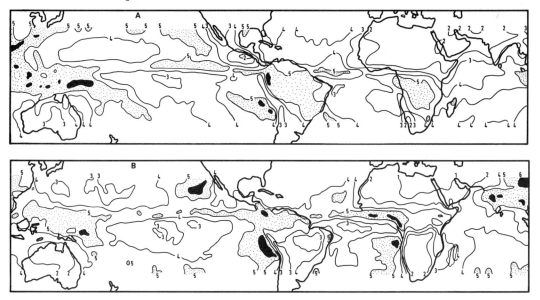

**Figure 3.9**   Analyses of average monthly cloudiness. A : January 2–year average. B : July 2–year average. Units : oktas. (*After Sadler, 1968.*)

patterns correlate well with the average seasonal rainfall. If the limited areas with relatively cold ocean currents are ignored, it can be assumed that there is a general relationship between average cloud amount and average atmospheric flow patterns.

Almost clear skies prevail throughout the year (as should be expected) over the deserts of northern Africa and the Middle East, southwestern Africa, central and western Australia, west central South America, and western Mexico, which are all areas dominated by subtropical anticyclones. The low cloudiness associated with the subtropical anticyclones is not restricted completely to the subtropics, because throughout the year in the Atlantic and eastern Pacific it tends to extend northwards from the southern highs towards the equator, and in some months the minimum is found at the equator itself.

Cloudiness maximums persist throughout the year to the north of the equator over the central and eastern Pacific, and over the Atlantic. They are embedded, according to Sadler in the mean surface easterlies, except over the eastern Pacific in the northern summer when they lie in westerly flow to the south of surface troughs of low pressure.

Large-scale variations in cloudiness occur with the changing seasons and especially in the monsoon belt stretching from West Africa eastward to the western Pacific. In January, cloud is restricted to the area near the equator while southern Asia has low cloud amounts similar to those found over the Sahara Desert; but in July a belt of dense cloud extends from western India to China, with a continuation eastward past the Philippines. In July, a virtually continuous zone of disturbed weather exists from the monsoon westerlies of India and southeast Asia across the South China Sea to the Philippines, where the monsoon westerlies and Pacific trades converge into the equatorial west Pacific. In January, greatest weather activity and cloud exist over equatorial Africa, equatorial South America, and over Indonesia. Indonesia is of interest because in January it lies within confluent low-level westerlies which are fed partly from the northeast trades of the northern hemisphere and partly by the winds of the southern Indian Ocean. By July this cloudy zone over Indonesia has moved northwards to join with the monsoon clouds over south China and the Philippines, leaving the area relatively cloud-free.

From the discussion in this chapter, it will have become clear that the tropical world can be divided into a number of climatic zones. One very obvious broad climatic type is formed under the subtropical anticyclones with their clear skies, which give rise to near desert conditions. As the western parts of the oceans experience a rather different climate from the high pressure-dominated eastern halves, they warrant separate treatment. The monsoon climate of southern Asia is really an extreme example of this last type, but because of its complexity it is discussed in a separate chapter. Equatorial trough climates are also examined separately mainly because of the unique properties of the atmosphere within a few degrees of the equator. This very broad classification is not meant to cover all of the tropics, but it does describe the main environments found within the tropical world.

# References

CHANG, C. P. 1970: Westward propagating cloud patterns in the tropical Pacific as seen from time-composite satellite photographs. *Journal Atmospheric Science* **27**, 133.

CHARNEY, J. G. 1967: The intertropical convergence zone and the Hadley circulation of the atmosphere. In *Proceedings of WMO/IUGG Symposium on numerical weather prediction in Tokyo. Japan Meteorological Agency, Technical Report* **67**.

CHARNEY, J. G. and ELIASSEN, A. 1964: On the growth of the hurricane depression. *Journal Atmospheric Science* **21**, 68.

CROWE, P. R. 1971: *Concepts in climatology.* London: Longman.

FRANK, N. L. 1969: The inverted V cloud pattern—an easterly wave? *Monthly Weather Review* **97**, 130.

FUJITA, T. T., WATANABE, K. and IZAWA, T. 1969: Formation and structure of equatorial anticyclones caused by large-scale cross-equatorial flows determined by ATS-1 photographs. *Journal Applied Meteorology* **8**, 649.

HOLTON, J. R., WALLACE, J. M. and YOUNG, J. A. 1971: On boundary layer dynamics and the ITCZ. *Journal Atmospheric Science* **28**, 275.

HUBERT, L. F., KRUEGER, A. F., and WINSTON, J. S. 1969: The double intertropical convergence zone—fact or fiction? *Journal of Atmospheric Science* **26**, 771.

JOHNSON, D. H. 1969: The role of the tropics in the global circulation. In Corby, G. A. (editor), *The global circulation of the atmosphere.* London: Royal Meteorological Society, p. 113.

MALKUS, J. C. 1962: Large-scale interactions. In Hill, M. N. (editor), *The sea.* Vol. **1**. New York: John Wiley.

MANI, A. CHACKO, O., KRISHNAMURTHY, V. and DESIKAN, V. 1965: Radiation balance of the Indian Ocean. *Proceedings Symposium Meteorological Results International Indian Ocean Expedition,* Bombay, July 1965, p. 165.

MANI, A., CHACKO, O., KRISHNAMURTHY, V., and DESIKAN, V. 1967: Distribution of global and net radiation over the Indian Ocean and its environment. *Archiv Meteorologie Geophysik und Bioklimatologie, Ser. B,* **15**, 82.

NEWELL, R. E., VINCENT, D. G., DOPPLICK, T. G., FERRUZZA, D. and KIDSON, J. W. 1969: The energy balance of the global atmosphere. In Corby, G. A. (editor), *The global circulation of the atmosphere.* London: Royal Meteorological Society, p. 42.

PALMEN, E. 1948: On the formation and structure of tropical hurricanes. *Geophysica* **3**, 26.

PALMER, C. E. 1951: Tropical meteorology. In Malone, T. F. (editor), *Compendium of meteorology*. Boston: American Meteorological Society, p. 859.

PIAZZI-SMYTH, C. 1858: An astronomical experiment on the Peak of Tenerife. *Transactions Royal Society*, London, **148**, 465.

PIKE, A. C. 1971: Intertropical convergence zone studied with an interacting atmosphere and ocean model. *Monthly Weather Review* **99**, 469.

REED, R. J. 1970: Structure and characteristics of easterly waves in the equatorial western Pacific during July–August, 1967. *Proceedings of Symposium Tropical Meteorology*, Honolulu. *American Meteorological Society* **EII**–**1** to **EII**–**8**.

RIEHL, H. 1954: *Tropical meteorology*. New York: McGraw-Hill.

RIEHL, H. 1969: On the role of the tropics in the general circulation of the atmosphere. *Weather* **24**, 288.

SADLER, J. C. 1968: Average cloudiness in the tropics from satellite observations. *International Indian Ocean Expedition Meteorological Monograph No. 2*. Honolulu: East–West Center Press.

SAWYER, J. S. 1970: Large-scale disturbance of the equatorial atmosphere. *Meteorological Magazine* **99**, 1.

SHEPPARD, P. A. 1969: The atmospheric boundary layer in relation to large-scale dynamics. In Corby, G. A. (editor), *The global circulation of the atmosphere*. London: Royal Meteorological Society, p. 91.

SIKDAR, D. N. and SUOMI, V. E. 1971: Time variation of tropical energetics as viewed from a geostationary altitude. *Journal Atmospheric Science* **28**, 170.

WILLIAMS, K. 1970: Characteristics of the wind, thermal and moisture fields surrounding the satellite observed mesoscale trade wind cloud clusters of the western North Pacific. *Proceedings of Symposium Tropical Meteorology*, Honolulu. *American Meteorological Society* **D** N–**1** to **D** N–**8**.

WORLD METEOROLOGICAL ORGANIZATION 1966: International meteorological vocabulary. *WMO No.* **182**, *Technical Paper* **91**, Geneva.

# 4
# Tropical arid and semi-arid climates

*1   Arid climates*

The important arid zones of the world are situated around latitudes 30°N and °s, where large areas are dominated by dynamic anticyclones, but it must not be assumed that arid areas cannot occur elsewhere, because aridity is normal and it is atmospheric precipitation which needs special explanation. In figure 2.16 which indicates the generalized pattern of precipitation over the earth, the arid zones in the subtropics stand out very clearly, and it is interesting that these zones extend over the subtropical oceans as well as over the subtropical land-masses. This indicates that a lack of surface water is not the prime cause of lack of rainfall, for indeed the actual cause of general aridity in the subtropics is the widespread subsidence of the atmosphere in the descending limbs of the Hadley cells. The subsidence effectively suppresses extensive low-level convergence, and it is therefore almost impossible for extensive cloud systems to form and rain to fall.

In figure 4.1 which explores the water balance over a small land area, rainfall (r) within the unit area is partly formed of moisture advected ($r_A$) into the area by the winds and partly of moisture which has evaporated ($r_E$) within the area:

$$r = r_A + r_E$$

The local evapotranspiration ($E_t$) partly supplies moisture for local rainfall and is partly advected (C) out of the unit area by the winds:

$$E_t = r_E + C$$

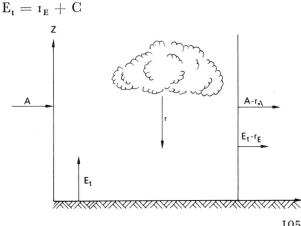

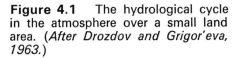

**Figure 4.1**   The hydrological cycle in the atmosphere over a small land area. (*After Drozdov and Grigor'eva, 1963.*)

Finally, the rainfall must either flow out of the unit area as rivers or ocean currents, or be used in local evapotranspiration.

A study of atmospheric moisture content (figure 2.14) shows that the air over the great deserts is normally very dry. This is further illustrated by table 4.1 which lists mean relative

**Table 4.1**   Mean relative humidity at various isobaric levels for radiosonde stations in the Sahara and Arabia

| Station | Isobaric Level (mb) | Month January | July |
|---|---|---|---|
| Fort Trinquet (25°14′N, 11°35′W; 360 m) | 850 | 33 | 16 |
| | 700 | 22 | 22 |
| | 500 | 15 | 31 |
| Kolon-Beshar (31°37′N, 2°13′W; 780 m) | 850 | 44 | 17 |
| | 700 | 33 | 22 |
| | 500 | 22 | 28 |
| Aulef-el′-Arab (27°04′N, 1°06′E; 275 m) | 850 | 28 | 14 |
| | 700 | 22 | 17 |
| | 500 | 15 | 21 |
| Cairo (30°08′N, 31°24′E, 68 m) | 1,000 | 47 | 38 |
| | 850 | 49 | 36 |
| | 700 | 40 | 24 |
| | 500 | 37 | 15 |
| Habbaniya (33°22′N, 43°34′E, 45 m) | 1,000 | 62 | — |
| | 850 | 53 | 15 |
| | 700 | 33 | 18 |
| | 500 | 26 | 7 |

*After Drozdov and Grigor'eva (1963).*

humidity at various isobaric surfaces for stations in the Sahara and Arabia. This dryness of the air is partly a result of the lack of local evapotranspiration and partly due to the lack of horizontal moisture advection. Subsidence within anticyclones does not extend right to the surface, since normally the warm, dry subsiding air is insulated from the surface by a shallow layer of relatively cool air. The properties of this surface layer, which may be several thousand metres thick, are often completely different from those of the subsiding air, and it is normally maintained by horizontal advection from a source region outside of the main anticyclone. If the air forming the surface layer originates over the sea, it may be moist and even contain layer cloud which can give rise to light rain or drizzle. Within the tropics, the surface moist layer has already been identified with the layer below the trade-wind inversion.

Most of the moisture advection within the tropical atmosphere takes place in the shallow layer below the trade-wind inversion. The high-level winds over the subtropical anti-cyclones are extremely strong, but they transport little water vapour because of their extreme cold, and therefore do not help in the formation of cloud. At low levels, winds are light and hence inefficient in transporting water vapour; the low level winds in the anti-cyclonic cores are also generally divergent. The dryness of the air above the major deserts results partly, therefore, from a lack of advection of water vapour.

The atmospheric moisture-balance model (figure 4.1) suggests that local evapotranspi-ration has to be considered as well as general moisture advection, and this means that the nature of the local surface has to be taken into account. If the surface consists of the open ocean, then evaporation will be high because of the high net radiation; the surface layer will become moist and this moisture will be available for use by occasional synoptic dis-turbances. In contrast, if the underlying surface is completely arid, there will be no local evapotranspiration and no moisture gain by the surface layers of the atmosphere. Thus it is possible to distinguish two general types of arid climate; a maritime arid climate and a continental arid climate. The maritime anticyclones are the source regions of the trade winds and the evaporated water is carried towards the equator as described in previous chapters. The moisture in the layer below the anticyclonic inversion can be made use of by any synoptic disturbances which may occur. On the other hand, over large arid continents all the net radiation is available for heating the air and soil, and as a result surface tempera-tures are high and humidity low. If surface convergence does occur over an arid land-mass, there is not normally enough water vapour in the air to form cloud and rainfall; moisture must be imported from outside the desert areas if rain is to be formed. Conditions are there-fore more extreme in large deserts than over the oceans.

## 2 Deserts

No desert area is completely rainless, though in extreme cases several years may elapse between individual storms. The subtropical anticyclones undergo only minor seasonal variations, so the desert cores are liable to remain almost rainless, but rainfall is apt to in-crease and become more seasonal towards the poleward and equatorial limits of the deserts. Rainfall within desert regions normally results from disturbances arising outside the true desert areas. For instance, upper cold pools and troughs from the middle latitudes bring rainfall to the poleward margins of the deserts; similarly, disturbances forming near the equator can bring rainfall to the equatorward margins. As middle latitude troughs are most intense and nearest to the equator in winter, the poleward margins of the deserts usually experience a winter rainfall maximum. The equatorial trough tends to move north–south with the sun; in contrast to the poleward margins, the equatorial limits of the desert areas normally experience summer rainfall maximums. It is therefore possible to recognize two distinct rainfall regimes in desert areas; these are discussed in more detail later.

On the equatorward side of the subtropical highs there are generally easterly winds, and where these originate over land-masses they are extremely dry. This dryness is carried over the oceans to leeward of the land-masses. The trade winds pick up moisture by local evapo-ration from the ocean surface as they travel westward across the oceans and because of this, deep surface moist layers are most likely to be found in the western parts of the tropical oceans. As tropical disturbances depend on the release of latent heat for their maintenance, they tend to become more frequent towards the western ends of the subtropical oceans where deep moist layers exist. The western edges of the oceans receive large amounts of rainfall from disturbances and therefore form a break in the normal subtropical arid belt; they are discussed in the chapter on trade wind climates.

It is clear from the atmospheric moisture balance model (figure 4.1), that where topo-graphical factors are added to those caused by the general circulation, the aridity is greatly increased and in these areas the most severe desert conditions in the world can be found. It has already been noted that in the tropics the moisture available for rain formation is trapped in a shallow layer below an inversion. The depth of this moist layer varies, but it is usually between 1 and 2 km. If a mountain barrier projects through the moist layer, it will interrupt the surface flow and the surface moist layer will not penetrate behind the moun-tain range. Even if the mountains do not completely block the moist layer, the reduction in moisture advection to the lee of the range can still be substantial. For example, under tropical conditions an air-stream with a surface moist layer 2 km deep, and which has its lowest kilometre blocked by mountains, will lose between 60 and 70 per cent of its precipi-table water content. Dryness can be increased by subsidence of air from near the inversion down the lee slopes of the range, and for this reason mountain-enclosed inland basins are often extremely arid; Death Valley, USA is a good example.

Northeastern Brazil provides a further instance of an inland dry area, where, as sug-gested by figure 4.2, annual rainfalls of up to 2,000 mm are found along the coast; inland they fall to about 400 mm. The major cause of this dry zone appears to be the presence of the south Atlantic subtropical anticyclone which throughout most of the year directs rela-tively dry southeasterly trade winds towards northeast Brazil. The coastal area around Recife receives a high rainfall from the trade winds, but the mountains behind the coast rise to 1,000 m and effectively stop the surface moist layer, which is only about 1,500 m deep, from penetrating inland. During the southern winter almost all of Brazil south of the equator is under the influence of a greatly enlarged subtropical anticyclone and rainfall

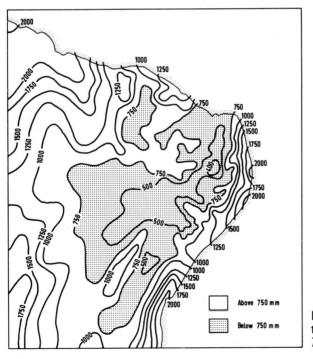

**Figure 4.2** Average annual precipi-tation (mm) over eastern Brazil. (*After Trewartha, 1962.*)

over much of the country is at a minimum. The coastal region near Recife receives much of its rainfall during this season from weak troughs in the trade-wind circulation. In the southern summer, the South Atlantic anticyclone weakens over the interior and humid equatorial air invades much of southern Brazil; but even in summer the northeast of Brazil remains under the influence of the subtropical anticyclone, and therefore the interior remains dry. This is an example of a semi-arid region resulting as much from the local relief as from the general atmospheric circulation.

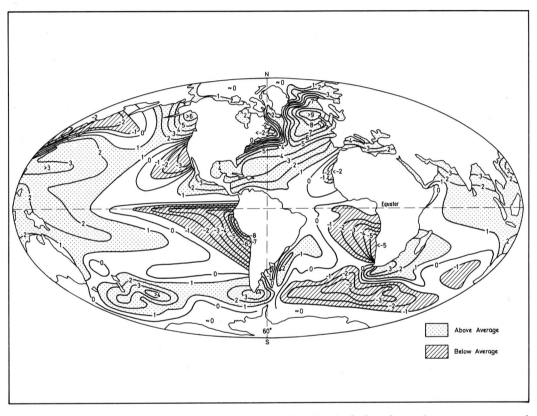

**Figure 4.3** Sea-surface temperature represented as a deviation from the average at each latitude. (*After Dietrich and Kalle, 1957.*)

*3   Ocean temperatures*
Dietrich and Kalle (1957) have mapped the difference between the sea-surface temperature and its average global value along the part of each latitude circle situated over the oceans. The results are illustrated in figure 4.3, where cold water regions are defined as ocean areas with sea-surface temperatures below the global average. By far the most extensive is the South Pacific cold water, for it stretches westwards from the coast of South America to about 85° longitude; in contrast the South Atlantic cold water continues westward from the coast of Africa to only about 40° latitude. The temperature difference along most of the coast of Peru exceeds −8°c, and the cold water reaches almost to the equator

where off Equator, differences of $-3.5°c$ are found. The corresponding coastal cold water off southwest Africa has an equally large negative temperature difference, but it does not extend very far to the north. The subtropical cold water areas in the North Atlantic and North Pacific are relatively small, and the north Indian Ocean has only a minor area off Somalia; there is also a minor patch off Australia in the south Indian Ocean.

Because much of the moisture content of the trade winds in their core region is of local origin, the sea-surface temperature will greatly influence the local rate of evaporation. This arises both from the direct relationship between water temperature and evaporation rate, and also from the effect of sea-surface temperature on atmospheric stability. If the ocean is warmer than the atmosphere, convection takes place with towering cumulus clouds forming, and under these conditions there is a maximum in the upward transfer of moisture. In contrast, if the ocean is colder than the atmosphere, then stability is increased and the upward transfer of moisture is at a minimum. Considerable rainfall could fall in the form of showers from cumulus clouds in the first case, but in the second there will be only slight drizzle from stratus cloud.

Bjerknes (1969) has studied the relationship between rainfall and air-sea temperature differences at Canton Island, which lies on the edge of the south Pacific cold water. Upwelling cold water reaches Canton Island for most of the year, but occasionally the upwelling process ceases and the sea surface becomes warmer than the atmosphere. Figure 4.4 shows the air and sea temperatures from 1950–1967 at Canton Island, together with the rainfall; it can be seen that big monthly totals of rain only occur during periods when the ocean is warmer than the atmosphere.

The Canton Island type of rainfall regime—that is, normal aridity but occasional wet summers—is known to prevail throughout most of the south Pacific cold water region. Along the coast of Peru the surface moist layer is less than 800 m deep and normally only occasional drizzle falls from stratus cloud. However, during the southern summer the southeasterly trades are sometimes replaced by northerly winds which in turn induce a southward flow of warm equatorial water, displacing the cool upwelling water of the Peruvian coastal current. The warm water and unstable air result in heavy rainfall in an otherwise almost completely arid desert—a phenomenon known locally as the El Niño effect.

Bjerknes states that when the cold ocean water along the equator is well developed, the air above will be too cold to take part in the ascending motion of the Hadley cell circulations. Instead, the equatorial air flows westward between the Hadley cell circulations of the two hemispheres to the warm west Pacific, where having been heated and supplied with moisture from the warmer waters, the equatorial air can take part in large-scale, moist-adiabatic ascent. This suggests general subsidence over the equatorial eastern Pacific and ascent over the equatorial western part of that ocean—a picture that fits the known rainfall distribution rather well. Bjerknes considers that the axis of this circulation is located at about 160°E, with the sinking air occupying about 120° longitude to the east and the rising air about 30° longitude to the west. The equatorial Atlantic is analogous to the equatorial Pacific in that the warmest part is in the west, at the coast of Brazil, but the west–east contrast of water temperature is much smaller than in the Pacific. However, in January a thermally driven equatorial circulation may operate from the Gulf of Guinea to the Andes, with the axes of the circulation near the mouth of the Amazon and sinking air over the equatorial Atlantic and the dry northeast corner of Brazil.

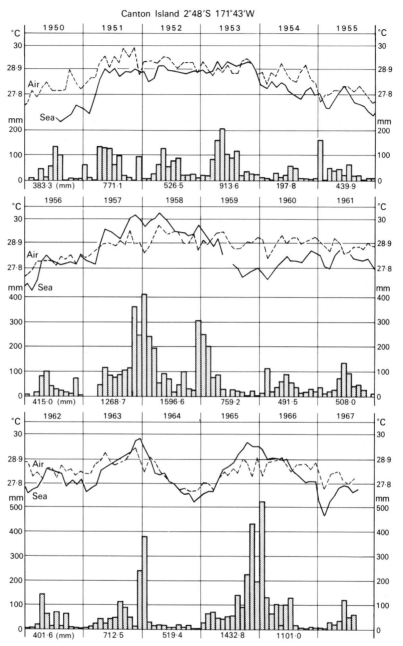

**Figure 4.4** Time series of monthly air and sea temperatures and of monthly precipitation at Canton Island from 1950–1967. (*After Bjerknes, 1969.*)

## Definitions of aridity

Wallén (1967) suggests that there are three approaches in attempting to define aridity and in classifying the arid or semi-arid areas of the world. These may be termed the classical approach, the index approach, and the water-balance approach.

The classical approach includes the fundamental study of the various climatic elements in their relation to vegetational or agricultural conditions, and the applying of various statistical methods until those elements or parameters are found which have particular significance in defining aridity. Rainfall dependability parameters have been found to be particularly useful and an example is discussed later in this chapter.

Classical studies have often been developed to the stage where they can be applied in an objective manner in the form of standard indexes. For example, after considerable comparisons between temperature and precipitation data, and vegetation conditions, Köppen (1931) concluded that the boundaries of arid regions could be established in terms of the mean annual precipitation P and the mean annual temperature T. He used temperature as a substitute for evaporation data, which are too scanty to be used on a world basis, and found that generally the limit of semi-arid regions in the subtropics could be expressed by

$$P = 2(T + 7)$$

where P is in centimetres and T in °c. For arid conditions he wrote

$$P = T + 7$$

Köppen was aware that aridity differences exist in areas of summer and winter rainfall, and he therefore amended his formula for semi-aridity to read

$$P = 2(T + 14)$$

in areas with summer rainfall and

$$P = 2T$$

in areas with winter rainfall. If a typical mean annual temperature of 18°c is assumed, then these formulas give a maximum precipitation of 640 mm for semi-aridity with summer rainfall, and 360 mm with winter rainfall. According to Wallén, these values are confirmed by agricultural experience.

Köppen's index has been applied by botanists, mostly in the Mediterranean area. For instance, Gaussen (1963) has established a system of aridity based on the number of dry months in the year. The definition of a dry month is analogous to the last equation, i.e., $p \geqslant 2t$. It has proved possible to establish a scale of aridity on this basis which seems to fit vegetation conditions in the Mediterranean region fairly well.

There are numerous other aridity indexes, the best known of which are probably those due to de Martonne (1926) and Emberger (1955). The fundamental weakness of the above-mentioned indexes was first clearly seen by Ångström (1936). He wrote that from a purely scientific point of view it is embarrassing to accept values based on a relationship between precipitation expressed in millimetres and temperatures expressed in degrees Celcius or Fahrenheit. Such indexes have no physical meaning, tell us nothing about the physical processes involved, and may easily be misleading in the hands of inexperienced people. Aridity should be expressed in terms of a relationship between precipitation and evapotranspiration $(E_t)$:

$$I_{arid} = f_1 (E_t, P)$$

However, since temperature is an approximate measure of the energy available for evapotranspiration through net radiation and sensible heat advection, indexes using it have proved useful for the description of climates on a world scale.

Thornthwaite (1933, 1948) was probably the first investigator to point out, with sufficient emphasis, the need for a water-balance approach in order to reach a more complete understanding of aridity. The theory of potential evapotranspiration and water balance were discussed in detail in chapter 1.

## Radiation and temperature

Generalized maps of the components of the heat balance were given in chapters 2 and 3 and the annual heat balance of Aswan (figure 1.22) was discussed in chapter 1. Since Aswan has a typical continental desert location, the soil is dry and the average annual rainfall is below 2 mm. Most of the available net radiation is used in heating the air, leading to extremely high temperatures. At Aswan, summer afternoon temperatures regularly exceed 38°C and temperatures above 48°C have been recorded on a number of occasions.

Annual heat balances for the eastern and western peripheries of the oceanic anticyclones are indicated in figures 4.5 and 4.6. The example for the eastern periphery pertains to a portion of the southeast Atlantic affected by the cold Benguela current, while the western example refers to the island of Trinidad (Brazil) in the South Atlantic. The net radiation at both sites varies in accordance with the annual march of total radiation, but the expenditure patterns for sensible heat and evaporation differ greatly. In the warm current area around Trinidad there is a transfer of energy from the ocean surface to the atmosphere. This is particularly strong in the southern winter when the expenditure of heat for evaporation and sensible heat transfer to the atmosphere exceeds the net radiation. At this time of year there is a transfer of heat from the deep layers of the ocean to the surface. In contrast, the eastern periphery is affected by cold sea currents, and the annual march of the heat balance components is altered accordingly. The flow of sensible heat, which is largest in summer, is from the atmosphere to the cold ocean surface. The expenditure of heat for evaporation is greatly reduced as compared with Trinidad, which is in the same latitude. The gain of heat by the ocean surface from the net radiation and the sensible heat flux is greater than the losses by evaporation, and therefore a great deal of heat is transferred to the deeper layers of the ocean.

The highest values of the diurnal temperature range are experienced in deserts, where the soil is generally dry and the atmosphere contains little water vapour. The hotter parts of the Sahara experience mean maximums of 45°C in the hottest month, and values nearly as high are probable in the Great Sandy Desert in western Australia, while Death Valley, California has a mean maximum temperature of 47°C in July. As explained earlier, these very high temperatures occur because nearly all the net radiation is available for heating the air and soil. The low water vapour content of the air allows relatively large long-wave radiation losses from the surface, leading to low temperatures at night. Deacon (1969) has illustrated this for the Sahara Desert, where in April the average cloud amount over considerable areas is less than one okta and the total precipitable water content only about 1·2 cm, while the global average for 20°N is nearly 4 cm. As a result, the mean daily range of temperature reaches 20°C over wide regions and as much as 22·5°C over the sandier areas. In July, a month with an equally small cloud amount, greater amounts of dust and moisture in the atmosphere cause the diurnal range to be nearly 5°C smaller than in April.

Priestley (1966) has considered the limitation of temperature by evaporation in hot

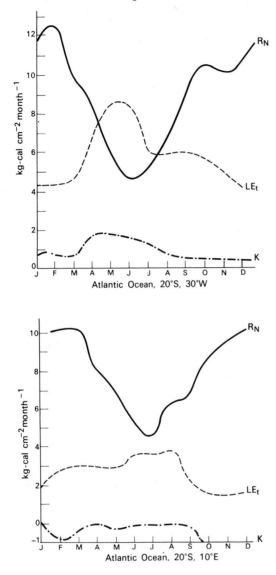

Atlantic Ocean, 20°S, 30°W

Atlantic Ocean, 20°S, 10°E

**Figures 4.5 and 4.6** Seasonal march of the major heat-balance components of the sea surface (K = sensible heat transfer to the atmosphere; $LE_t$ = latent heat transfer to the atmosphere; $R_N$ = net radiation).
**Figure 4.5** Tropical climate of the western periphery of oceanic anticyclones (Atlantic 20°S, 30°W).

**Figure 4.6** Tropical climate of the eastern periphery of oceanic anticyclones (Atlantic 20°S, 10°W). (*After Budyko, 1956.*)

climates, and concluded that with extremely few exceptions 33°c is the highest mean maximum temperature (from monthly statistics) attained by the air over any extensive freely-evaporating surface. A study by Deacon (1969) of temperature and rainfall data at Alice Springs, which is in the desert near the centre of the Australian land-mass, provides a good illustration of this effect. At Alice Springs the rainfall occurs mainly as brief heavy showers, so its main effect in reducing temperature will be via evaporation, since the influence of cloud amount on radiation can be neglected. Deacon showed that the January mean maximum temperature at Alice Springs is not only influenced by the January rainfall but also by that of the previous month. Working with rainfalls on a logarithmic scale, he obtained partial correlation coefficients from 50 years of data as follows:

January mean maximum temperature and January rainfall; December rainfall constant
—0·66.

January mean maximum temperature and December rainfall; January rainfall constant
—0·40.

Net radiation values in the poleward regions of the subtropical deserts undergo large seasonal variations (see figure 1.22), and lead to large seasonal variations in temperature. The high summer temperatures at Aswan have been mentioned, but in January and February the night temperature can fall as low as 5°C. Indeed, large areas of the Sahara have recorded absolute minimum temperatures of 0°C or below, indicating that frost is not unknown. At Alice Springs night temperatures regularly fall below 0°C in June, July, and August, and the absolute minimum temperature is about —7°C.

## Rainfall

The nature of desert rainfall, mentioned in the introductory section, can be explored further by a study of North Africa. The Mediterranean and North Africa lie on the southern flank of the middle latitude westerlies and are strongly influenced by them. Kirk (1964) states that, particularly during the colder months, the weather is essentially determined by the large-scale fluctuations of the main westerly flow pattern, and that the long waves and their modes of behaviour are of fundamental importance. There are two main jet streams in the westerlies of the northern hemisphere: the northern jet stream is related to the polar front, while the southern subtropical jet is not associated with surface fronts on most occasions. The northern jet stream undergoes irregular quasi-cyclic variations connected with index cycles, and the associated weather patterns are illustrated in figures 8.4 and 8.5. With a high zonal index, there is a strong zone of flow in middle and high latitudes and bad weather tends to be confined to this fairly narrow band; but at low zonal index, wave amplitude in the westerlies is large and blocking activity is at a maximum, extending bad weather into the subtropics. The middle latitude and subtropical jet streams combine or approach each other in periods of low zonal index, thus complicating the upper wind and vorticity fields. Index cycles, blocking activity, and related flow patterns are discussed in detail in chapter 8.

In the Mediterranean, Middle East, and adjacent areas, cyclones appear that are completely different from the warm-sector depressions formed along the polar front. These Mediterranean cyclones are associated with domes of cold air which have been cut off from the main cold air to the north and are surrounded by warm air. Cut-off cold pools, as these domes are known, form mostly at times of low index flow in middle latitudes and are most frequent in winter. Once a cut-off cold pool is formed in the Mediterranean, a corresponding surface cyclone soon appears, and convergence into the low leads to shower activity. The cold pools have a life period of from 2–12 days and usually remain stationary or drift slowly westward.

The distribution of the mean annual amounts of rainfall over Egypt (figure 4.7) shows a maximum on the Mediterranean Coast with a rapid decrease inland, for indeed Egypt south of 28°N is practically rainless. Ali (1953) has found that generally wet periods in Egypt or the Middle East are connected with the development of upper lows over the eastern Mediterranean which affect mainly the coastal regions. This is clearly shown by table 4.2, where Soliman (1953) has analysed the rainfall from various types of disturbance

**Table 4.2** Analysis of rainfall at three stations in Egypt during the period December 1951–November 1952

|  | Alexandria | Port Said | Cairo |
|---|---|---|---|
| Rain due to cold fronts (mm) | 9·8 | trace | trace |
| Rain due to instability in absence of upper troughs or lows (mm) | 8·4 | 1·7 | trace |
| Rain due to upper cold troughs or lows (mm) | 173·2 | 105·2 | 69 |
| Number of days of cold-front rain | 3 | 3 | 2 |
| Number of days of mere instability rain | 6 | 5 | 3 |
| Number of days of rain associated with upper cold troughs or lows | 70 | 33 | 23 |

*After Soliman (1953).*

at Alexandria, Port Said, and Cairo for the period December 1951–November 1952. The importance of upper cold troughs is well illustrated by this last table. Eastern Mediterranean lows originate as upper lows at the 500 mb level north of 30°N and gradually move southwards to the northern coast of Africa. This is probably brought about by cold outbreaks from the north which push the polar front jet into amalgamation with the subtropical jet.

The role of upper tropospheric divergence in the formation of North African depressions has been mentioned, but without water vapour such systems will give little rainfall. A study

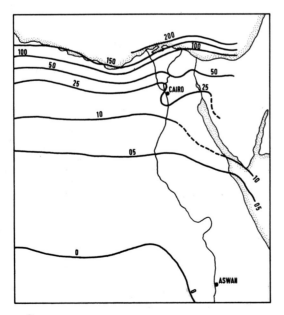

**Figure 4.7** Mean annual precipitation (mm) over Egypt. (*After Soliman, 1953.*)

of the climatological distribution of autumn thunderstorms suggests a maximum frequency in the middle Egypt and Sinai area, and it is considered by several investigators that this can be attributed to the existence of the Red Sea, which is an important source of water vapour and also forms a local convergence zone.

According to Solot (1943) the summer rains in the transition zone along the southern margins of the Sahara may be of two kinds, the one possibly associated with local random convection, and the other with organized shower activity occurring in connection with extensive atmospheric disturbances. The nature of the atmospheric disturbances is not clear, but they appear to be related to perturbations in the equatorial westerlies.

It is characteristic of rainfall in North Africa that much of it comes in the form of showers. Although these showers are normally localized, on some occasions they are accompanied by such heavy precipitation that they cause destructive floods in otherwise arid areas. Sometimes the mean monthly or mean annual rainfall will be precipitated by one storm, and numerous examples of this are contained in the meteorological literature. No desert area is completely immune from destructive rainfall; a good example of this in Tunisia and Algeria has been described by Winstanley (1970). On 23 September 1969, a cyclonic storm developed over the central Mediterranean and during the next few days it slowly retrogressed westwards, moving across northwestern Libya and southern Tunisia, and finally dissipating over northeastern Algeria. It was accompanied by widespread and heavy rains, which caused such devastation in Tunisia and northeastern Algeria that nearly 600 people were killed and a further quarter of a million were made homeless. Figure 4.8 shows that the rainfall during the 10-day period, 20–29 September, was several times greater than the mean for the whole month over large areas of central and southern Tunisia and northeastern Algeria. At Biskra in Algeria, the rainfall on 27 and 28 September alone was over twice the mean annual fall of 148 mm; the annual fall has only once

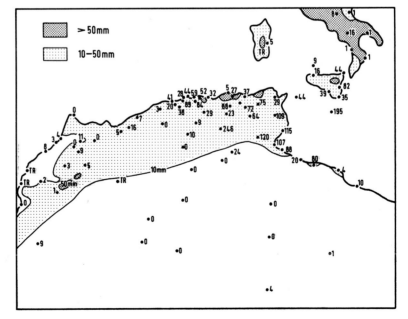

**Figure 4.8** Ten-day rainfall totals (mm) over North Africa for the period 20–29 September 1969. Mean September rainfall is shown for comparison. (*After Winstanley, 1970.*)

exceeded the monthly total for September 1969, namely 329.8 mm in 1934. Winstanley comments that at many places in North Africa, September, rather than in the middle of the wet season, is the month with the greatest mean rainfall intensity. This is probably because in the early autumn the vapour content of the air over the Mediterranean is at a maximum and conditions very suitable for heavy rainfall give the right synoptic conditions. This particular cyclonic storm appears to have resulted from the amalgamation of mid-latitude and subtropical circulation systems.

Similar intense rainfalls are recorded in desert regions elsewhere in the world. In central Australia, for example, Alice Springs has an average annual rainfall of about 267 mm, the highest daily fall recorded being of the order of 145 mm, while the probable maximum daily rainfall in the Alice Spring's region is about 450 mm (Wiesner, 1970). The intensity and irregularity of desert storms tend to make mean annual rainfall statistics for desert areas rather meaningless.

*Rainfall variability*
Apart from its bearing on rates of erosion, the extreme variability of rainfall is of import-ance in dry-land farming. Wallén (1967, 1968) has discovered that the minimum annual rainfall at which dry-land farming is possible in the Middle East varies from region to region, mainly because the variability of annual rainfall (P) is higher in some than in others. For simple investigations of the variability of precipitation, Wallén applied the following parameter:

$$1AV_{rel} = \frac{100 \sum\limits_{r=2}^{r=n} (P_{r-1} - P_r)}{\overline{P}(n-1)}$$

where $1AV_{rel}$ is the interannual variability relative to the mean value. He calculated this parameter for a variety of stations in the Middle East and then plotted the calculated values against mean annual rainfall. A scatter of points was obtained, but it was possible to fit a curve indicating the general decrease of variability with the increase of annual precipitation. Wallén fitted further curves to his diagram separating stations where dry-land farming is possible from those where it is not possible. The intersection of these two curves and the primary curve gave the approximate values of the precipitation conditions at the limit for dry-land farming in the Middle East, namely, 240 mm of annual rainfall with an interannual variability of 37 per cent.

Landsberg (1951) has published a diagram showing the probability of obtaining, at various places, certain amounts of rainfall in an individual year in relation to the mean annual rainfall. By using this diagram as a basis for his calculations and allowing for a crop failure because of insufficient rain in 2 years out of 10, Wallén found that the minimum amount of rainfall in an individual year that would permit dry-land farming was different in the various parts of the Middle East. He calculated that the amount was as low as 180 mm in the area of regular rainfall in Jordan, southwestern Syria, and the 'Fertile Crescent', and as high as 210 mm in the Zagros mountains and 230 mm in the agricultural region of northern Iran. Further, by making use of these minimum values, it is possible to estimate the corresponding average annual rainfall and thus draw a boundary (figure 4.9) on the map of mean annual rainfall of those areas with no possibilities of dry-land farming.

In trying to define a wet season or a season of reliable rainfall it is necessary to consider

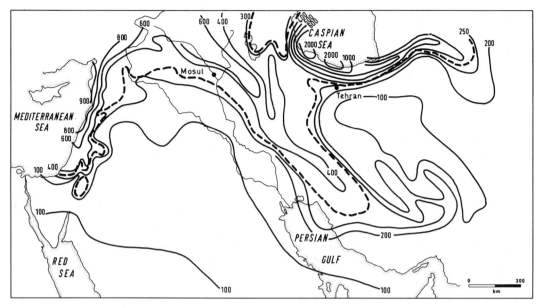

**Figure 4.9** Annual rainfall and limits for regular dry-land farming in part of the Middle East. Dashed line indicates theoretical outer limit for dry-land farming; rainfall amounts are in mm. (*After Brichambaut and Wallén, 1963 and Wallén, 1967, 1968.*)

not only the amount of rainfall but also its variability. In the Middle East, which has winter rainfall, Wallén considers that the start and end of the reliable rainfall season can be defined to correspond to a monthly rainfall of 25 mm and a variability of 100 per cent. The following parameter was then calculated for each station:

$$q = \frac{\bar{p}}{1AV_{rel}}$$

and the value q = 0.25 used to establish the dates of the beginning and end of the wet season.

For the Middle East, Wallén discovered only small variations in the start of the wet season, which is usually in November or early December. The end of the wet season is, however, extremely variable. It comes as early as the beginning of April in the semi-arid parts of Jordan, while in the 'Fertile Crescent' the season lasts until the end of April or the beginning of May, and in Iraq and Iran it may last until the end of May or even the middle of June. Some of Wallén's results are summarized in table 4.3.

## Evapotranspiration and water balance

Soil water balance which was briefly discussed in chapter 1, is clearly of great importance in desert areas and therefore needs to be explored more fully.

The soil water balance equation is generally written in the form

$$P - O - U - E_a + \Delta W = 0$$

where $\Delta W$ is the change in soil water storage (initial minus final) during the period, and P,

**Table 4.3** Agroclimatic subregions in the Middle East

| | Agroclimatic subregion | | Key station | Rainfall conditions | | | Water balance conditions | | | |
|---|---|---|---|---|---|---|---|---|---|---|
| | | | | Annual rain-fall | Rel* (IaV) in % | Date end reliable rainfall season | Winter season** | | | Summer*** season |
| | | | | | | | $\bar{p}$ | $E_t$ | $\dfrac{\bar{p}}{E_t}$ | $E_t$ |
| 1 | Arid and irrigated parts of Jordan Valley | | Deir Alla | 289 | 41 | Mid III | 272 | 447 | 0·61 | 1510 |
| 2 | Semi-arid zones with dry-land farming in east Jordan and southwest Syria | | Irbid | 482 | 41 | 13.IV | 476 | 469 | 1·01 | 1120 |
| | | | AmmanRAF | 271 | 41 | 2.IV | 264 | 487 | 0·54 | 1190 |
| | | | Amman Land Dept | 451 | 43 | 10.IV | 439 | 487 | 0·91 | 1190 |
| 3 | Semi-arid zones with dry-land farming in northern part of Lebanon, Syria, Iraq | a | Shlifa | 402 | 25 | 24.IV | 389 | 343 | 1·13 | 1150 |
| | | | Hama | 358 | 28 | 25.IV | 317 | 332 | 0·95 | 1260 |
| | a Hama district | b | Aleppo | 364 | 35 | 23.IV | 327 | 322 | 1·01 | 1120 |
| | b Aleppo and Hassekieh districts | | Hassekieh | 268 | 26 | 30.IV | 262 | 328 | 0·80 | 1130 |
| | | c | Kamishlieh | 430 | 34 | 12.V | 307 | 350 | 1·16 | 1210 |
| | c Kamishlieh and Mosul districts | | Mosul | 389 | 29 | 9.V | 371 | 362 | 1·02 | 1220 |
| 4 | Arid zones of the Syrian steppe | | Damascus | 220 | 35 | 15.III | 204 | 383 | 0·53 | 1230 |
| | | | Deir-ez-Zor | 148 | 36 | 24.III | 144 | 509 | 0·35 | 1300 |
| | | | Rutba | 129 | 52 | 15.III | 122 | 527 | 0·23 | 1330 |
| 5 | Semi-arid zones with dry-land farming in the Piedmont of northern and central Zagros | | Khanaquin | 325 | 47 | 30.IV | 291 | 325 | 0·90 | 1330 |
| 6 | Arid zones of central Iraq | | Baghdad | 151 | 52 | 18.III | 133 | 358 | 0·37 | 1650 |
| 7 | Arid zones of south Iraq and lower Khuzistan in Iran | | Ahwaz | 194 | 37 | 26.III | 188 | 365 | 0·52 | 1860 |
| 8 | Semi-arid valleys with dry-land farming in south Zagros | | Shiraz | 371 | 42 | 14.IV | 365 | 532 | 0·69 | 1250 |
| 9 | Semi-arid zones with dry-land farming in Azerbaidjan and valleys of northern Zagros | | Rezayeh | 376 | 22 | 8.VI | 349 | 393 | 0·89 | 810 |
| 10 | Semi-arid zones with dry-land farming in Kurasan | | Meshed | 252 | 33 | 1.VI | 240 | 342 | 0·70 | 890 |
| 11 | Arid zones of western part of high plateau of Iran | | Teheran | 213 | 37 | 2.V | 205 | 469 | 0·44 | 1110 |
| | | | Ispahan | 121 | 55 | 10.III | 116 | 488 | 0·24 | 1100 |
| 12 | Arid zones along the Persian Gulf | | Bushire | 278 | 52 | | 247 | 404 | 0·61 | 1840 |

*Relative interannual variability (%)

**Defined as starting by 1 November and ending 15 days after heading date (approximated to the 1 or 15 of the month)

***Defined as starting at the date when mean temperature passes +15°C in spring and ending when the same mean temperature is passed in autumn

$\bar{p}$ = mean precipitation

*After Wallén (1967).*

O, U, and $E_a$ are the precipitation, runoff, deep drainage, and actual evapotranspiration respectively.

This equation can be more specifically written as

$$\int_{t_1}^{t_2} [(P - O) - E_a - V_z]\, dt = \int_{t_1}^{t_2} \int_{o}^{z} \frac{\partial \theta}{\partial t}\, dz\, dt$$

where $(t_2 - t_1)$ is the time interval over which the measurements are made; z is the depth to the lowest point of measurement; $V_z$ is the net downward flux of water at depth z (cm sec$^{-1}$), and $\theta$ is the volumetric soil water content (cm³ water/cm³ soil).

The amount of water in the soil available for plant growth is generally considered to be that retained at soil water potentials lower than 15 atmospheres, and the difference in volumetric water content at 0·3 atmospheres and 15 atmospheres is commonly referred to as the available water range. The significance of 15 atmospheres as a lower limit of soil water availability depends on the type of plant under consideration, but it appears reasonable for most crops. Water extraction in arid plant communities does not cease when the available soil water storage is depleted but continues until a fairly consistent minimum value is reached, which represents an additional 'survival' storage capacity, estimated to be approximately 0·025 cm/cm depth. Slatyer (1968) has suggested for Alice Springs that the available water range of the soils is approximately 0·075 cm water/cm soil depth and that the rooting depth of the vegetation approximates 150 cm, giving a water storage capacity of approximately 112·5 mm. If the survival storage capacity is added to this amount, the total storage capacity in the range available for plant growth rises to 150 mm.

Following rainfall and soil water recharge in an arid region, plant growth recommences in the case of a dormant perennial vegetation, or commences in the case of an annual pasture or sown crop. If the soil water supply is adequate, leaf area increases progressively with growth rate and community evapotranspiration may reach values dependent primarily on atmospheric conditions. At this stage the rate of evapotranspiration is commonly referred to as potential evapotranspiration and under many conditions it can be predicted, with a fair degree of accuracy, from meteorological data alone. In communities with widely spaced plants, the limited extent of the transpiring leaf surface may prevent this stage from being reached. Progressive reductions in actual to below potential rates will occur as soil water storage is reduced.

## *I   Alice Springs*

The nature of water balance in a semi-arid area is well illustrated by work carried out at Alice Springs in central Australia. At Alice Springs the long-term average annual rainfall is 267 mm with 191 mm falling in the summer months (October–March inclusive). The rainfall is described by Winkworth (1970) as sporadic, with occasionally up to 90 per cent of the annual total being recorded in 2 months of the year. In the average year 16 groups of 1–3 wet days supply 171 mm of rain and 2 groups exceeding 3 days account for the remaining 96 mm, with long dry periods separating the wet periods. Over 80 per cent of the wet periods have rainfalls of less than 13 mm, and so it can be expected that an appreciable proportion of rain is directly evaporated, since the evaporation from the free water surface of a standard Australian tank varies from 90 to 300 mm per month.

Winkworth found that after significant rainfall, the soil profile at Alice Springs was wetted rapidly down to 30 or 40 cm, reaching field capacity at these depths in 3–4 days.

Downward movement to deeper layers continued for a further 7–10 days, so that water contents at 41 cm and below were increasing or constant while the upper layers had begun to lose water. Depletion of stored water normally started 6–12 days after rain in the upper 30 cm of soil and one week later at greater depths. The duration of periods of available soil water ranged from 25–141 days.

Typical soil moisture drying profiles are illustrated in figure 4.10. Moisture extraction usually continues until all available water has been removed from the whole of the wetted profile, and the soil then remains at low moisture content until the process is repeated at the next effective rainfall. As mentioned earlier, during the dry periods the soil water potential became greater than 15 atmospheres and at 10 cm reached 550 atmospheres at times, while at other depths it ranged from 70 to 170 atmospheres.

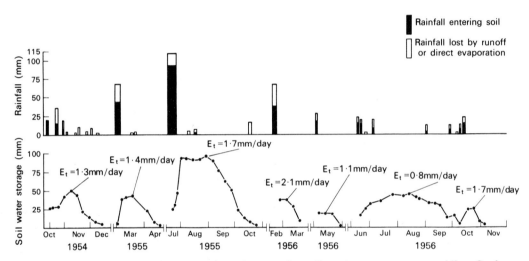

**Figure 4.10**  Rainfall and progressive changes in soil moisture storage at Alice Springs, Australia, for the 6 periods of significant recharge and extraction from October 1954–November 1956. (*After Winkworth, 1970.*)

Fitzpatrick, Slatyer, and Krishnan (1967) have estimated, for Alice Springs, the likelihood of commencement of growth or non-growth periods of various duration, and their results are contained in table 4.4. At Alice Springs, growth periods of 30 days duration or longer can only be expected three times in 2 years, and of 90 days only once in 4 years. The probability of a year without plant growth is extremely small.

## 2  Southern Sahara

Cochemé and Franquin (1967) have studied the agroclimatology of the semi-arid region south of the Sahara in detail, and their work illustrates another approach to water-balance problems in such regions. The zone of study was bounded to the north by the limits of dry-land farming corresponding more or less to the isohyets of 400 to 500 mm of mean annual rainfall, while the approximate southern limits were where the length of the wet season exceeds that of the dry season. The belt of territory thus investigated in West Africa was 6,000 km long and about 1,000 km wide, bounded to the west by the Atlantic Ocean and

**Table 4.4** Expected number of growth and non-growth periods of specified duration occuring seasonally and annually at Alice Springs

| Expected number of growth periods of duration equalling or exceeding | Oct–Mar | Apr–Sept | Annual |
|---|---|---|---|
| 5 days | 2·34 | 1·45 | 3·79 |
| 15 days | 1·39 | 1·09 | 2·48 |
| 30 days | 0·74 | 0·70 | 1·44 |
| 90 days | 0·09 | 0·18 | 0·27 |
| **Expected number of non-growth periods of duration equalling or exceeding** | | | |
| 30 days | 1·32 | 1·22 | 2·54 |
| 90 days | 0·45 | 0·54 | 0·99 |
| 180 days | 0·77 | 0·10 | 0·17 |
| 365 days | 0·00 | 0·00 | 0·00 |

*After Fitzpatrick, Slatyer, and Krishnan (1967).*

to the east by the border between Sudan and Chad. This zone, shown in figure 4.11, is situated between the tropical anticyclonic belt and the equatorial convergence. It is virtually free from maritime influence from the west, and shows very little geographical relief.

In this particular region, the climate tends to be latitudinally uniform, but to vary along the meridians. As a result of this climatic zonation, the native vegetation is also arranged zonally and consists of sparsely wooded steppe in the north changing southwards to relatively dry woodlands and savannahs. Mean annual rainfall is depicted in figure 4.11 and as suggested, the pattern is basically zonal with a slight dip southward from the Atlantic coast to Lake Chad, the pattern being reversed further east. The rains come in the summer, are heaviest on average in August, with their duration varying from about 6 months in the south to 2 in the north.

For the period 1953–1962, monthly mean potential evapotranspiration was calculated assuming a reflection coefficient of 0·25 for 35 stations in or near the area by Cochemé and Franquin using Penman's formula. As noted before, the meteorological variables in Penman's formula are distributed zonally except for the wind, which shows a maximum in north Mali and on the coast of Senegal, and a minimum in south Chad. Therefore, mean

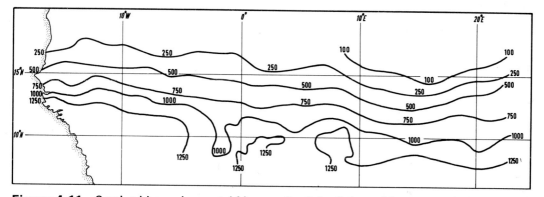

**Figure 4.11** Semi-arid area in west Africa south of the Sahara. Mean annual rainfall (mm). (*After Cochemé, 1968.*)

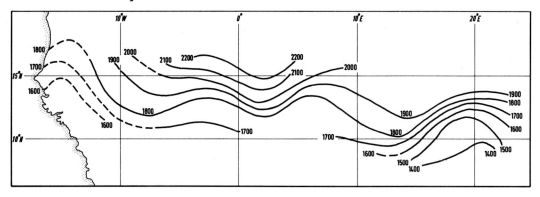

**Figure 4.12** Semi-arid area in west Africa south of the Sahara. Mean annual evapotranspiration (mm). (*After Cochemé, 1968.*)

annual potential evapotranspiration is also distributed zonally (figure 4.12), and is higher in the north where it reaches 2,200 mm, than in the south where it is below 1,500 mm. Comparison of figures 4.11 and 4.12 suggests that the demand for water increases from south to north faster than the rainfall decreases, and thus the ratio rainfall/potential evapotranspiration varies from about 0·8 in the south to 0·2 in the north.

The annual march of the water balance at a typical station is shown in figure 4.13. There are two maximums to the potential evapotranspiration curve, before and after the wet season (the former being higher), and two minimums, in August and winter. During the dry season the soil moisture is almost completely exhausted and therefore the actual evapotranspiration will be considerably less than the estimated potential values. Cochemé and Franquin made some assumptions in their study about the relationship of the actual to the potential evapotranspiration. During the humid period (figure 4.13), when rainfall exceeds

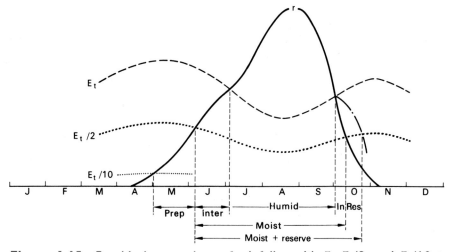

**Figure 4.13** Graphical comparison of rainfall r, with $E_t$, $E_t/2$, and $E_t/10$ to define preparatory, intermediate, moist, moist with reserve, and humid periods. (*After Cochemé, 1968.*)

potential evapotranspiration, it is assumed that actual and potential evapotranspiration are equal and that there is a water surplus; clearly this is the period of maximum vegetative growth. Because at the beginning of the wet season the soil will be dry and actual evapotranspiration will approach the potential as the soil moisture increases, two periods are defined: the preparatory period when the actual evapotranspiration increases from 0·1 to 0·5 of the potential, and the intermediate period when it varies from 0·5 to the full potential rate. A similar intermediate period occurs at the end of the wet season when the soil starts to dry and actual evapotranspiration falls to 0·5 of the potential. The humid period plus the two intermediate periods are defined as the moist period. At the end of the humid period it is assumed that up to 100 mm of water are stored in the ground and that this water is available for use in evapotranspiration. It is therefore possible to define a third period: the moist period plus reserve. The length of the various periods, for a selection of stations varying from the wettest to the driest, is shown in figure 4.14. The two intermediate periods are of fairly constant duration, 25 days for the prehumid and 15 for the posthumid; but the humid period on the other hand increases from zero on the monthly scale in the north to 140 days in the south. To the moist period may be added up to 20 days for the soil moisture reserve, giving a range of about 55 to 200 days for the moist period, including reserves. Since temperature is always adequate, the length of the moist period with reserve seems to be the most important agroclimatological factor, because the main adaptation of semi-arid crops to their environment involves the matching of the growing cycle and its main biological events to the moist period with reserve and its main climatic events.

### 3   North Africa and Middle East

Similar studies have been undertaken for North Africa and the Middle East by Wallén (1967, 1968), the major difference from the previous study being that these are winter rainfall areas. Potential evapotranspiration was calculated using Penman's method and values were found to be surprisingly similar over large parts of the Middle East, the main exceptions being in low latitudes where the incoming radiation was higher than further north. Water-balance diagrams were constructed and theoretical dates calculated when there would be no more water available in the soil for active plant growth. The limits for dry-land farming established by the water-balance method agree well with those established from rainfall conditions only (figure 4.9). Indeed, Wallén concludes that in large areas of similar winter evapotranspiration and soil conditions the degree of aridity is mainly determined by the amount of rainfall and its variability.

Basic details of the water balance for a variety of areas are given in table 4.3. The values of the ratio rainfall/potential evapotranspiration for the winter season are considered by Wallén to give a reasonable idea of the degree of aridity throughout the region. For instance, Hassekieh, a station which agricultural experience shows is on the limit of dry-land farming, has a ratio of 0·80, while Kamishlieh, situated in one of the best agricultural districts of the whole area, has a ratio of 1·16. There are similarities between the basic subregions in table 4·3 and other parts of the world. The climate of the semi-arid zones with dry-land farming in east Jordan and southwest Syria is like that of south and southwest Australia. Likewise, the climate of the semi-arid zones of the Syrian wheat belt and of the semi-arid valleys in southern Zagros of Iran is related to that of the San Joaquin Valley of California. There are several common features between the climate of lower Mesopotamia and that of the Imperial Valley in California. Also, the Azerbaidjan area of northern Iran

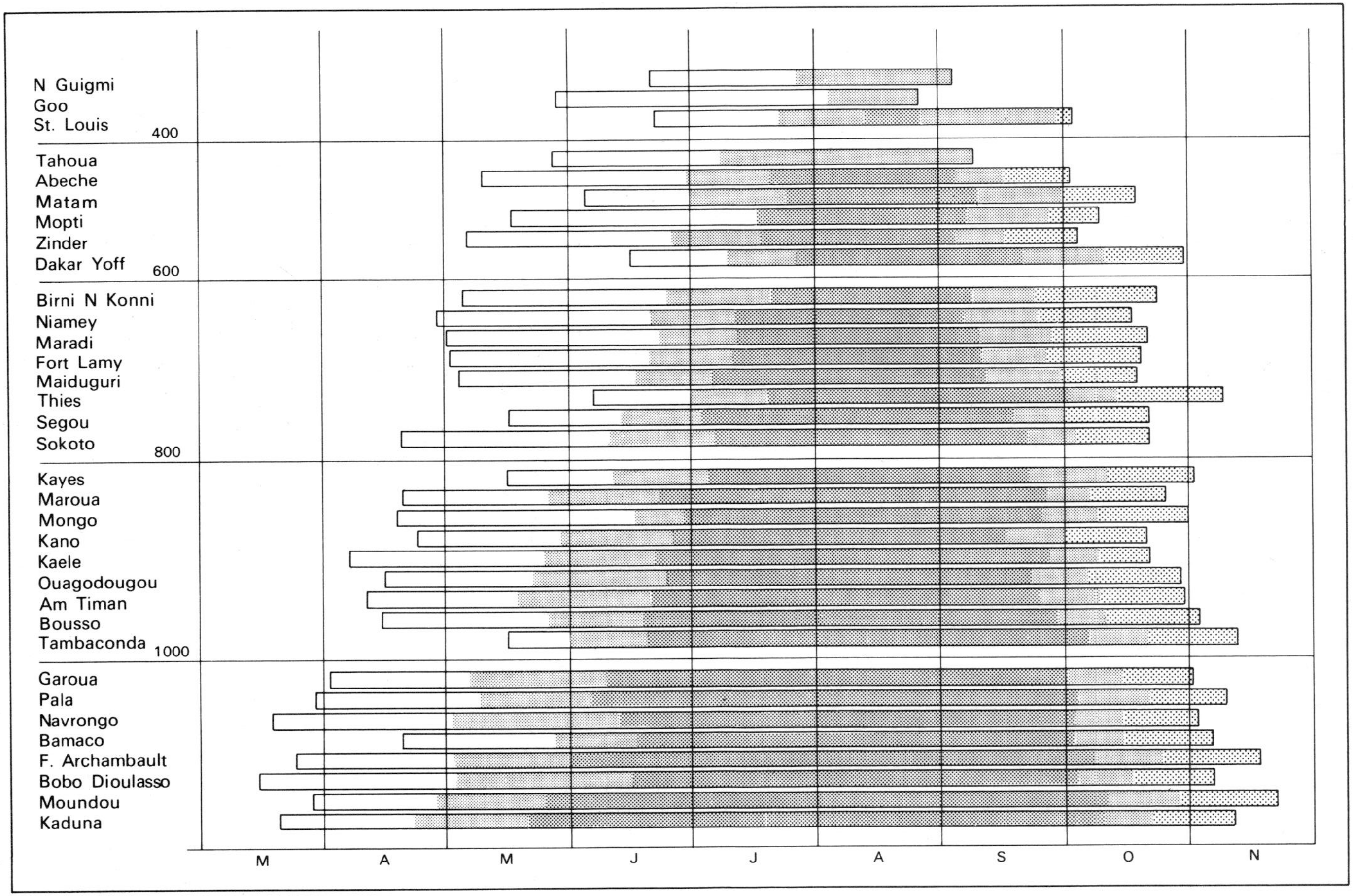

**Figure 4.14** Length of moisture periods defined in figure 4.13 at a selection of stations in west Africa. Stations arranged in order of increasing rainfall and in classes 200 mm apart. The preparatory period is blank, the humid period is darkest, and the intermediate shaded; the shading at extreme right represents reserve addition to moist period. (*After Cochemé, 1968.*)

corresponds with the Anatolian Plateau throughout the year and with the high plateaux of Spain and North Africa in summer. Wallén suggests that these climatic similarities could be important when considering the exchange of crop varieties between areas.

# References

ALI, F. M. 1953: Prediction of wet periods in Egypt four to six days in advance. *Journal Meteorology* **10**, 478.

ÅNGSTRÖM, A. 1936: A coefficient of humidity of general applicability. *Statens Met-Hydrogr. Anstalt Meda.* **11**, Stockholm.

BJERKNES, J. 1969: Atmospheric teleconnections from the equatorial Pacific. *Monthly Weather Review* **97**, 163.

BRICHAMBAUT, G. P. de and WALLÉN, C. C. 1963: A study of agroclimatology in semi-arid and arid zones of the Near East. *WMO Technical Note* **56**, Geneva.

BUDYKO, M. I. 1956: The heat balance of the earth's surface. Translated by N. I. Stepanova. Washington: US Weather Bureau.

COCHEMÉ, J. 1968: Agroclimatology survey of a semi-arid area in West Africa south of the Sahara. In *Agroclimatological methods. Proceedings of the Reading Symposium.* Paris: UNESCO, p. 235.

COCHEMÉ, J. and FRANQUIN, P. 1967: An agroclimatology survey of a semi-arid area in Africa south of the Sahara. *WMO Technical Note* **86**, Geneva.

DEACON, E. L. 1969: Physical processes near the surface of the earth. In Flohn, H. (editor), *General climatology*, **2**. Amsterdam: Elsevier Publishing Co., p. 39.

DIETRICH, G. and KALLE, K., 1957. *Allgemeine Meereskunde: eine Einführung in die Ozeanographie.* Berlin: Gegrüder Borntraeger.

DROZDOV, O. A. and GRIGOR'EVA, A. S. 1963: *The hydrologic cycle in the atmosphere. (Vlagooborot v atmosfere.)* Leningrad: Gidrometeorologicheskoe Izdatel'stvo.

EMBERGER, L. 1955: *Une classification biogéographique des climats. Recueil des travaux.* Faculté Sciences de L'Université de Montpellier. Fascicule **7**.

FITZPATRICK, E. A., SLATYER, R. O. and KRISHNAN, A. I. 1967: Incidence and duration of periods of plant growth in central Australia as estimated from climatic data. *Agricultural Meteorology* **4**, 389.

GAUSSEN, H. 1963: Bioclimatic map of the Mediterranean zone. *UNESCO Arid Zone Research* **21**.

KIRK, T. H. 1964: Discontinuities with reference to Mediterranean and North African meteorology *WMO Technical Note* **64**, Geneva, 35.

KÖPPEN, W. 1931: *Die Klimate der Erde.* Berlin: Walter de Gruyter.

LANDSBERG, H. 1951: A study of the Hawaiian rainfall. *Meteorological Monographs* **3**.

MARTONNE, E. de 1926: L'indice d'aridité. *Bulletin Association de Géographes Français* **9**.

MCALPINE, J. R. 1970: Estimating pasture growth periods and droughts from simple water balance models. *Proceedings of International Congress on Grassland* **11**, 484.

MEIGS, P. 1953: World distribution of arid and semi-arid homoclimates. *UNESCO Arid Zone Research* **1**, 203.

MINISTERIO DA AGRICULTURA 1948: *Atlas pluviometrico da Brasil (1914–1938).* Rio de Janeiro.

PRIESTLEY, C. H. B. 1966: The limitation of temperature by evaporation in hot climates. *Agricultural Meteorology* **3**, 241.

SLATYER, R. O. 1962: Climate of the Alice Springs area. CSIRO *Land Research Division* **6**, 109.

SLATYER, R. O. 1967: *Plant-water relationships.* London: Academic Press.

SLATYER, R. O. 1968: The use of soil water balance relationships in agroclimatology. In *Agroclimatological methods. Proceedings of the Reading Symposium.* Paris: UNESCO, p. 73.

SOLIMAN, K. H. 1953: Rainfall over Egypt. *Quarterly Journal Royal Meteorological Society* **79**, 389.

SOLOT, S. B. 1943: The meteorology of central Africa. *Report of Weather Research Center*. Accra, Gold Coast: 19th Weather Region.

THORNTHWAITE, C. W. 1933: The climates of the earth. *Geographical Review* **23**.

THORNTHWAITE, C. W. 1948: An approach toward a rational classification of climate. *Geographical Review* **38**, 55.

TREWARTHA, G. T. 1962: *The earth's problem climates*. Madison: University of Wisconsin Press.

WALLÉN, C. C. 1967: Aridity definitions and their applicability. *Geografiska Annaler* **49**, 367.

WALLÉN, C. C. 1968: Agroclimatological studies in the Levant. In *Agroclimatological Methods. Proceedings of the Reading Symposium*, Paris: UNESCO, p. 225.

WIESNER, C. J. 1970: *Hydrometeorology*. London: Chapman and Hall Ltd.

WINKWORTH, R. E. 1970: The soil water regime of an arid grassland community in central Australia. *Agricultural Meteorology* **7**, 387.

WINSTANLEY, D. 1970: The North African flood disaster, September 1969. *Weather* **25**, 390.

WORLD METEOROLOGICAL ORGANIZATION, 1964: High-level forecasting for turbine-engined aircraft operations over Africa and the Middle East. *WMO Technical Note* **64**, Geneva.

# 5
# The disturbed trade-wind climates

The subtropical anticyclones are most strongly developed over the eastern parts of the oceans, and so cause extreme desert conditions along the western margins of the subtropical continents. Atmospheric disturbances are frequent over the western parts of the subtropical oceans and since these disturbances drift westward, they bring ample rainfall to the eastern coasts of the subtropical land masses. The most frequent disturbances are the easterly waves and tropical storms (cyclones) discussed in chapter 3, which are best studied over the northwestern parts of the tropical Atlantic and Pacific Oceans. Since the occurrence of tropical storms is the essential and basic characteristic of this climatic type, it is necessary to identify storm development areas.

## 1  Tropical storm development regions

Figure 5.1 designates the various tropical storm development regions, tropical storms and cyclones being defined according to the WMO definitions mentioned in chapter 3. Table 5.1 lists the areas shown in figure 5.1 and adds further details. The $26\frac{1}{2}$°C sea-surface isotherm for the summer months is also shown in figure 5.1. Figure 5.2 gives the latitudinal

**Table 5.1**  Tropical storm development areas

| Area Number | Area Location | Average number of tropical storms per year | Per cent of world-wide total |
|---|---|---|---|
| I | NE Pacific | 10 | 16 |
| II | NW Pacific | 22 | 36 |
| III | Bay of Bengal | 6 | 10 |
| IV | Arabian Sea | 2 | 3 |
| V | South Indian Ocean | 6 | 10 |
| VI | Off NW Australian coast | 2 | 3 |
| VII | South Pacific | 7 | 11 |
| VIII | NW Atlantic (including W Caribbean and Gulf of Mexico) | 7 | 11 |

(Tropical storms are defined by WMO as a warm-core vortex circulation with sustained maximum winds of at least 20 m/sec.)

*After Gray (1968).*

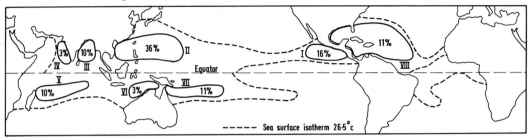

**Figure 5.1** Designation of various tropical storm development regions and percentage of tropical storms occurring in each region relative to the global total. Further details of the development regions are given in table 5.1. The 26·5°C sea-surface isotherm for August in the northern hemisphere and January in the southern hemisphere is also shown. (*After Gray, 1968.*)

distribution of sea-surface temperatures in the regions where disturbances and storms develop. It has been suggested by Palmer (1951) that a sea-surface temperature of at least $26\frac{1}{2}$°c is necessary for storm development and it is clear that this particular isotherm encloses all the development areas.

Since tropical storms and cyclones consist largely of cumulonimbus and cumulus clouds,

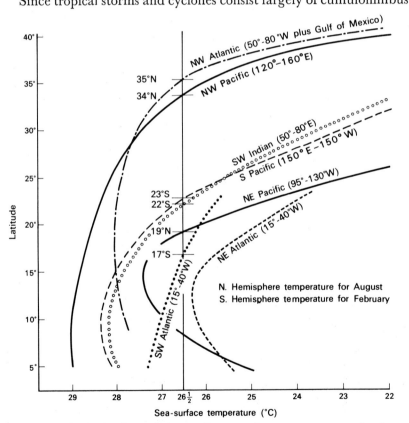

**Figure 5.2** Latitude variations of sea-surface temperatures in the various storm development regions during the warmest summer month. (*After Gray, 1968.*)

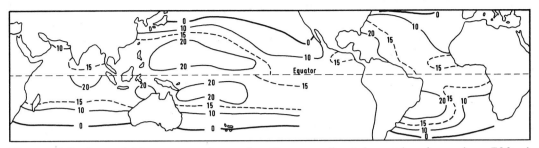

**Figure 5.3** Potential buoyancy of cumulus. Summer climatology of surface minus 500 mb level equivalent potential temperature difference. Northern hemisphere is for an August–September average; southern hemisphere is for a January–February average. (*After Gray, 1968.*) The equivalent temperature ($T_e$) of a moist air sample is the temperature that would be attained on the assumption of condensation at constant pressure of all the water vapour in the sample, all the latent heat released in the condensation being used to raise the temperature of the sample. The equivalent potential temperature is found on an aerological diagram by progressing along a dry adiabatic line from $T_e$ to the 1000 mb level.

it is of interest to study the cumulus potential buoyancy in the lower half of the tropical troposphere. Figures 5.3 and 5.4 portray the average potential buoyancy over the tropical oceans in summer and winter respectively. In these figures the potential buoyancy is defined as the difference of equivalent potential temperature from the surface to 500 mb. Gray (1968) states that it is controlled mainly by the surface air temperature and humidity, which in turn are controlled by the sea-surface temperatures. The potential buoyancy is greatest in the summer season over the warm western oceans, while in the subtropics and over the cold water areas of the eastern oceans it falls to low values. Tropical storms and cyclones are most frequent in the summer (see figure 5.5).

The subtropical anticyclones are best developed over the eastern ends of the subtropical oceans (see figure 5.6, where the summer averages of surface relative vorticity are reproduced). Large areas of negative (anticyclonic) vorticity are seen in the subtropics and over the cold water eastern sectors of the oceans. On the whole, there is a strong correlation between the areas of large-scale positive relative vorticity and the areas of initial disturbance formation.

A detailed study of the locations of the tropical storm development regions gives further insight into the causes of their geographical distribution. The South Pacific is largely free of

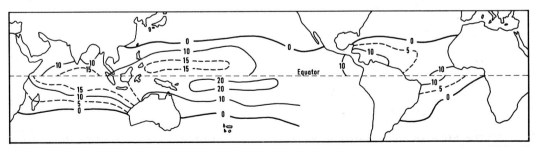

**Figure 5.4** Potential buoyancy of cumulus. Winter climatology of surface minus 500 mb level equivalent potential temperature difference. Northern hemisphere is for a January–February average; southern hemisphere is for an August–September average. (*After Gray, 1968.*)

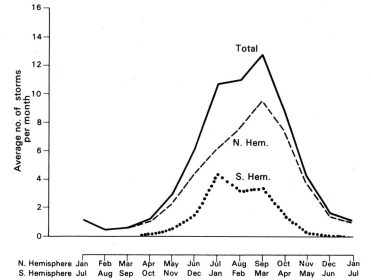

**Figure 5.5** Monthly totals of tropical storms relative to solar year. (*After Gray, 1968.*)

storms except for a narrow zone off Queensland over relatively warm ocean water, and comment has already been made in chapter 4 on the general subsidence over the equatorial eastern Pacific. An interesting area of storm development is located off western Mexico, which is a region of high rainfall, in marked contrast to the arid equatorial region to the south. The heavy precipitation is probably partly a result of disturbances from the Atlantic passing over the narrow neck of land which forms Central America and developing fully in the Pacific. It is curious that these storms do not normally move very far into the Pacific, presumably because of the strong anticyclonic circulation just to the west. Major tropical storms do not normally develop in the South Atlantic Ocean, mainly because of the persistence of the subtropical anticyclone. This is clearly shown in figure 5.6 where the only positive relative vorticity is near the African coast. It would seem reasonable that a storm development area should exist off the coast of southwest Asia between the entrances to the Red Sea and the Persian Gulf. Tropical storms occur in the northern summer around the coasts of India, but they do not move far across the Arabian Sea. In summer, relatively cold water is found in the western Arabian Sea, and the area is one of marked subsidence,

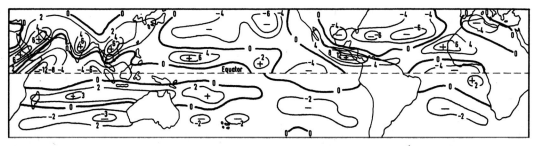

**Figure 5.6** Monthly averages of surface relative vorticity in units of $10^{-6} sec^{-1}$ for the northern hemisphere in August and the southern hemisphere in January. (*After Gray, 1968.*)

probably resulting from the ascent over India. The Indian subcontinent is a special case in the study of tropical climates and is discussed further in chapter 6 on monsoon climates.

Though on the whole the ocean islands experiencing climates influenced by tropical storms all enjoy a similar type of climate, there are some differences. The Philippine area is significantly different from the other regions in that it enjoys a seasonal reversal of the surface winds, which occurs under the influence of the Indian monsoon. This has a slight influence on the distribution of precipitation but appears to be environmentally of little consequence.

The particular climate of the disturbed trade winds is distinctly maritime because major tropical storms require large amounts of water vapour which can only be supplied by warm oceans. Nevertheless, the climate of these tropical seas does impose itself completely on small tropical islands and also on neighbouring coastal areas. The distinctive features of the climate are the extremely high rainfalls and wind speeds associated with tropical storms. The highest daily rainfalls in the world are found in the subtropics, intensities often being greater than those found nearer to the equator. Tropical cyclone winds have little influence on the long-period wind averages according to Watts (1969), because cyclones rarely affect any particular location more than once during each season and because winds of hurricane force (33 m/sec) are rarely maintained at any one place for as long as two hours. Gale force winds (17 m/sec) occur in Hong Kong on an average of only once per year (Heywood, 1950), with many gale-free years and a maximum of five in 1964. Even if high winds are infrequent, they are important because of the vast amount of damage and loss of life they can cause in coastal areas from storm-surges.

Information on peak wind speeds within tropical cyclones is difficult to obtain, but gusts of 75 and 79 m/sec have been recorded at Hong Kong, and short-period winds of 103 m/sec have been reported during an intense typhoon over the China Sea. The greatest storm-surge (recorded tide-height minus predicted tide-height) at Hong Kong is a little over 2 m, but surges up to 6 m have probably occurred in narrowing bays and estuaries along the China coast (Watts, 1959); storm-surges are considered in detail in chapter 6. The loss of life associated with tropical cyclones can be enormous: along the China coast, 300,000 people died at Haiphong during a typhoon in 1881 (Tannehill, 1927), 60,000 at Swatow in 1922 (Heywood, 1950), and 11,000 in Hong Kong in September 1937 (Jeffries, 1938).

## 2 *Extreme wind speeds*

Extreme-value wind distributions are determined by two parameters, a shape parameter, y, which defines the shape of the frequency distribution, and a scale parameter, $\beta$, which determines the wind speeds themselves. Thom (1968a and b) has found that for storms in the USA, Caribbean, and Hong Kong, the distribution of the extreme wind speeds, v, is of the Frechet type and is given by

$$F(v) = \exp\left[-(v/\beta)^{-y}\right]$$

where y equals 9 for extratropical storms and thunderstorms and 4·5 for tropical storms. He further concluded for the southern North Atlantic and the Caribbean that the annual extreme wind population, $G_v$, will be a mixture of extremes from tropical and extratropical storms, giving a mixed extreme-value distribution function:

$$G(v) = p_E \exp\left[-(v/\beta)^{-9}\right] + p_T \exp\left[-(v/\beta)^{-4·5}\right]$$

where the first term allows for extratropical storms and the second for tropical storms; $p_E$ and $p_T$ are the probabilities of an annual extreme being produced by an extratropical and a tropical storm respectively, and $p_E = 1 - p_T$; and the parameter $\beta$ is related to the highest average monthly mean wind speed $V_m$:

$$\beta = (320 \cdot 5\ V_m + 248 \cdot 7)^{\frac{1}{2}} - 15 \cdot 7$$

Thom also found that $p_T$ could be estimated from the mean number of tropical storm passages per year, f, through the area concerned:

$$p_T = 1/[1 + 99\ \exp\ (-\ 3 \cdot 0\ f)]$$

**Table 5.2**  Fastest mile speeds for return periods of 10, 50, and 100 years

| Location | Return Period (years) | | |
|---|---|---|---|
| | 10 | 50 | 100 |
| | | mile/h | |
| San Juan, Puerto Rico | 72 | 94 | 105 |
| Palisadoes, Jamaica | 71 | 93 | 105 |
| Coolidge, Antigua | 78 | 102 | 113 |
| Seawell, Barbados | 79 | 100 | 110 |
| Pearls, Grenada | 67 | 85 | 94 |
| Piarco, Trinidad | 48 | 63 | 69 |
| East Coast, Trinidad | 63 | 78 | 86 |
| Crow Point, Tobago | 67 | 85 | 93 |

(The speeds given in mile/h can be converted to m/sec by multiplying by 0·447.)

*After Shellard (1971).*

Shellard (1971) has studied extreme wind speeds in the Caribbean by using the above equation. He suggests that the fastest mile speeds for selected islands are as listed in table 5.2; corresponding maximum (3-second) gust speeds are shown in table 5.3. The British

**Table 5.3**  Maximum (3-second) gust speeds for return periods of 10, 50, and 100 years

| Location | Return Period (years) | | |
|---|---|---|---|
| | 10 | 50 | 100 |
| | | m/sec | |
| Palisadoes, Jamaica | 38 | 49 | 55 |
| Coolidge, Antigua | 42 | 54 | 59 |
| Seawell, Barbados | 42 | 52 | 57 |
| Pearls, Grenada | 36 | 45 | 50 |
| Piarco, Trinidad | 27 | 34 | 37 |
| East Coast, Trinidad | 34 | 42 | 46 |
| Crown Point, Tobago | 36 | 45 | 49 |

*After Shellard (1971).*

*Plate 9*  Typhoon damage in Hong Kong. The photograph was taken while the signposting on the right was collapsing; seconds later it was carried away by the wind. *Reproduced by permission of the Controller of Her Majesty's Stationery Office; Crown Copyright reserved.*

Standard Code of Practice on Wind Loading has adopted as the basic wind speed for design purposes, the maximum 3-second gust speed at 10 m above the ground with a return period of 50 years. On this basis, Shellard proposes that basic design wind speeds for the Caribbean Islands range from 45 to 60 m/sec, values which may also be applicable to the Far East. He further suggests that an engineer who wishes to design a structure to withstand the full force of a major tropical storm may choose a return period of perhaps 1,000 years and a corresponding wind speed of 80 m/sec or more. Tropical cyclones do not come within 5 degrees of the equator, so design wind speeds fall equatorwards.

## Temperature

The general radiation characteristics of tropical climates have been discussed in earlier chapters. The typical temperature characteristics of tropical cyclone regions can be illustrated by a study of the south China coast and the Philippines. In January average temperatures along the south China coast vary between about 14° and 16°C, but in July they reach about 28°C. This temperature regime is completely different from that found nearer the equator where uniform temperatures tend to prevail throughout the year. During the winter an intense anticyclone becomes established over Mongolia and cold air flows south over China, leading to extremely cold conditions in north China and a steep meridional temperature gradient. Along the coast the 0°C January mean isotherm reaches 35°N and inland it comes south of 30°N, bringing almost polar conditions into the tropics. A similar steep temperature gradient is found in winter, northwards from the gulf coast of the USA The cold air normally warms very quickly over the sea, and therefore cold air rarely reaches off-shore islands. Over most of China in summer the temperatures are uniformly high, and similar comments can be made about the southern USA.

At oceanic islands temperatures are higher in winter than those of the coast. In the Philippines, which are a good example of oceanic islands within the typhoon zone, the coldest month is January with a lowland average of 25·6°C; the hottest month is May with an average of 28·2°C; at 42 stations located there the average annual range is 2·6°C. These temperature conditions are very similar to those found in the equatorial trough region. The southern island of Mindanao can be considered to lie within the trough zone, but even Basco, the northern-most station in the Philippines, has only an annual range of 6·0°C.

## Rainfall

Most stations within the tropical storm regions show summer rainfall maximums, but the rainfall regimes can be greatly influenced by local conditions. The summer rainfall maximum is correlated with the increased frequency of disturbances in that season, but as much of it comes from waves in the easterlies and other weak disturbances, the influence of tropical cyclones may be small. Annual rainfall totals can be very high and often exceed 2,000 mm, particularly on windward-facing hill slopes.

As in the case of temperature, the rainfall characteristics of this particular climatic type are illustrated by reference to south China and the Philippines. Annual rainfall is greatest in the southeastern provinces of China with totals exceeding 2,000 mm over large areas; it decreases towards the northeast. Hong Kong may be taken as typical of the south with, over 2,000 mm/year, 25 mm or more each month, more than 25 mm in each summer month, and a maximum in June. Minor disturbances can be equally as important as major ones in bringing rainfall, but Lu (1944) considers that typhoons provide 30–35 per cent of the rainfall on the south China coast. Chang (1958) estimates that the average contribution of typhoons to annual rainfall is 21 per cent at Canton, 17 per cent at Foochow, 11 per cent at Shanghai, 4 per cent at Nanking, 3 per cent at Peking, and only 2 per cent at Wuhan. At Hong Kong, Watts (1969) judges that typhoons contribute 24 per cent of the total rainfall, and in July, September, and October about 40 per cent. According to Kao and Tsang (1957), on average three typhoons strike the China coast between Shanghai and Canton each year, but there are still many records of subnormal rainfall on this coast in years when successive storms have either passed along 20°N latitude to the south of Kwangtung or recurved towards Japan before entering the China Sea. The annual rainfall is therefore characterized by irregularity and the standard deviation of annual rainfall along the south China coast reaches 400 mm.

Flores and Balagot (1969) state that the average annual rainfall for the Philippines is 2,533 mm, with individual totals ranging from less than 1,000 mm to over 4,000 mm. Taking an average for the whole country, August has the highest rainfall with 304 mm, while April has the least with 100 mm. June to December may be considered the rainy months and coincide with the typhoon season, leaving January to May as the dry months. There are many local rainfall variations within the Philippines since the eastern coasts of the islands derive their rainfall not only from tropical cyclones and the southwest monsoon, but also from the northeast trades, while the western coasts do not receive appreciable rain from the northeast trades which bring dry weather. The driest localities are found in valleys or plains which are shielded by high mountain ranges.

The world's greatest observed point rainfalls are listed in table 5.4; those above 9 hours

**Table 5.4**  World's greatest observed rainfall depths (as of 1965)

| Duration | Depth (mm) | Location | Date |
|---|---|---|---|
| 1 min | 31·2 | Unionville, Maryland | 4 July 1956 |
| 8 min | 126·0 | Fussen, Bavaria | 25 May 1920 |
| 15 min | 198·1 | Plumb Point, Jamaica | 12 May 1916 |
| 20 min | 205·7 | Curtea-de-Arges, Romania | 7 July 1889 |
| 42 min | 304·8 | Holt, Missouri | 22 June 1947 |
| 2 hr 10 min | 482·6 | Rockport, W Virginia | 18 July 1889 |
| 2 hr 45 min | 558·8 | D'Hanis, Texas | 31 May 1935 |
| 4 hr 30 min | 782·3 | Smethport, Pennsylvania | 18 July 1942 |
| 9 hr | 1,086·9 | Belouve, La Réunion | 28–29 Feb 1964 |
| 12 hr | 1,340·1 | Belouve, La Réunion | 28–29 Feb 1964 |
| 18 hr 30 min | 1,688·8 | Belouve, La Réunion | 28–29 Feb 1964 |
| 24 hr | 1,869·9 | Cilaos, La Réunion | 15–16 Mar 1952 |
| 2 days | 2,500·0 | Cilaos, La Réunion | 15–17 Mar 1952 |
| 7 days | 4,110·0 | Cilaos, La Réunion | 12–19 Mar 1952 |

*After Paulhaus (1965).*

in duration all occurred in the tropics. The occurrence of extreme rainfalls is due to a combination of exceptional conditions such as the presence of very moist air together with an efficient storm mechanism, and perhaps favourably located topography. At the time of writing, the world's highest rainfalls for durations of 1 to 8 days were recorded at Cilaos, La

Réunion. Here 1,870 mm of rain fell in one 24 hour period during 1952, the result of a combination of an intense tropical cyclone together with the funnelling of the winds up a steep valley which rises to about 3,000 m. The previous world record 24 hour fall was 1,168·1 mm, during July 1911, at Baguio on the western slope of the Cordillera Central Ranges in northern Luzon, Philippine Islands. This was associated with the passage of a typhoon, the rainfall being greatly intensified by the relief.

To exploit the full flood potential of a river catchment, the duration of the storm should exceed the time of concentration of the catchment, that is to say, the time taken for water to travel from the most distant part of the catchment to the exit. Obviously, as the catchment increases in size so too will the time of concentration. Under uniform rainfall conditions over the catchment, the river discharge will increase until the time of concentration is exceeded, after which it becomes constant, and is proportional to the rainfall intensity.

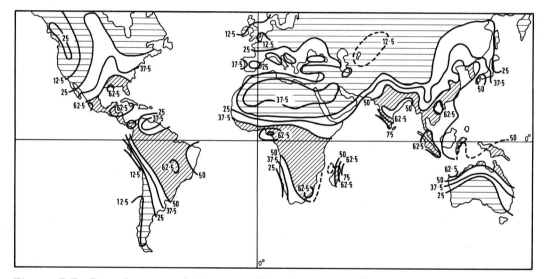

**Figure 5.7** Tentative map of 1-hour precipitation value likely to be equalled or exceeded once every 2 years (mm). (*After Reich, 1963.*)

Under extreme conditions the soil moisture content and rate of infiltration will be of little importance, since most of the rainfall will run off directly over the soil surface. The type of storm capable of causing an extreme flood varies with the size of the catchment. For example, thunderstorms are important causes of flooding in catchments with times of concentrations of the order of about 15 to 30 minutes, while tropical storms are more significant in those with times from 12 to 36 hours.

The intensity of thunderstorms is partly controlled by the supply of moist air, which is greatest near the equator, and it is likely therefore that the greatest short-term rainfalls will be found in the humid equatorial zone. Reich (1963) has produced a tentative map (figure 5.7) of 1 hour rainfalls likely to be equalled or exceeded only once every 2 years, and this clearly shows that the greatest intensities are near the equator. Above the thunderstorm scale, the organization of storms within 5° of the equator is poor and so the maximum extreme rainfalls with durations above about 6 hours are found in the subtropics.

Within the subtropical regions, the probable maximum precipitation will be caused by a tropical cyclone, though other types of less severe storm can produce very heavy rainfalls. Schwarz (1963) suggests that the probable maximum point precipitation resulting from a tropical cyclone without any orographic influence is between 1,000–1,100 mm in 24 hours. This is the highest daily rainfall which will result from the passage of a tropical cyclone over a completely flat island. The extreme daily rainfalls quoted earlier result from a combination of a severe storm and relief suitable for intensifying the rising air currents. Therefore, the daily probable maximum point precipitation over land areas can only be arrived at by adding a suitable orographic component to the probable maximum storm precipitation.

## Evapotranspiration and water balance

Potential evapotranspiration values for Canton were computed by Penman's method, while mean values for Hong Kong were taken from data published by Watts, and the results, shown in table 5.5, are typical for the continental coastal regions in this climatic

**Table 5.5** Average annual water balance (mm)

| Month | Rainfall | $E_t$ | $E_a$ | Soil Moisture | Runoff |
|---|---|---|---|---|---|
| | | Canton | | | |
| Jan | 27·1 | 56·7 | 27·1 | — | — |
| Feb | 65·0 | 54·9 | 54·9 | 10·1 | — |
| Mar | 100·7 | 67·2 | 67·2 | 43·6 | — |
| Apr | 184·7 | 81·6 | 81·6 | 100·0 | 46·7 |
| May | 256·1 | 119·1 | 119·1 | 100·0 | 137·0 |
| Jun | 291·5 | 126·0 | 126·0 | 100·0 | 165·5 |
| Jul | 264·4 | 157·2 | 157·2 | 100·0 | 107·2 |
| Aug | 248·5 | 149·4 | 149·4 | 100·0 | 99·1 |
| Sept | 149·1 | 135·6 | 135·6 | 100·0 | 13·5 |
| Oct | 48·5 | 115·8 | 115·8 | 32·7 | — |
| Nov | 50·5 | 78·9 | 78·9 | 4·3 | — |
| Dec | 34·1 | 62·4 | 38·4 | — | — |
| | | Hong Kong | | | |
| Jan | 32·5 | 79·0 | 32·5 | — | — |
| Feb | 45·0 | 76·0 | 45·0 | — | — |
| Mar | 72·5 | 86·0 | 72·5 | — | — |
| Apr | 135·0 | 97·0 | 97·0 | 38·0 | — |
| May | 287·5 | 128·0 | 128·0 | 100·0 | 97·5 |
| Jun | 387·5 | 127·0 | 127·0 | 100·0 | 260·5 |
| Jul | 375·0 | 146·0 | 146·0 | 100·0 | 229·0 |
| Aug | 355·0 | 150·0 | 150·0 | 100·0 | 205·0 |
| Sept | 252·5 | 138·0 | 138·0 | 100·0 | 114·5 |
| Oct | 112·5 | 137·0 | 137·0 | 75·0 | — |
| Nov | 42·5 | 113·0 | 113·0 | 5·0 | — |
| Dec | 30·0 | 89·0 | 35·0 | — | — |

$E_t$—Potential evapotranspiration
$E_a$—Actual evapotranspiration
Soil moisture calculated assuming a field capacity of 100 mm.

*Computed from data in Watts (1969).*

type. The annual water balances for Canton and Hong Kong are also shown in table 5.5. Since the rainfall is greatest in summer, there is a large water surplus and considerable run-off between May and September. The winter months have only slight rainfall and normally this falls short of the potential evapotranspiration, suggesting a water deficit. Indeed, the water-balance calculations indicate that the soil must become almost completely dry by the end of the northern winter and that drought must prevail.

From botanical and pedological indications (Frenzal, 1930; Hou, Chen, and Wang, 1956; Grant, 1968), it appears that tropical evergreen rain-forest, similar to that on Hainan Island, is the natural vegetation of the Hong Kong–Canton area, but that poor soil and human interference have prevented the natural regeneration of the forests after intensive deforestation in the tenth and eleventh centuries. During the Sung Dynasty (960–1279 AD) the coastal plains and larger river valleys were colonized by farmers who embarked on an effective policy of deforestation to eliminate troublesome animals and to provide timber for the new settlements. Under normal conditions of decay and natural replacement, the soil surface had not been subject to the direct erosional effect of rainfall, but the felling was followed by erosion of the poorly cohesive organic and podzolic topsoil. The local Hong Kong pine (*Pinus massoniana*) has shown remarkable powers of regeneration and colonization in clear felled or eroded areas (Frenzel, 1930), so it is probable that the original forest was replaced, after a period of severe erosion, by the existing pine and grassland vegetation. Unfortunately, the acid pine-needle litter accelerates and intensifies podzolization processes so that there has been a progressive deterioration in the soil structure and in the nutrient status of the soil.

The data available from Hainan Island are poor and do not allow satisfactory water-balance diagrams to be constructed, but Yulin does appear to have a climate which is at least as dry as that at Hong Kong. Allowing for a field capacity of 200 mm within the rooting zone of forest trees growing on a deep soil, the period of insignificant soil moisture at Yulin probably lasts for about 4 months. A similar field capacity at Hong Kong creates a dry period of 3 months, while the soil always contains some moisture at Canton. From these considerations there is no reason why the natural vegetation of Hainan should not also exist in the Hong Kong area.

Much of the effective rainfall comes within a very short season, most of the year having various degrees of water deficit. The amount of water which can be stored and so carried over to the dry season is therefore of more importance than in less extreme climates. This is illustrated in table 5.6, which is constructed assuming field capacities ranging from 50 to 200 mm. The length of the period with an almost completely dry soil increases with decreasing field capacity, and can reach 5 months at Hong Kong with a field capacity of 50 mm. Similarly, at Canton the dry period ranges from almost zero with a field capacity of 200 mm, up to 4 months at 50 mm. Clearly, vegetation which can survive the dry period while growing on a deep soil with a large field capacity may be completely unable to grow on a thin soil which is dry for a significantly longer time. Thus the original tropical evergreen rain-forest of the Canton area survived because it was growing on deep soil with a large moisture-holding capacity, but the soils are now considerably thinner because of erosion and therefore the effective dry season has been increased in length and the original forest no longer exists.

An important crop of many oceanic islands in the disturbed trade-wind area is sugar-cane, and the influence of water balance on this crop has been studied in Barbados by

**Table 5.6**  Soil moisture at end of month assuming various field capacities

| Month | Field capacities of | | | | Field capacities of | | | |
| | 50 mm | 100 mm | 150 mm | 200 mm | 50 mm | 100 mm | 150 mm | 200 mm |
| | | Hong Kong | | | | Canton | | |
|---|---|---|---|---|---|---|---|---|
| Jan | — | — | — | — | — | — | — | 46·5 |
| Feb | — | — | — | — | 10 | 10 | 10 | 56·5 |
| Mar | — | — | — | — | 43·5 | 43·5 | 43·5 | 90 |
| Apr | 38 | 38 | 38 | 38 | 50 | 100 | 146·5 | 193 |
| May | 50 | 100 | 150 | 198 | 50 | 100 | 150 | 200 |
| Jun | 50 | 100 | 150 | 200 | 50 | 100 | 150 | 200 |
| Jul | 50 | 100 | 150 | 200 | 50 | 100 | 150 | 200 |
| Aug | 50 | 100 | 150 | 200 | 50 | 100 | 150 | 200 |
| Sept | 50 | 100 | 150 | 200 | 50 | 100 | 150 | 200 |
| Oct | 25·5 | 75·5 | 125·5 | 175·5 | — | 32·5 | 82·5 | 132·5 |
| Nov | — | 5·0 | 55·0 | 105 | — | 4·5 | 54·5 | 104·5 |
| Dec | — | — | — | 46 | — | — | 26 | 76 |

Oguntoyinbo (1966). Here, the annual rainfall varies from 1,000 mm to over 2,000 mm, with the wet season mainly between July and November. Actual and potential evapotranspiration values through a crop cycle are reproduced in figure 5.8, showing that periods of soil moisture deficiency occur during the dry season. Investigations in Hawaii (Chang, *et al.*, 1963) have shown that the availability of moisture to sugar-cane is a major factor in determining the size of the yield to be expected. When the ratio of the actual to the potential evapotranspiration falls below 0·73, the size of the yield will depend to a great extent on the period when the drought occurred, the formative stages being the most critical. Oguntoyinbo has discovered that 60 per cent of the crop cycles in the low rainfall

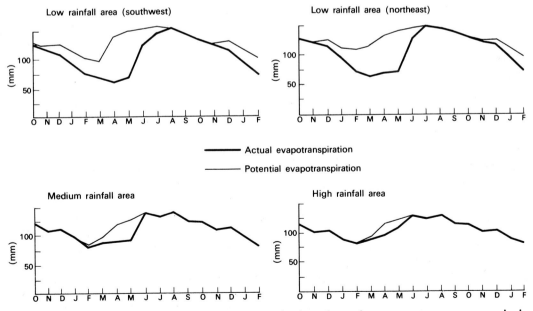

**Figure 5.8**  Actual and potential evapotranspiration through a sugar-cane crop cycle in Barbados. (*After Oguntoyinbo, 1966.*)

areas (below 1,400 mm year) of Barbados have periods when the actual to potential evapotranspiration ratio is below 0·73. He concludes that crops grown over about 50 per cent of the island will be critically affected by moisture deficiency, first because of insufficient moisture supply, and second because most of the droughts occur within the critical first 8 months of the crop cycle.

## References

ARAKAWA, H. (editor). 1969: *Climates of northern and eastern Asia. World survey of climatology*. Vol. **8**. Amsterdam: Elsevier Publishing Co.

CHANG, J., CAMPBELL, J. and ROBINSON, F. E. 1963: On the relationship between water and sugar-cane yield in Hawaii. *Agronomy Journal* **55**, 450.

CHANG, KEE-KAR 1958: The relation between long-period fluctuations of the frequency of occurrence of typhoons in the Pacific and long-period fluctuations of the general circulation. *Acta Meteorologica Sinica* **29**, 135.

FLORES, J. F. and BALAGOT, V. F. 1969: Climate of the Philippines. In Arakawa, H. (editor), *Climates of northern and eastern Asia*. Amsterdam: Elsevier Publishing Co.

FRENZAL, G. 1930: Problems of reforestation in Kwongtung with respect to climate. *Linnan Science Journal* **9**, 97.

GRANT, C. J. 1968: Gullying and erosion characteristics of Hong Kong soils. In Misra, R. and Gopal, B. (editors), *Proceedings of the Symposium on Recent Advances in Tropical Ecology, International Society Tropical Ecology*, Varanasi, India, p. 94.

GRAY, W. M. 1968: Global view of the origin of tropical disturbances and storms *Monthly Weather Review* **10**, 669.

HEYWOOD, G. S. P. 1950: Hong Kong typhoons. *Hong Kong Royal Observatory Technical Memorandum* **3**.

HOU, CHEN and WANG. 1956: *Vegetation of China with special reference to soil types*. Peking: Soil Science Society of China.

JEFFRIES, C. W. 1938: *Meteorological Results, 1937, Appendix* **11**. Hong Kong: Royal Observatory.

KAO, Y. H., and TSANG, Y. Y. 1957: *Typhoon tracks and some related statistical analyses*. Peking: Scientific Publishers.

LU, A. 1944: *Chinese climatology. Collected scientific papers (meteorology)*. Peking: Academia Sinica, 1954, p. 441.

MISRA, R. and GOPAL, B. 1968: *Proceedings of the Symposium on Recent Advances in Tropical Ecology*. Varanasi, India: International Society Tropical Ecology.

OGUNTOYINBO, J. S. 1966: Evapotranspiration and sugar-cane yield in Barbados. *Journal Tropical Geography* **22**, 38.

PALMER, C. E. 1951: Tropical meteorology, In Malone, T. F. (editor), *Compendium of meteorology*. Boston: American Meteorological Society, p. 859.

PAULHAUS, J. L. E. 1965: Indian Ocean and Taiwan rainfalls set new records. *Monthly Weather Review* **93**, 331.

REICH, B. M. 1963: Short-duration rainfall-intensity estimates and other design aids for regions of sparse data. *Journal Hydrology* **1**, 3.

SCHWARZ, F. K. 1963: Probable maximum precipitation in the Hawaiian Islands. *Hydrometeorological Report No.* **39**, Washington.

SHELLARD, H. C. 1971: Extreme wind speeds in the Commonwealth Caribbean. *Meteorological Magazine* **100**, 143.

TANNEHILL, I. R. 1927: Some inundations attending tropical cyclones. *Monthly Weather Review* **55**, 453.

THOM, H. C. S. 1968a: New distribution of extreme winds in the United States. *Journal Structural Division American Society Civil Engineers* **94**, 1787, New York.

THOM, H. C. S. 1968b: Toward a universal climatological extreme wind distribution. National Research Council of Canada, International Research Seminar: Wind effects on buildings and structures, Ottawa, September 1967. *Proceedings Vol.* **1**. Toronto: University of Toronto Press, p. 669.

WATTS, I. E. M. 1959: The effect of meteorological conditions on tide height at Hong Kong. *Hong Kong Royal Observatory Technical Memorandum* **8**.

WATTS, I. E. M. 1969: Climates of China and Korea. In Arakawa, H. (editor), *Climates of northern and eastern Asia, World survey of climatology* Vol. **8**. Amsterdam: Elsevier Publishing Co.

# 6
# The monsoon climates of southern Asia

Originally the term 'monsoon' was applied to the surface winds of southern Asia which reverse between winter and summer, but the word is now used for many different types of phenomena, e.g., the stratospheric monsoon, the European monsoon, etc. It is therefore necessary to use a definition of the term 'monsoon' that depends neither on cause nor geographical location. A suitable definition has been suggested by Chang Chia-Ch'eng (1959): 'the monsoon is a flow pattern of the general atmospheric circulation over a wide geographical area, in which there is a clearly dominant wind in one direction in every part of the region concerned, but in which this prevailing direction of wind is reversed (or almost reversed) from winter to summer and from summer to winter'. Such a wind system is found over much of southern Asia and the northern Indian Ocean.

The characteristics of the monsoon climate are to be found mainly in the Indian sub-continent, where over much of the region the annual changes may conveniently be divided as follows:

1  The season of the northeast monsoon
  a  January and February, winter season
  b  March to May, hot weather season
2  The season of the southwest monsoon
  a  June to September, season of general rains
  b  October to December, post-monsoon season

January is a fine sunny month over most of India, since the generally northeasterly winds give little or no rain. The winter season merges gradually into the hot weather season, for in April and May the sun is nearly overhead, and the days become hotter and hotter, giving the maximum annual temperatures over much of the area. In June the southwest monsoon sets in, or to use the more common term, 'bursts'. It is a change not so much in the direction of the wind as in its force, and indeed is a transformation in the whole aspect of the weather as the result of the arrival of a new-air mass undergoing new dynamic influences. The monsoon rains are not continuous and the breaks can be lengthy. In October and November the southwest monsoon retreats, the sky clears, the sun shines again, and the temperature rises for a few weeks before falling to the winter minimum.

From an environmental viewpoint, monsoon climates are very similar to those of the disturbed trade winds, and it will be shown later that the causes of rainfall in the two regimes are comparable. The difference is really a matter of degree, for although much of north India would have a desert climate if it were not for the summer monsoon rains, the same cannot be said about areas in the disturbed trade winds. Seasonal changes in parts of the Indian subcontinent are extreme, ranging from a desert-type climate in the winter season to plentiful rain in the summer, whereas, though the disturbed trade winds usually have a summer rainfall, the winter season is not normally completely dry. The transition from the true monsoon climate to that of the disturbed trade winds is often gradual. China and the Philippines have seasonal reversals of surface winds, often referred to as monsoons, and yet their climates are discussed under the disturbed trade-wind climates since they are not so seasonal as those of India and are considered to be environmentally similar to the Gulf of Mexico. It would appear, therefore, that the rather extreme climates of parts of the Indian subcontinent warrant separate discussion.

## Atmospheric circulation

Hadley proposed the first explanation of the Asiatic monsoon in a memoir presented to the Royal Society in 1686. His theory is general and can be applied to all continental regions, not only Asia. Over the cold continents during winter the intense cooling leads to the establishment of thermal high-pressure systems, while pressure remains relatively low in the warmer lighter air over the oceans. The surface air flow is therefore from the highs over the land towards the lows over the oceans. During summer, the land is warmer than the sea and a reverse current circulates from the relatively cool sea towards the heated land. The average sea-level pressure charts for the two extreme seasons (figure 2.11) show a reversal of pressure over Asia and its maritime surroundings. In winter the Siberian anticyclone accumulates subzero masses of air ($-40°$C to $-60°$C) under a pressure of between 1,040 and 1,060 mb, while in summer the torrid heat (about $50°$C) reduces the pressure to 950 mb over northwest India.

Flohn (1955) has suggested a rather different explanation of the monsoon, for he has noted that two quite different circulation patterns are discernible in the tropics according to oceanic or continental conditions. They are represented by the central Pacific and, on the other side of the globe, by central Africa and the Indo-Australian region. Over the sea only small annual temperature variations are discerned and therefore the seasonal shiftings of the thermal and pressure belts amount to only a few degrees. Over land, large annual temperature changes are perceived and these changes cause large seasonal shifts of the temperature and pressure belts; for example, over India in summer the equatorial trough travels more than 30 degrees from the equator. This migration of the equatorial trough leads to a pressure gradient from the equator to the region of maximum heating and lowest surface pressure, which is associated with a quasi-geostrophic westerly current deflected by surface friction towards the low-pressure area in the subtropics. Thus, the existence of the Asian monsoon is apparently not due to the thermal differences between land and sea, but primarily to the seasonal shifting of thermally-produced planetary belts of pressure and winds under continental influences. On a homogeneous oceanic earth, only tropical easterlies with frictional trade-wind components would be evident, but on a homogeneous continental earth, a zone of westerlies reaching from the equator towards the zone of lowest pressure would occur. On this basis Flohn explains the occurrence of the monsoon as being due to the thermal response of the tropical continental atmosphere

145

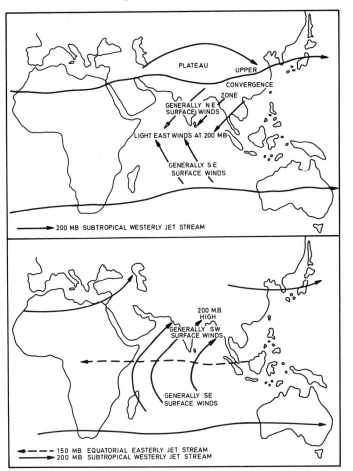

**Figure 6.1** Schematic illustrations of the major features of Asian monsoon. *Upper:* northern winter. *Lower:* northern summer. (*After Lockwood, 1965a.*)

to the annual variation of solar radiation. In the introduction to chapter 7 on equatorial trough climates it is suggested that there are dynamic reasons for expecting westerly winds to be rainy and cloudy and easterly winds to be generally fine and dry.

When upper winds are taken into account, it is found that the Asian monsoon is a fairly complex system. During the northern winter season (figure 6.1), the subtropical westerly jet stream lies over southern Asia, with its core located at about 12 km altitude. It divides in the region of the Tibetan Plateau, with one branch flowing to the north of the plateau, and the other to the south. The two branches merge to the east of the plateau and form an immense upper convergence zone over China. In May and June the subtropical jet stream over northern India slowly weakens and disintegrates, causing the main westerly flow to move north into central Asia. While this is occurring, an easterly jet stream, mainly at about 14 km, builds up over the equatorial Indian Ocean and expands westward into Africa (figure 6.1). The formation of the equatorial easterly jet stream is connected with the formation of an upper-level high-pressure system over Tibet. In October the reverse process occurs; the equatorial easterly jet stream and the Tibetan high disintegrate, while the subtropical westerly jet stream reforms over northern India.

## *1  Seasonal changes*

Various authors have found relationships between the seasonal changes which take place over Asia. It was first demonstrated by Yin (1949) that the 'burst' of the Indian southwest monsoon is connected with the disintegration of the subtropical jet stream over northern India, and the formation of a new current to the north of the central Asiatic highlands. Sutcliffe and Bannon (1954) have demonstrated, for the years 1948–1953, a time-relationship between the shift of the 200 mb winds from west to east over Aden and Bahrain, the ending of the polar tropopause over Habbaniya (Iraq), and the onset of the southwest monsoon on the Malabar coast of India. Yeh tu-cheng, Dao Shih-Yen, and Li Mei-Ts'un (1959) have claimed to show that there is an abrupt change in the pattern of the upper tropospheric circulation over the northern hemisphere in June.

An examination of the rather more plentiful data now available suggests that while the conclusions of the above authors are in general correct, the changes they describe are not as sharp as originally thought. The southwest monsoon does not sweep across India in a short period of time, for in some years it first appears at the Malabar coast and then becomes stationary, or even retreats, and does not start to advance across India until a later date. Similarly, the breakdown of the subtropical westerly jet stream over northern India takes place gradually, with the current sometimes reforming after a partial breakdown.

In general, the significant changes in the upper troposphere during May and June may be considered as taking place in the following order:

1  the first notable change is the movement of the high wind speeds associated with the subtropical jet stream from the southern to the northern side of the highlands of central Asia;
2  then the equatorial easterly jet stream expands rapidly across southern Asia, central Africa, and the western Pacific;
3  and sometime after the subtropical jet stream has left India, and before the upper level easterlies become definitely established over the Indian Ocean, the southwest monsoon reaches the Malabar coast of India.

Changes in October are less distinct than those in the spring, but it is still possible to detect some order in the sequence of events. The main easterly jet stream leaves the western Indian ocean before the southwest monsoon retreats from northern India, and only then do winds of jet stream strength become noticeable over north India. This seems to be the reverse of the sequence of events occurring in May and June.

The southwest monsoon does not flow continuously over India from June to October, but instead there are frequent breaks in the monsoon rains which appear to be connected with changes in the upper troposphere. When the southwest monsoon is active, there is a marked anticyclone at 500 mb and above over Tibet; at the 200 mb level westerlies are well to the north of the plateau; and at 150 mb strong easterlies are found over India. During breaks in the southwest monsoon, a large-amplitude trough in the upper troposphere extends from middle latitudes to subtropical Indo-Pakistan, destroying the Tibetan upper anticyclone. Wind speeds and shears characteristic of a westerly jet stream are occasionally found in these troughs.

## 2 The Tibetan plateau

Emphasis has been placed in recent years on the influence of the Tibetan Plateau on the climate of Asia. The Tibetan Plateau is an enormous block, more than 4,500 m above sea level except in the valley bottoms of the east. Its length is about 2,000 km and its width about 600 km in the west and 1,000 km in the east. Atmospheric pressure on the surface of the plateau varies between 700 mb and 500 mb, and on the highest peaks it approaches 300 mb. The Tibetan Plateau can affect the atmosphere by acting in two ways, either separately or in combination: as a mechanical barrier, or as a high-level heat source.

The earlier monsoon theories involving the plateau considered it as a purely mechanical barrier, and one such theory is due to Yin (1949). At the beginning of June the westerly subtropical jet stream passes from the southern to the northern side of the plateau. Yin considers that this shift corresponds to the slowing down of the westerlies over the whole of Eurasia, and that the plateau accentuates the normal retreat of the jet stream towards the north. In the same way the barrier causes it to advance far to the south in the middle of October. His conclusion is that the sudden irruption of the surface equatorial westerly flow at the beginning of June is set off by the hydrodynamic effect of the Himalayas and not by the thermal depression of northwest India.

Rather more recent theories have stressed the influence of the Tibetan Plateau as a high-level heat source. The atmosphere is transparent to short-wave radiation, which is mostly absorbed by the soil surface. It is therefore possible that a high mountain will receive as much incoming radiation as the neighbouring lowlands. In the tropics, as incoming short-wave radiation exceeds outgoing long-wave radiation, there will be a sensible heat transfer from the mountain to the atmosphere, and the air near the mountain will be at a higher temperature than that found in the free air at the same altitude. Since warm air is less dense than cool air, pressure will fall less rapidly with height in the relatively warm air near the mountain, than in the cooler air at some distance away. The vertical distribution of equal pressure and density lines is shown in figure 6.2. The resulting solenoid field sets up a vertical circulation with rising air in the relatively high pressure over the plateau and sinking air over the plains. An anticyclone appears in the upper troposphere over Tibet during the southwest monsoon, and this high appears to be closely related to the strength of the surface winds. When Flohn (1964b) compared the

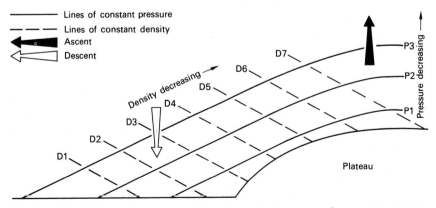

**Figure 6.2** Schematic cross-section through the atmosphere over a warm tropical plateau. (*After Lockwood, 1965a.*)

August heat balances of the Pamir (Kara-Kul, 3990 m) and the adjacent low-level Kara-kum desert, he found the values (shown in table 6.1) to be very similar. He comments that it is only the larger contribution of advection and vertical mixing in the elevated regions which prevents the temperatures from rising to the same values as at sea-level. Similarly, the Staff Members of the Academia Sinica, Peking (1958), having considered the radiation balance of the Tibetan Plateau and also the latent heat released by rainfall, deduced that the plateau acts as a heat source in summer, and that the southeast part plays the role of a warm source in winter, while the remaining parts most probably form a cold source.

**Table 6.1** Energy budget for the Pamir and the Karakum Deserts during August (units : g cal cm$^{-2}$ day$^{-1}$)

|  | Pamir | Karakum |
|---|---|---|
| Incoming short-wave radiation absorbed by soil surface | 546 | 525 |
| Outgoing long-wave radiation from soil surface | −305 | −244 |
| Net radiation | 241 | 281 |
| Sensible heat transfer surface to air | −240 | −274 |
| Sensible heat transfer surface to deep soil layers | 17 | −7 |
| Latent heat of evaporation | −18 | 0 |

Energy fluxes towards the soil surface are positive, while those away are negative.

*After Flohn (1964a).*

Recent calculations by Godbole and Kelkar (1969) suggest that in July the radiative heating close to the top of the Himalayas is about five times that at the corresponding height for the Gangetic plain. Flohn (1964a and b) is of the opinion that part of the energy for the equatorial easterly jet stream is provided by the remarkable heating of the middle and upper troposphere above the Tibetan Plateau. Because this increases the meridional pressure and temperature gradients towards the equator, it leads to a marked baroclinic state in the tropical atmosphere. To this persistent transfer of sensible heat from elevated surfaces must be added the large amounts of latent heat released by the monsoon rains over India.

As investigations over other continents reveal no comparable phenomena to the equatorial easterly jet stream nor to the southwest monsoon, the origin of these changes must therefore lie in the heat balance of the very extensive plateau of central Asia, for no other tropical continent has so much elevated land. Even so, the complete explanation of the monsoon is far more complex. The moist southwesterly winds over India are a necessary but not a sufficient condition for summer rainfall. Also required are monsoon depressions, which by increasing convergence in the monsoon current and its speed, enhance rainfall through dynamic and mechanical lifting (Sarker, 1966). For example, synoptic disturbances which bring rain to northern Burma and to eastern and northeastern India may be moderately vigorous, resembling weak typhoons. They normally form over the northern part of the Bay of Bengal and then move slowly west–north–westward along the general line of the Ganges Valley. Equally heavy rains fall in the vicinity of Bombay through the agency of subtropical cyclones, which according to Miller and Keshavamurthy (1968) usually form, intensify, and decay just off the west coast. Unlike the Bay of Bengal disturbances, these cyclones are most intense at around 4,500 to 6,000 m above the surface.

Large amounts of latent heat released in monsoon disturbances over India are available to increase the temperature of the upper troposphere, and therefore increase the temperature gradient northwards from the equator. A crude model of the energy exchanges in a warm core disturbance is shown in figure 6.3, where it can be seen that the disturbance will intensify the upper level warming from the Tibetan Plateau. The possibility that the heavy rainfall is a cause of the subtropical easterly jet is further suggested by the work of Raman and Remanathan (1964), who found a high correlation between heavy monsoon rains over Indo-Pakistan and the subsequent (several hours) development to the south of a sharply defined jet in the upper tropospheric easterlies. Heavy monsoon rains and previous development of a jet to the south were found to be uncorrelated.

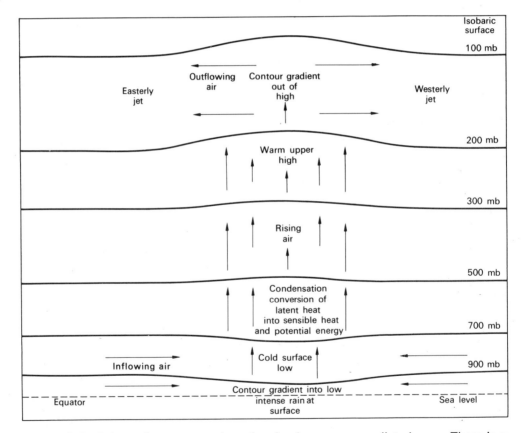

**Figure 6.3** Schematic representation of a simple warm-core disturbance. There is a cold low at the surface which is overlain by a warm upper high. The system contains active cumulonimbus clouds and is largely maintained by the latent heat released by condensing water vapour. Actual systems can have vertical structures which are more complex than the basic simple structure illustrated. (*After Lockwood, 1965b.*)

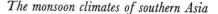

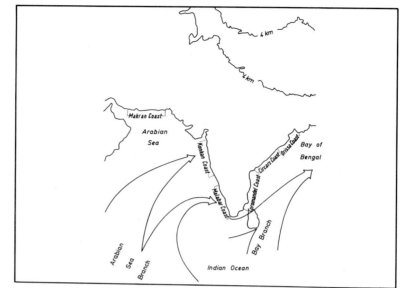

**Figure 6.4** The main branches of the south-west monsoon current in the lowest 5 km. (*After Miller and Kesh-avamurthy, 1968.*)

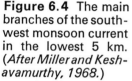

## 3   *Monsoon currents*

According to Miller and Keshavamurthy (1968), the southwest monsoon current in the lower 5 km near India consists of two main branches (figure 6.4), which are the Bay of Bengal branch, influencing the weather over the northeast part of India and Burma, and the Arabian Sea branch, dominating the weather over the west, central, and northwest parts of India. The latter branch can be further divided into two separate currents. One part of the main low-level current flows northward across the equator between 38°E and 55°E and then, turning northeast, flows over the Arabian Sea, crossing the Konkan coast north of 15°N. The southern current separates near 10°N over the Arabian Sea and flows eastward across the south Malabar coast to join the Bay of Bengal branch in the vicinity of Ceylon.

The low-level flow across the equator is not evenly distributed between latitudes 40°E and 80°E as was previously thought, but has been found by Findlater (1969a, b 1969b, 1971, 1972) to take the form of low-level high-speed southerly currents which are con-centrated between about 38°E and 55°E. A particularly important feature (Figure 6.5) of this flow is the strong southerly current with a mean wind speed of about 15 m/sec observed at the equator over eastern Africa from April to October. The strongest flow occurs near the 1·5 km level, but it often increases to more than 25 m/sec and occasionally to more than 45 m/sec at heights between 1·2 and 2·4 km. According to Findlater, this high-speed current flows intermittently during the southwest monsoon from the vicinity of Mauritius through Madagascar, Kenya, eastern Ethiopia, Somalia, and then across the Indian Ocean towards India.

Recent aircraft observations (Bunker, 1965) suggest that the depth of the southwest monsoon current in the Arabian Sea is also fairly shallow west of longitude 65°E, having a depth of only about 1·5 km. Eastwards of 65°E, however, the depth of the monsoon cur-rent increases rapidly reaching about 6 km near the Indian coastline. It appears that there is a discontinuity in the moisture content and other properties of the monsoon air along

65°E. It is also interesting to note that a similar discontinuity is not observed in the Bay of Bengal branch which is uniformly about 6 km deep. The cause of this discontinuity is not well understood but it may be due to the current striking the Western Ghats.

According to Bunker, the skies are generally clear over the western section of the Arabian Sea in summer with only a few scattered cirrus and occasional patches of lower cloud.

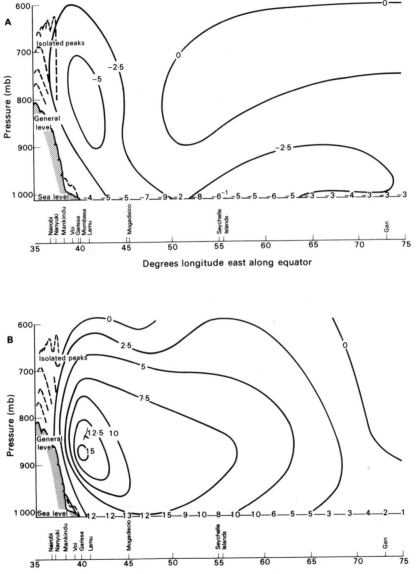

**Figure 6.5** Mean meridional flow at the equator over the western Indian Ocean (units: m sec⁻¹, surface winds in knots). A: mean January flow; B: mean July flow. (*After Findlater, 1969b.*)

East of 60°E variable amounts of cumulus occur, as well as increasing altostratus and cirro-stratus clouds. As the Indian coast is approached from the west, the amounts of cumulus cloud increase and groups of cumulonimbus are formed, while over the Indian coast showers are widespread. The heat budgets of the air east and west of 60°E are shown in table 6.2, which indicates that in summer the air is accumulating sensible and latent heat nearly twice as fast west of 60°E as east of the boundary line.

Table 6.2  Summer monsoon heat budgets for areas west and east of 60°E Meridian (units: cal day $^{-1}$)

| Layer | Sensible heat flux | Latent heat flux | Infra red cooling in layer | Absorption of short-wave radiation by layer | Net accumulation of heat |
|---|---|---|---|---|---|
| West of 60°E meridian | | | | | |
| 600–700 mb | −13 | 21 | −5 | 1 | 4 |
| 700–850 mb | 33 | 144 | −53 | 20 | 144 |
| 850–1000 mb | 109 | 248 | −63 | 36 | 334 |
| Total | 129 | 413 | −121 | 57 | 478 |
| East of 60°E meridian | | | | | |
| 600–700 mb | 17 | 29 | −33 | 12 | 25 |
| 700–850 mb | 10 | 46 | −45 | 24 | 35 |
| 850–1000 mb | 10 | 167 | −47 | 34 | 164 |
| Total | 37 | 242 | −125 | 70 | 224 |

Heat gain by layer shown by positive sign.

*After Bunker (1965).*

Traditionally, meteorologists have linked the Arabian Sea summer monsoon circulation with a heat low over Arabia, West Pakistan, and northern India (Ramage, 1965). This heat low which develops during May, establishes the low-level westerly monsoon wind regime a full month before heavy monsoon rains start over western India. In mid-July the heat low is deepest, the southwest monsoon is strongest, and the west Indian rains are heaviest. The question then arises as to why the heat low remains cloud-free while rain falls to the south? Simple reasoning would lead to the conclusion that moist air and associated rain and cloud should penetrate deep into the low pressure area.

The passage of cyclones in the vicinity of Bombay and over the Ganges Valley during the summer months leads to the release of massive amounts of latent heat into the upper troposphere and the generation of high-level easterlies with a climatological speed maximum roughly coincident with the west coast of India. Thus, above the heat low and over most of the Arabian Sea, northeast Africa, and Arabia, the upper easterlies are convergent and subsidence prevails through the middle troposphere (Flohn, 1964).

Meteorologists contend that the resulting subsidence affects the heat low in two ways. First, radiosonde ascents at Karachi and Jodhpur indicate the frequent presence of a sharp inversion at around 850 mb with moist air originating over the Arabian Sea below the inversion (Ramage, 1965). Subsidence limits the height to which surface air from the Arabian Sea can ascend, restricting cloud development, and thus favouring strong insolational heating. Second, while Dixit and Jones (1965) were comparing monsoon rain periods with monsoon lulls along the west coast of India, they located the greatest middle and upper tropospheric temperature differences above the heat low, with the rain situation 2°–6°C warmer than the lull situation. It is therefore concluded that subsidence over the heat low, by raising the temperature of the middle and upper tropospheric air, reduces surface pressure below what is observed in heat lows elsewhere. Both these effects operate to maintain the heat low circulation and consequently the strength of the low-level monsoon flow. Further, it appears that subsidence into the heat low waxes and wanes in unison with the monsoon rains. When the rains are heavy the heat low is vigorous, and stronger than normal surface winds prevail over most of the Arabian Sea.

Ramage considers that there exists a symbiotic relationship between the heat lows of northwest India and subtropical cyclones. He is of the opinion that the summer regime over the Arabian Sea is a strange one whose character seems basically determined by the vast desert area extending from northeast Africa to northwest India and by coastline orientation. The heat low develops over the deserts but then is maintained by the subtropical cyclones near India.

The distribution of radiation over southern Asia was discussed in chapter 3 and charts of net radiation produced in figure 3.2. The associated temperature distribution over southern Asia is such that temperature is always adequate for plant growth and rarely a limiting factor in human endeavour.

## Monsoon rainfall and storms

*I  Seasonal variations*

The southwest monsoon season comprising the 4 months of June, July, August, and September accounts for more than 75 per cent of the annual rainfall for most parts of India, except for peninsula India south of 15°N, which experiences its rainy season during the northeast monsoon months of October and November. The northern and northeastern parts of India also experience precipitation associated with pre-monsoon thunderstorms which progressively increase from March to May. Similarly travelling disturbances in the extra-tropical westerlies give rise to light precipitation from December to May in the northwestern parts of India. Figure 6.6 shows the precipitation received over different parts of the area in the four main seasons expressed as a percentage of the annual rainfall which is shown in figure 6.7.

Rainfall is, by and large, the criterion adopted by the India Meteorological Department for fixing the dates of onset and withdrawal of the southwest monsoon over different parts of the country (Ananthakrishnan and Rajagopalachari, 1964). The Department chooses the middle dates of the five-day periods during which the characteristic rise or fall occurs in the average pentad rainfall curve. The advance of the monsoon rainfall over India and the neighbouring seas is from the southeast towards the northwest, while the retreat is in the opposite direction (figure 6.8).

Monsoon rainfall over India presents a variety of patterns, which are partly due to orography and partly to dynamic factors which have been described by Ananthakrishnan and Rajagopalachari. Stations on the west coast show a steep increase in rainfall followed by a gradual decline, in marked contrast to the gradual rise and fall revealed by the north Indian stations. This is clearly illustrated in figure 6.9 by Mangalore and Marmagoa. Mangalore has a sharp increase in rainfall commencing about the 20 May and reaching a peak value by the middle of June. Indeed the steep rise in the rainfall curves of stations

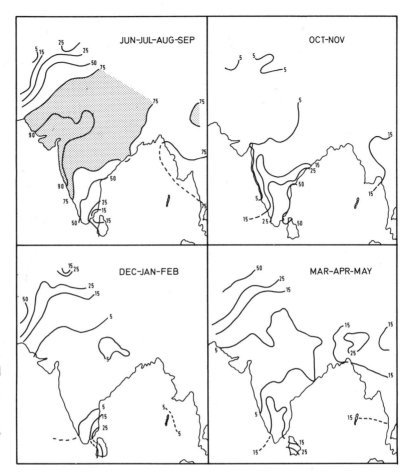

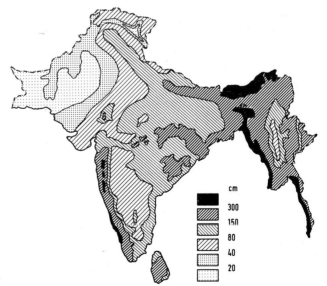

**Figure 6.6** Seasonal distribution of rainfall as percentage of annual for Indian sub-continent. (*After Ananthakrishnan and Rajagopalachari, 1964.*)

**Figure 6.7** Average annual precipitation over the Indian sub-continent. (*After Carter, 1954.*)

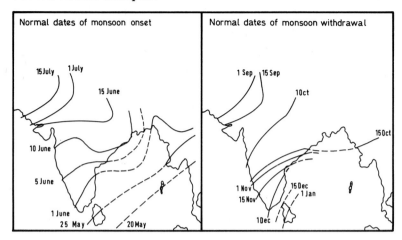

**Figure 6.8** Normal dates of onset and withdrawal of the southwest monsoon. (*After Ananthakrishnan and Rajagopalachari, 1964.*)

on the west coast of India explains the origin of the expression 'burst of the monsoon on the Malabar coast', for by taking rainfall as the criterion, the date of onset of the southwest monsoon can be fixed with greater precision here than for the rest of the subcontinent. Stations on the west coast, south of latitude 12°N, such as Trivandrum (08°29′N, 76°57′E), reveal two rainfall peaks, one in June and the other in October, with the October maximum being lower than the one in June. On the east coast of India, stations between latitudes 19° and 15°N have their maximum rainfall in the month of October, those south of 15°N in November, and those north of 20°N in July. Stations in the interior of the peninsula show a number of interesting features; for example, Sholapur receives its maximum rainfall in September while Hyderabad has two rainfall peaks in July and September. In north India the rainfall decreases from east to west, and the monsoon current does not reach Srinagar in the Kashmir Valley, since this station experiences more rainfall in the winter months in association with western disturbances than in the southwest monsoon months.

*2 Distribution*

The distribution of the average annual precipitation of India and vicinity is illustrated in figure 6.7. Maximums of precipitation coincide with the higher elevations in the Ghats, the eastern Himalaya, and the Arakan Yoma but a general decrease from southeast to northwest is characteristic of the region between the Ghats and the Himalayas. Minimums of precipitation are found mainly in the northwest but subsidiary centres appear in the lee of the Ghats, the Khasi Hills, and the Arakan Yoma. The region of large precipitation amounts in India has annual totals comparable to the greatest amounts in central Africa; but this zone in India extends to latitude 30° whereas its African counterpart is distinctly equatorial.

In India the southwest monsoon and tropical storms or cyclones are the principal causes of rainfall. It is of interest to determine what proportion of the annual rainfall is due to tropical storms. In general, July and August are the most rainy months, and on average about two tropical storms develop in each of these months and last for about 4–5 days. Most of them form in the Bay of Bengal, north of latitude 18°N, and their subsequent movement across India varies little from year to year. Based on data for stormless days, average total rainfalls for July and August have been computed by Raghavan (1967), and then compared with the normal monthly rainfalls inclusive of days with tropical storms. The difference between the two sets of data, one representing the month without storms and the other including storms, are expressed as a percentage of the normal and reproduced in figure 6.10. This diagram indicates that, although a tropical storm causes widespread rainfall, it may actually inhibit rainfall over certain areas. Raghavan's study suggests that in the month of July, tropical storms deprive nearly one third of India of about 10 per cent of its share of monsoon rainfall, while an almost equal area receives a similar increase of rainfall. In August, however, only a small part of the country benefits from the storms, while a major part is deprived of monsoon rainfall. In July there are two distinct areas where storms are mainly responsible for rainfalls of about 20 per cent above normal; one of these covers a part of coastal Orissa where storms usually strike from the Bay of Bengal, and the other is in Saurashtra–Katch where storms usually strike from the Arabian Sea. Both peninsula India east of the Western Ghats south of about 17°N, and the northern districts of Uttar Pradesh, Bihar, West Bengal, and Assam experience a rainfall increase during July of 10–20 per cent when the area is free from tropical storms. It appears therefore that tropical storms create conditions unfavourable for these regions to receive rainfall, for instead their rainfall comes mostly from minor upper troughs. The area where tropical storms exercise maximum influence detrimental to the production of rainfall is found in the southeast of Madras State during August, when the departure amounts to 40 per cent.

Variations in rainfall caused by tropical storms can be explained by assuming that precipitation is concentrated within certain sectors of the storms. Lockwood (1968) has mapped the air flow in a tropical storm over India and found that the above assumption was correct in this particular case.

During early July 1964, a tropical storm developed in the Bay of Bengal and moved westwards into northern India. At 0000 GMT on 4 July 1964, the storm was centred to the north of Calcutta, while a minor disturbance existed over West Pakistan. The pattern of air flow in the tropical storm is illustrated in figure 6.11 by the flow along the surface with a constant total energy content ($Q$) value of 83 cal/g. Over periods of a few days, the total energy content ($Q$) of air particles in the free atmosphere remains approximately constant, and it is, therefore, possible to use the $Q$ content as a 'label' on air particles and thereby trace their motions in three dimensions. The most convenient method of doing this is by drawing surfaces of constant $Q$ and constructing streamlines on these surfaces. To construct figure 6.11 the heights, in millibars, of the 83 cal/g surface were computed for each of the available radiosonde ascents. At some distance from the storm centre each ascent passed through the 83 cal/g surface twice, once near the ground and once at about 200 mb; but nearer to the storm centre all the $Q$ values were greater than 83 cal/g; therefore the shape of the surface is of the nature of a hyperboloid.

From figure 6.11 the picture that emerges is one of air flowing into the storm at low levels, rising rapidly near the storm centre, and flowing out at high levels. The main ascent along the 83 cal/g surface is over Orissa, where the streamlines rise from 900 mb to 400 mb in a relatively short distance. Over much of Orissa between 1200 GMT on 3 July and 0300 GMT on 4 July, the rainfall averaged about 60 mm (figure 6.11); this implies an average rate of approximately 4 mm/h. If the atmosphere over Orissa is assumed to be almost saturated with a surface dew-point of 25°C, then an ascent rate of 6 to 15 cms/sec in

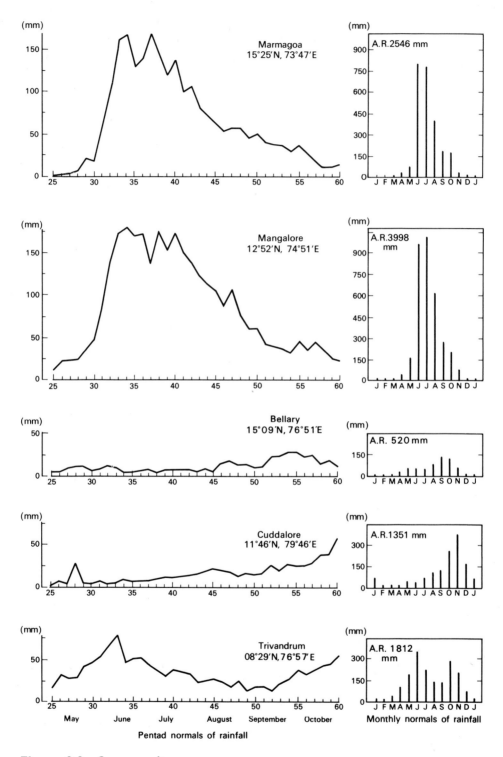

**Figure 6.9** *See opposite.*

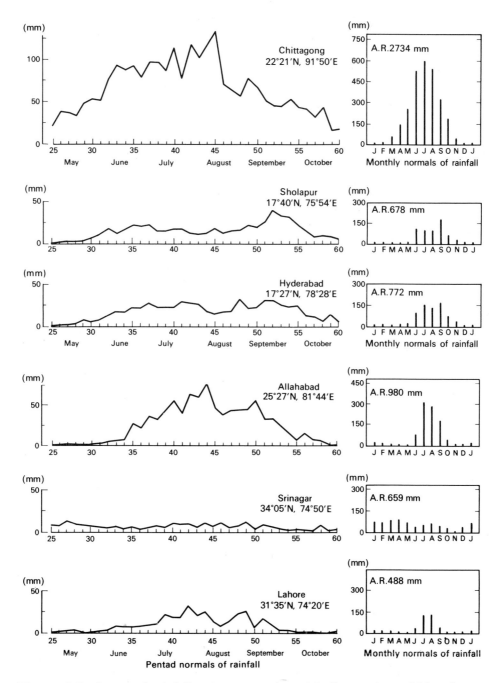

**Figure 6.9** Seasonal rainfall patterns at selected Indian stations. (*After Ananthakrishnan and Rajagopalachari, 1964.*)

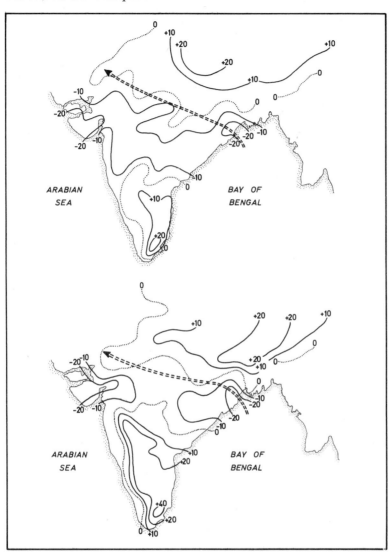

**Figure 6.10** Per cent departure from mean monthly rainfall over India for month without storms. A negative departure indicates below-normal rainfall in storm-free month while a positive departure indicates above-normal rainfall. Arrows indicate mean tracks of tropical storms. *Upper:* July. *Lower:* August. (*After Raghavan, 1967.*)

the lower atmosphere would be required to produce a rainfall rate of 4 mm/h. The calculated vertical component of the air flow at the 83 cal/g surface over Orissa was about 15 cm/sec; the values are close enough to indicate that the rainfall over east–central India was associated with the air flow along the marked upward curve of the 83 cal/g surface. On 4 July there was a minor disturbance over West Pakistan. Figure 6.11 indicates that there was no active ascent into this disturbance and that subsidence might have been occurring; rainfall amounts from the system were low. Similar comments can be made about the northern sector of the main storm, where the Himalayas appear to restrict low-level inflow. The flow along the 83 cal/g surface does not explain the rainfall along the west coast of India in figure 6.11, but investigation of the 81 cal/g surface shows clearly that air was rising from the Arabian Sea over the Western Ghats and then sinking over eastern India. It is therefore suggested that the rainfall deficits found by Raghavan result from the interaction of the relief of southern Asia with the tropical storm circulation patterns, and in particular from the lack of low-level flow from the north. The problem of the relationship between rainfall, surface conditions, and vertical motion is considered further in the discussion on the possible modification of the monsoon circulation.

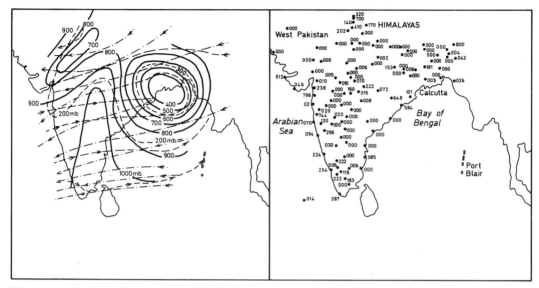

**Figure 6.11** *Left:* Contours and streamlines of the 83 calorie per gramme surface over India at 0000 GMT on 4 July 1964.

——————— Isobaric heights of the lower surface.
– – – – – – Isobaric heights of the upper surface.
— → — — → Flow along lower surface (direction shown by arrow).
— · ← — · ← Flow along upper surface (direction shown by arrow).

*Right:* Rainfall during the period from 1200 GMT on 3 July 1964, until 0300 GMT on 4 July 1964. (Rainfall amounts are plotted in tenths of millimetres.) (*Both maps after Lockwood, 1968.*)

*3   Storm-surges*

Tropical cyclones forming in the Bay of Bengal can be extremely destructive, both because of the associated high rainfalls giving rise to river floods and also because of storm-surges along the coasts. Storm-surges are not peculiar to the Bay of Bengal, since such waves are met with in the coastal regions of all the oceans which are subject to severe cyclonic storms, but the Bay does have the biggest storm-surges in the world. The magnitude and importance of storm-surges are indicated by some historic facts, which have been summarized by Naqvi (1960).

Naqvi has described particularly interesting storms occurring in 1864, 1876, and 1945. On 5 October 1864, a severe cyclonic storm arrived over the mouth of the Hoogly River about 2 hours before high tide. There was an enormous accumulation of water in the northwest angle of the Bay of Bengal, and the storm-surge rose suddenly and broke on the shore like a wall of water. It flooded the whole of the low-lying ground in the neighbourhood

*Plate 10* At about midnight 12/13 November 1970, a storm surge struck the coastal regions of the Ganges Delta in Bangladesh. The surge is estimated to have been about 5 to 10 m in height and according to one survivor it came with a hollow, rolling roar and a cold luminous glow. An international rescue operation was needed to help the survivors. The four pictures, which were made by the weather satellite ITOS1 on successive days between 9–12 November show the northward movement in the Bay of Bengal of the tropical cyclone which caused the storm surge. The spiral structure of the cyclone is clearly seen and on 11 November there is evidence of a central eye. *Above left to right:* 9 November 1970; 10 November 1970. *Below left to right:* 11 November 1970; 12 November 1970. *Reproduced by permission of the United States Department of Commerce, National Oceanic and Atmospheric Administration.*

of the Hoogly River and then moved slowly up to Calcutta where it arrived an hour before high tide. The total loss of life due to this storm surge was about 80,000 people. A severe cyclonic storm crossed the coast near the mouth of the Meghna River on 1 November 1876, just after the spring tide. The storm prevented the tidal and river water from dispersing into the Bay and eventually a storm-surge rushed inland covering the coastal districts with water 3 to 10 m, and at places even 13 m deep. In less than half an hour all standing crops were destroyed and 100,000 persons drowned. During the early hours of 18 October 1945, a severe cyclonic storm crossed the coast between Cocanada and Masulipatam, causing heavy rain and violent gales along part of the Circars Orissa coast. The surge associated with this storm inundated a coastal belt from about 50 km northeast of Cocanada to the neighbourhood of Masulipatam. The height of the wave is reported as varying between 4 and 10 m, and its rate of movement inland between 3 to 5 m/sec.

Pressure differences within tropical cyclones are only capable of producing differences in sea level of 30–50 cm, and therefore pressure differences alone cannot produce the huge storm-surges actually experienced. Instead the wave is produced by the wind stress on the ocean surface, as modified by the shallow waters of the continental shelf and the shape and topography of the coastline. The stress of the wind on the sea surface can cause the piling of water in the direction of that wind. The slope of the sea surface that can be maintained by a given wind is greater in shallow water than in deep, since the stress is balanced by the component of gravity acting upon the entire water body. For winds exceeding 6 m/sec, the slope of the sea surface in shallow water is given by the formula

$$\Delta P = 4 \cdot 8 \times 10^{-9} w^2 \Delta D / d$$

where $\Delta D$ is the horizontal distance of a section of the sea surface taken along a line perpendicular to the coast; d is the average depth of the section; w is the sustained on-shore component of the wind in m/s; and $\Delta P$ is the corresponding difference in the heights of the sloping surface.

Naqvi notes that in the storm of October 1945, the height of the wave at Cocanada was 3·46 m, out of which the contribution of the astronomical tide was 1·68 m, and the rest of the 1·78 m was due to piling up of water on the coast. This last value corresponds well with the calculated value from the above equation, suggesting that the storm-surge is partly due to the piling of water along the coast and partly due to the astronomical tidal wave which is added to it.

Storms which are associated with high storm waves in the Bay of Bengal, are of severe intensity and are not common (table 6.3). Only 27 storms produced storm-surges during the 165 year period from 1780 to 1945 (Naqvi). Although storms develop in the Bay at any time during the period from April to December, it is only those of May and September to December that normally are of extraordinary severity. The storms which form in the north of the Bay of Bengal at the peak of the monsoon season in July and August are generally of small extent and do not have a core of hurricane winds. Cyclonic storms are of frequent occurrence in May, and they form mostly in the central part of the Bay and more northwards or northeastwards. At least ten severe cyclonic storms with associated surges have been recorded in the month of October. These storms usually travel either towards the Godavary Delta or towards the Ganges Delta and the Chittagong coast. November appears to be almost as dangerous, and it was in November 1970 that a cyclone sent a storm surge up to 10 m high into the Ganges Delta drowning up to 500,000 people.

**Table 6.3** Monthly totals of cyclonic storms in the Bay of Bengal and the Arabian Sea, 1891–1960

| Month | Bay of Bengal | Arabian Sea |
|-------|---------------|-------------|
| Jan | 4 (1) | 2 (0) |
| Feb | 1 (1) | 0 (0) |
| Mar | 4 (2) | 0 (0) |
| Apr | 18 (7) | 5 (4) |
| May | 28 (18) | 13 (11) |
| Jun | 34 (4) | 13 (8) |
| Jul | 38 (7) | 3 (0) |
| Aug | 25 (1) | 1 (0) |
| Sept | 27 (8) | 4 (1) |
| Oct | 53 (19) | 17 (7) |
| Nov | 56 (23) | 21 (16) |
| Dec | 26 (9) | 3 (1) |
| Total | 314 (100) | 82 (48) |

The values within brackets indicate the number of storms which reached 'severe' intensity.

*After Das (1968).*

Dates of significant storm surges in the Bay of Bengal between 1787 and 1945 are contained in table 6.4.

**Table 6.4** Date of storm waves in the Bay of Bengal, 1787–1945

| | |
|---|---|
| Godavary delta | May 1787 |
| Godavary delta | Dec 1789 |
| Bakargarij | May 1822 |
| North of Cuttack | May 1823 |
| Godavary delta | Dec 1830 |
| Barisal | Oct 1831 |
| Godavary delta | Dec 1839 |
| False Point | Oct 1848 |
| Contai | Oct 1864 |
| Contai | Oct 1874 |
| Mouth of Meghna | Oct 1876 |
| Akyab | May 1884 |
| Near False Point | Sept 1885 |
| Balasore | May 1887 |
| A little south of False Point | Sept 1888 |
| Near Chaiya | Nov 1891 |
| Over Chittagong | Oct 1897 |
| Nellore | Nov 1915 |
| Masulipatam | Nov 1922 |
| Midnapur | May 1925 |
| Cox's Bazar–Akyab | May 1926 |
| Nellore–Masulipatam | Nov 1933 |
| Kyaukpu–Sundoway | Apr 1936 |
| Masulipatam | Oct 1936 |
| Comilla | May 1941 |
| Midnapur | Oct 1942 |
| Cocanada | Oct 1945 |

*After Naqri (1960).*

Rao and Mazumdar (1965) have calculated the maximum wind piling-up effect for a large number of locations along the coasts of the Bay of Bengal and the Arabian Sea. They assumed the most favourable direction from which the wind would blow for producing the maximum effect. With the help of these calculations they have classified the coastal belts into three categories: *a*, *b*, and *c* which indicate the degree of vulnerability (in increasing order) to storm-surge afflictions of any part of the coastal belt (figure 6.12). At coasts in category *a*, the value of the maximum piling-up effect is 2 m or less. These are comparatively safe areas where major storm-surges will not occur even when all other conditions are favourable. Coasts under category *b* are those where the maximum piling-up effect

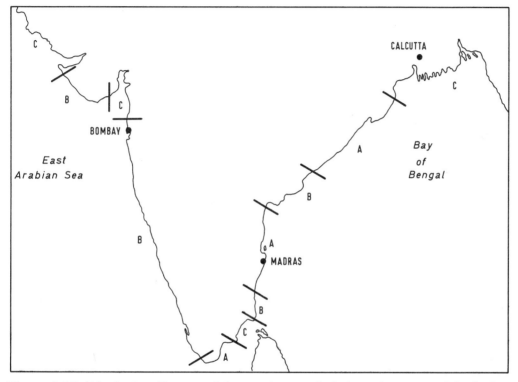

**Figure 6.12** Maximum piling-up of the sea due to wind along the coasts of the Indian subcontinent. A—2 m or less; B—2 to 5 m; C—over 5 m. (*After Rao and Mazumdar, 1965.*)

is between 2 and 5 m, and therefore major flooding will occur only in association with severe storms provided all the conditions are favourable. Under category *c* are placed those coastal areas where the maximum piling-up effect exceeds 5 m, and so these are highly vulnerable areas where major surges should always be anticipated in association with approaching cyclonic storms.

The classification of coasts by Rao and Mazumdar appears to be valid in that all the known major storm waves which caused losses of life in the past occurred in the *c*-type coastal areas. There are four coastal areas of the Indian subcontinent which belong to the *c* type. The first is the coastal area around the head of the Bay of Bengal, roughly to the north of 20°N, where some areas like the Sundarbans appear to be highly vulnerable;

the second is the coastal region around the Palk Strait and adjoining the south Coromandel Coast; the third is the Konkan coast to the north of 18°N and the coastal region around the Gulf of Cambay; the fourth lies between Dwarka and Karachi.

## Evapotranspiration and water balance

Average annual potential evapotranspiration as calculated by Carter (1954) according to the 1948 system of Thornthwaite is shown in figure 6.13. Nearly all the productive lands of the Indian subcontinent except the valleys of Kashmir have annual potential evapotranspiration values of at least 1140 mm and over the lowlands it reaches at least 1500 mm. Annual totals appear to exceed 1700 mm in only three small regions: the Irrawaddy Delta, the Coromandel coast, and the Malabar coast.

Wide variations in moisture deficit and water surplus occur in southern Asia. Virtually all the Indus Valley in Pakistan has a moisture deficit of rainfall below potential evapotranspiration of more than 1,000 mm per year, and the climate may be considered as arid. Besides the Indus Valley, there are three other areas of excessive deficit: the central Irrawaddy Valley, the Kistna Valley in the central Deccan, and a small area in northwest Ceylon opposite a similar area on the mainland. According to Carter, world record amounts of rainfall surplus occur in the Khasi Hills where Cherapundji has more than 10,000 mm of surplus in eight months. The coast of Burma near Gwa and the Mergui coast further south also have large surpluses (between 4,000 mm and 5,000 mm), while in the eastern Himalayas and in the central part of the Western Ghats some surpluses exceed 2,000 mm. In India south of about 20°N the water surplus decreases from west to east, while the opposite gradient exists in the north of the subcontinent.

Variation of water balance throughout the year tends to reflect the annual march of rainfall, and a selection of diagrams taken mostly from the work of Subrahmanyam and Murty (1968) are shown in figure 6.14. Mercara in the Western Ghats is typical of an extremely wet area, for it shows one rainfall peak with a massive water surplus during the southwest monsoon, but there is evidence of a dry season during the northeast monsoon with some use of stored soil moisture. Satkhira, in Bangladesh, is more typical of a rather drier type of climate with a marked soil moisture deficit at the end of the dry season. Some areas in the south of the subcontinent experience two rainfall maximums per year, and a good example of these is provided by Colombo.

Carter (1954) classifies large areas of India as having dry climates (figure 6.13), and the examples considered so far have been taken from areas with moist climates. In southern India aridity increases towards the east, and typical of this arid east are Hyderabad and Cuddalore. Both have rainfall maximums in the period from September to December and very marked dry seasons. At Cuddalore where there is, on average, a water surplus only for the two months of November and December, most of October is spent in recharging the soil moisture, while the soil appears to be almost completely dry from July until September. Conditions are even more extreme at Hyderabad where the soil is only moist on average in the months of August and September, and almost completely dry for the rest of the year. Bellary is typical of the extremely arid climate found in the central Kistna basin, where there is a water deficit throughout the year; a similar climate exists at Mandalay in the central Irrawaddy valley.

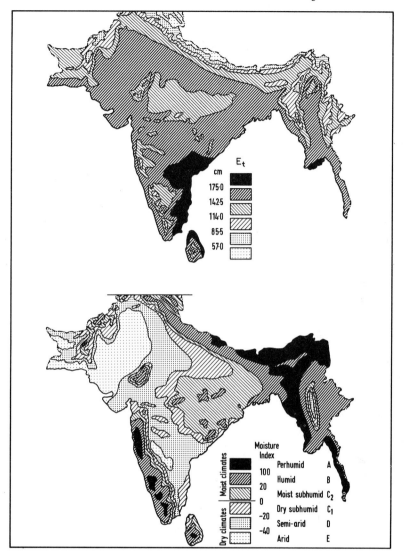

**Figure 6.13** *Upper:* average annual potential evapotranspiration over the Indian subcontinent according to Thornthwaite's method. *Lower:* distribution of Thornthwaite's moisture index over the Indian subcontinent.

$$\text{Moisture index} = \frac{100 \times \text{annual water surplus} - 60 \times \text{annual water deficit}}{E_t}$$

*(After Carter, 1954.)*

*Drought*

Water balance and drought are closely interconnected and in a largely agricultural country it is of some interest to investigate the incidence of drought. Table 6.5 showing the frequency of occurrence in northern India of dry years, is taken from the *Report of the Indian Irrigation Commission 1901–3*. Though the term 'drought' is not clearly defined, this

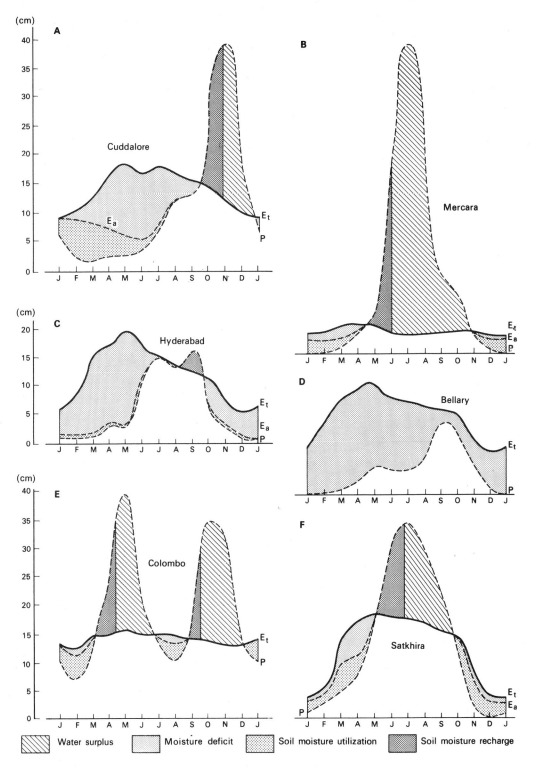

**Figure 6.14** *See opposite.*

**Table 6.5**   Frequency of occurrence in Northern India of dry years and of severe drought

| Place (Names as in British India) | Average Annual Rain (mm) | Number of dry or drought years that may be expected in 50 years | |
|---|---|---|---|
| | | Dry years incl. yrs of severe drought | Years of severe drought |
| Orissa | 1475 | 5 | 0 |
| Chota Nagpur | 1350 | 4 | 1 |
| Central Bengal | 1400 | 6 | 1 |
| Bihar | 1200 | 10 | 3 |
| Uttar Pradesh Submontane | 1125 | 10 | 3 |
| Uttar Pradesh east | 1025 | 10 | 3 |
| Punjab Submontane | 825 | 10 | 4 |
| British Bundelkhand | 900 | 10 | 5 |
| Uttar Pradesh west | 725 | 10 | 5 |
| Uttar Pradesh central | 925 | 12 | 5 |
| Punjab southeast | 575 | 13 | 5 |
| Punjab central | 575 | 13 | 6 |
| Ajmer Merwara | 500 | 11 | 6 |
| Punjab west | 250 | 14 | 9 |

*After Indian Irrigation Commission, 1901–3.*

table suggests that droughts are most frequent, as would be expected, in the low rainfall areas and rarely occur in high rainfall areas such as Orissa.

Most of the rainfall over the Indian subcontinent arrives during the southwest monsoon, and it is the rate of failure of the monsoon rains which determines the incidence of droughts. Figure 6.15, after Rao and Ramamoorthy (1958), indicates the seasonal variability of monsoon rainfall (June–September). It supports table 6.5, in showing that the variability is low in the high rainfall states of the northeast and increases towards the more arid regions. Agriculture is therefore not only more difficult in the arid regions but also more risky.

In most parts of India, the moisture stored in the soil in the wet period is almost entirely used up soon after the start of the dry season; as a result, a long period of pronounced dryness invariably precedes every rainy season. Consequently, not only does every deficiency mean disaster for the farmer, but the mere prolongation of the pre-monsoon dry season by a week or a fortnight may destroy all hopes of a normal harvest for that year. Subrahmanyam and Subramaniam (1964) have studied the water balance at three typical stations in southern India under drought conditions, and their results are given in table 6.6. Bijapur, situated in the arid zone, has its rainfall always below the potential evapotranspiration, except in September when there is a slight excess which is temporarily stored in the soil and used in later months. In the drought year of 1936, rainfall did not exceed potential evapotranspiration even in September and the climate became completely arid. Trichy has a semi-arid climate with a short rainy season in which enough rainfall is received to temporarily recharge the soil moisture reservoir, but the excess is hardly sufficient to produce runoff. During the drought year of 1927, the water surplus was very

**Figure 6.14**   Water-balance diagrams for selected stations in the Indian subcontinent. Potential evapotranspiration calculated by Thornthwaite's method. (*After Subrahmanyam and Murty, 1968.*)

**Table 6.6**  Water balance concepts in climatic study of droughts

Water balances (cm) according to the 1955 scheme of Thornthwaite

|       | J    | F    | M    | A    | M    | J    | J    | A    | S    | O    | N    | D    | Year  |
|-------|------|------|------|------|------|------|------|------|------|------|------|------|-------|
| BIJAPUR (1936) | | | | | | | | | | | | | |
| $E_t$ | 9·0  | 9·3  | 15·5 | 17·6 | 19·1 | 15·4 | 15·5 | 12·5 | 13·1 | 13·0 | 8·8  | 7·8  | 156·6 |
| P     | 0    | 0·2  | 0    | 1·4  | 4·4  | 1·8  | 0·1  | 3·4  | 9·8  | 0·3  | 5·6  | 0    | 27·0  |
| $E_a$ | 0    | 0·2  | 0    | 1·4  | 4·4  | 1·8  | 0·1  | 3·4  | 9·8  | 0·3  | 5·6  | 0    | 27·0  |
| WD    | 9·0  | 9·1  | 15·5 | 16·2 | 14·7 | 13·6 | 15·4 | 9·1  | 3·3  | 12·7 | 3·2  | 7·8  | 129·6 |
| WS    | 0    | 0    | 0    | 0    | 0    | 0    | 0    | 0    | 0    | 0    | 0    | 0    | 0     |
| TRICHY (1927) | | | | | | | | | | | | | |
| $E_t$ | 14·4 | 14·0 | 16·7 | 17·8 | 18·8 | 17·8 | 17·7 | 17·6 | 16·1 | 15·7 | 13·3 | 11·9 | 191·8 |
| P     | 0·4  | 0·4  | 0·7  | 2·1  | 2·0  | 7·3  | 4·2  | 3·2  | 18·8 | 10·2 | 10·2 | 4·3  | 63·8  |
| $E_a$ | 1·0  | 0·8  | 0·9  | 2·3  | 2·1  | 7·3  | 4·2  | 3·2  | 16·1 | 10·6 | 10·5 | 4·7  | 63·8  |
| WD    | 13·4 | 13·2 | 15·8 | 15·5 | 16·7 | 10·5 | 13·5 | 14·3 | 0    | 5·1  | 2·8  | 7·2  | 128·0 |
| WS    | 0    | 0    | 0    | 0    | 0    | 0    | 0    | 0    | 0    | 0    | 0    | 0    | 0     |
| CUDDALORE (1949) | | | | | | | | | | | | | |
| $E_t$ | 8·5  | 8·9  | 14·0 | 16·7 | 18·0 | 17·7 | 17·2 | 16·8 | 16·1 | 15·6 | 11·2 | 8·7  | 169·4 |
| P     | 0    | 0    | 0    | 0·4  | 10·6 | 10·6 | 19·5 | 10·2 | 7·1  | 5·2  | 10·8 | 0·8  | 75·2  |
| $E_a$ | 0·2  | 0·2  | 0·1  | 0·5  | 10·7 | 10·6 | 17·2 | 10·6 | 7·6  | 5·7  | 10·8 | 1·0  | 75·2  |
| WD    | 8·3  | 8·7  | 13·9 | 16·2 | 7·3  | 7·1  | 0    | 6·2  | 8·5  | 9·9  | 0·4  | 7·7  | 94·2  |
| WS    | 0    | 0    | 0    | 0    | 0    | 0    | 0    | 0    | 0    | 0    | 0    | 0    | 0     |

Climatic water balances (cm) according to the 1955 scheme of Thornthwaite

|       | J    | F    | M    | A    | M    | J    | J    | A    | S    | O    | N    | D    | Year  |
|-------|------|------|------|------|------|------|------|------|------|------|------|------|-------|
| BIJAPUR | | | | | | | | | | | | | |
| $E_t$ | 7·7  | 10·6 | 15·7 | 17·4 | 18·9 | 15·9 | 14·2 | 13·5 | 12·7 | 12·7 | 8·1  | 6·5  | 153·9 |
| P     | 0·5  | 0·2  | 0·5  | 2·2  | 2·8  | 7·7  | 6·0  | 6·1  | 14·1 | 7·3  | 3·2  | 0·6  | 51·2  |
| $E_a$ | 0·6  | 0·4  | 0·6  | 2·4  | 2·9  | 7·7  | 6·0  | 6·2  | 12·7 | 7·5  | 3·4  | 0·8  | 51·2  |
| WD    | 7·1  | 10·2 | 15·1 | 15·0 | 16·0 | 8·2  | 8·2  | 7·3  | 0    | 5·2  | 4·7  | 5·7  | 102·7 |
| WS    | 0    | 0    | 0    | 0    | 0    | 0    | 0    | 0    | 0    | 0    | 0    | 0    | 0     |
| TRICHY | | | | | | | | | | | | | |
| $E_t$ | 10·5 | 12·7 | 15·4 | 17·8 | 19·0 | 18·2 | 18·3 | 17·6 | 16·4 | 15·3 | 13·2 | 10·3 | 184·7 |
| P     | 2·5  | 1·2  | 0·9  | 4·6  | 8·4  | 4·1  | 3·4  | 9·7  | 12·0 | 18·3 | 14·8 | 7·1  | 87·0  |
| $E_a$ | 3·6  | 2·2  | 1·8  | 5·1  | 8·7  | 4·3  | 3·6  | 9·7  | 12·0 | 15·3 | 13·2 | 7·5  | 87·0  |
| WD    | 6·9  | 10·5 | 13·6 | 12·7 | 10·3 | 13·9 | 14·7 | 7·9  | 4·4  | 0    | 0    | 2·8  | 97·7  |
| WS    | 0    | 0    | 0    | 0    | 0    | 0    | 0    | 0    | 0    | 0    | 0    | 0    | 0     |
| CUDDALORE | | | | | | | | | | | | | |
| $E_t$ | 9·3  | 10·9 | 12·3 | 16·8 | 18·9 | 16·9 | 18·2 | 17·1 | 15·8 | 14·7 | 12·0 | 10·0 | 172·9 |
| P     | 6·1  | 2·3  | 1·8  | 2·5  | 2·5  | 3·6  | 6·6  | 12·2 | 13·2 | 29·2 | 39·4 | 19·0 | 138·4 |
| $E_a$ | 9·2  | 9·0  | 7·8  | 7·9  | 6·3  | 5·4  | 7·6  | 12·5 | 13·4 | 14·7 | 12·0 | 10·0 | 115·8 |
| WD    | 0·1  | 1·9  | 4·5  | 8·9  | 2·6  | 11·5 | 10·6 | 4·6  | 2·4  | 0    | 0    | 0    | 57·1  |
| WS    | 0    | 0    | 0    | 0    | 0    | 0    | 0    | 0    | 0    | 0    | 13·6 | 9·0  | 22·6  |

$E_t$ = potential evapotranspiration, P = precipitation, $E_a$ = actual evapotranspiration, WD = water deficit, and WS = water surplus

*After Subrahmanyam and Subramaniam (1964).*

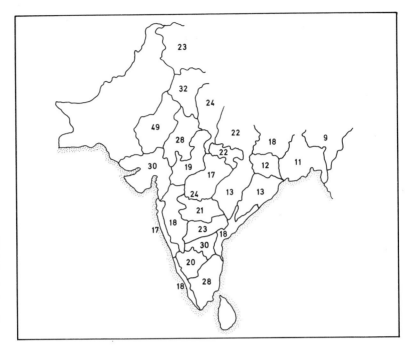

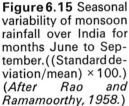

**Figure 6.15** Seasonal variability of monsoon rainfall over India for months June to September. ((Standard deviation/mean) × 100.) (*After Rao and Ramamoorthy, 1958.*)

small, not even adequate for the next month's water demand, and the water deficiency was about 30 per cent more than the climatological average. Cuddalore has a dry sub-humid climate possessing highly variable water balances, for in extremely dry years the moisture balance may resemble that of an arid climate, while in a wet year it may be superior to that of a humid station. The water balance for 1949 at Cuddalore shows an established drought year.

The main problem of water supply in India lies in the fact that a large part of the rainfall arrives during a relatively short period of time. Since soil moisture storage is limited, plants are bound to be under moisture stress during the following dry season. As much of the rainfall during the wet season forms runoff and escapes to the sea, the problem in India, as often elsewhere, is how to store the surplus so that it can be used in times of water deficit. Fortunately, although droughts producing severe economic strain are frequent in parts of India, they are not spread over all the country every year, and indeed historical records show no example of a serious drought affecting the whole of the subcontinent.

## Semi-arid India

Nearly half of the Indian subcontinent is arid or semi-arid, and by far the larger part of this dry area is located in northwest India and southwestern West Pakistan, with its centre on the Thar Desert. The larger area of deficient rainfall surrounding and including the Thar Desert is known as the Rajputana or Rajasthan Desert, and occupies about a fifth of the total land area of India.

## 1  The Rajasthan Desert

During the months of the summer monsoon, the Rajasthan Desert is under the influence of the Arabian Sea branch of the monsoon. The moisture content of the air over Rajasthan is consequently fairly high, but the rainfall totals are disappointingly low and observations suggest the existence of an inversion in the lower troposphere at about 1·5 km above sea level. Clearly, the low-level inversion suppresses convection and therefore inhibits the formation of rainfall. If the inversion could be removed, there would be an increased tendency for rising air currents to occur over Rajasthan, and as the supply of water vapour is already there, this might lead to an increase in monsoon rainfall.

The Rajasthan Desert was once the centre of a flourishing civilization, which would suggest that there was at that time a higher rainfall than at present. According to Das (1968) the climatic history of the area may be divided into four stages:

1  8,000 BC: Moist, wet, and cool climate.
2  8,000–3,000 BC: A dry climate, but less arid than at present. There is evidence to suggest the beginning of agriculture.
3  3,000–1,700 BC: Period of higher rainfall.
4  1,700–1,500 BC: Period of dry conditions. There is evidence of fresh-water lakes drying up.

Following the last period, the indications are that very arid conditions prevailed over Rajasthan.

Bryson and Baerreis (1967) and Ghosh (1952) consider that three major cultures have existed in the region. From 2,000 to 1,700 BC, when the Rajasthan area was very favourable for human habitation, the region was the centre of the Harappan civilization. Ghosh reports that remains of the Harappan culture have been identified in what were once tributaries of the Indus, but are now dessicated river valleys.

The people of the Harappan civilization were followed in the area by the Painted Grey Ware culture at about 500 BC, approximately 1,000 years after the termination of the previous culture. Ghosh comments that the sites of this new culture are small in extent and present few features. The Grey Ware sites were not superimposed upon the Harappan sites but established in new locations. The Harappan sites were located overlooking the valleys as though the flood-plains were farmed, while the later sites are on the river channels proper where the last bit of a dwindling water supply might be used.

A final group of sites has been classed by Ghosh as the Rangmahal culture, which flourished in the early centuries AD. The history of this period appears to have been one of extreme fluctuations, for there were periods of intense cultivation followed by years of dry and semi-arid conditions.

There seems to be little doubt that the Rajasthan Desert is more hostile to human occupation at the present time than it has been in the past. There is also evidence that the increasing aridity is purely local and not of a world-wide nature. Indeed, the consensus of many experts is that the Rajasthan Desert is largely a man-made desert, being the result of the work of man in cutting down and burning forests and causing the deterioration of the soils. It has also been suggested that the decline in soil texture has led to an abnormal increase in the dust content of the atmosphere.

## 2 Dust effect

Bryson and Baerreis (1967) have reported the results of meteorological reconnaissance flights which found a layer of dust extending over much of southern Asia. In the spring, deep dense dust over North Africa and Arabia appears to thin eastward along the southern coast of Iran and Baluchistan, but becomes very deep and dense over the Rajasthan Desert. Eastward and southward the dust diminishes, though it is still present in considerable quantities over northeast India, and observable as layers over Burma, Thailand, and Cambodia. The major source of dust for the Indian subcontinent would seem to be the Rajasthan Desert. The dust particles which are extremely small with diameters of about a micron, are often carried up to at least 5 km and remain in suspension for several days. The particles are formed mostly of quartz, although small quantities of silicates, such as mica and felspar, have also been observed.

Dust is carried aloft by vertical currents, which also cause rainfall. Das (1962) has studied the mean vertical motion over India during July using a ten-layer numerical model and has published his results for 900, 500, and 200 mb levels. His computations revealed two well-defined zones of vertical motion: one an area of marked ascent over northeast India and the other a zone of subsidence along the western periphery of the mountains, with the rate of subsidence of the same order of magnitude as the rate of ascent. He found no other regions where the vertical motion was of comparable intensity. The mean July vertical motion charts agree with the vertical motion picture suggested by Lockwood for an individual cyclone in July 1964, which was discussed in the section on monsoon rainfall.

## 3 Air circulation

Miller and Keshavamurthy (1968) have computed by Bellamy's (1949) triangle method the mean July divergence at 900 m, 700 mb, 500 mb, and 200 mb (figure 6.16). The divergence patterns agree fairly well with the suggested distribution of vertical motion. At low levels (900 m) the flow patterns appear complex, and particularly interesting is the convergence over the deserts of the northwest. This convergence into the heat low is only of limited depth because it has vanished at 700 mb; it probably consists of moist air from the sea which is trapped below an inversion.

Normally, mean vertical motions imply mean non-adiabatic changes and therefore Das converted his computed vertical velocities into mid-tropospheric cooling rates. His calculations revealed that non-adiabatic sources must be capable of producing an average warming at the rate of 1·6°C per 12 hours in northeast India and an average cooling at the rate of 1·2°C per 12 hours in northwest India. Latent heat released by precipitation can easily account for the warming over northeast India but the outgoing radiation does not appear to be sufficient to explain the maximum cooling over northwest India, for the latter, calculated from the observed water vapour and carbon dioxide distribution, is only 0·9°C per 12 hours.

Bryson, Wilson, and Kuhn (1964) have measured the radiation divergence in the troposphere over northwest India, and discovered that the observed rate of cooling is very nearly the same as that required to match the subsidence calculated by Das. They ascertained that the observed rate of cooling differed from the calculated rate by 50 per cent, due to the presence of atmospheric dust over the desert, and thus it is concluded that the dust increases the mid-tropospheric subsidence rate by perhaps 50 per cent.

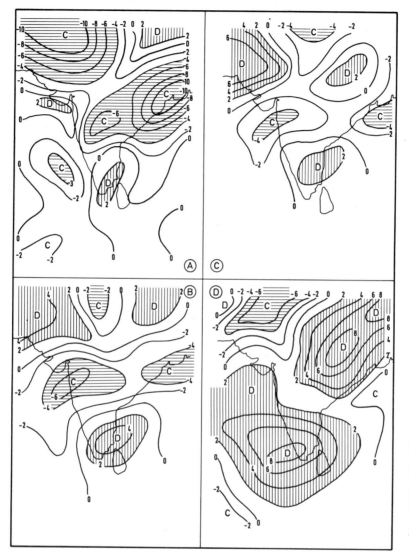

**Figure 6.16** Mean July divergence $(10^{-6} \, sec^{-1})$ over Indian subcontinent compiled by Bellamy's triangle method. A : 900 m level. B : 700 mb level. C : 500 mb level. D : 200 mb level. (*After Miller and Keshavamurthy, 1968.*)

## 4  Climate modification

It is suggested by Bryson and Baerreis (1967) that the atmospheric dust over northwest India could be reduced by stabilizing the surface of the Rajasthan Desert. This could perhaps be achieved by planting suitable grasses. The two authors consider that with an adequate grass cover there would be little blowing dust, in which case the average subsidence over the desert would be cut down by perhaps one-third, because the infra-red cooling rate in mid-troposphere would be decreased. Since a shallow inflow of moist air already exists over the desert, a reduction in the average rate of subsidence could create more favourable conditions for occasional upward motion, which in turn would increase the chances of more precipitation. The removal of the dust would also allow more solar energy to reach the surface during the day and cause the surface to cool more quickly at night. This would result in a larger diurnal temperature range at the surface, and could lead to an increase in the formation of dew, which is an important source of moisture in semi-arid and arid lands. Therefore, a favourable climatic modification may be possible in parts of India.

The initial growing of the grass is an agricultural problem, but experiments at the Central Arid Zone Research Institute, Jodhpur, indicate that it is only necessary to exclude animals for a year of two so that a fine stand of native grass may spring up (Bryson and Baerreis, 1967). The original environmental deterioration in Rajasthan was probably, as suggested earlier, caused by over-grazing which led to soil erosion and increased deflation. At the present time the desert is said to be advancing into agricultural lands at 0·5 to 1·0 km per year, which suggests that farming practices are probably not adjusted to the environment. As the monsoon circulations are extremely complex, the analysis of the circulation over Rajasthan may not be entirely correct, but it does at least seem possible, by adjusting farming practices, to stop the advance of the desert and probably win some land for agriculture. It has already been pointed out that the subsidence over West Pakistan and northwest India is partly a result of monsoon rains elsewhere. As it is therefore unlikely that changing the desert surface will produce widespread heavy rainfall, the best that can be hoped for is that the areas will become slightly less arid.

# References

ANANTHAKRISHNAN, R. and RAJAGOPALACHARI, P. J. 1964: Pattern of monsoon rainfall distribution over India and neighbourhood. *Proceedings of WMO/IUGG Symposium Tropical Meteorology*, Rotorua, New Zealand, p. 192.

BASU, S., RAMANATHAN, K. R., PISHAROTY, R. R. and BOSE, U. K. (editors). 1960: *Monsoons of the world.* Symposium, New Delhi, 1958. New Delhi: Hindi Union Press.

BELLAMY, J. C. 1949: Objective calculations of divergence, vertical velocity, and vorticity. *Bulletin American Meteorological Society* **30**, 45.

BRYSON, R. A. and BAERREIS, D. A. 1967: Possibilities of major climatic modifications and their implications: northwest India, a case study. *Bulletin American Meteorological Society* **48** (3), 136.

BRYSON, R. A., WILSON, C. A. and KUHN, P. M. 1964: Some preliminary results of radiation sonde ascents over India. *Proceedings of WMO/IUGG Symposium Tropical Meteorology*, Rotorua, New Zealand, p. 737.

BUNKER, A. F. 1965: Interaction of the summer monsoon air with the Arabian Sea. *Proceedings of Symposium on meteorological results of International Indian Ocean Expedition*, Bombay, July 1965, p. 3.

CARTER, D. B. 1954: Climates of Africa and India according to Thornthwaite's 1948 classification. *John Hopkins University Publications in Climatology* **7** (4).

CHANG CHIA-CH'ENG. 1959: Certain views on the nature of Chinese monsoons. *Acta Meteorologica Sinica* **30**, 350.

DAS, P. K. 1962: Mean vertical motion and non-adiabatic heat sources over India during the monsoon. *Tellus* **14**, 212.

DAS, P. K. 1968: *The monsoons.* New Delhi: National Book Trust.

DIXIT, C. M. and JONES, D. R. 1965: A kinematic and dynamical study of active and weak monsoon conditions over India during June and July, 1964. Bombay: International Meteorological Centre.

FINDLATER, J. 1969a: A major low-level air current near the Indian Ocean during the northern summer. *Quarterly Journal Royal Meteorological Society* **95**, 362.

FINDLATER, J. 1969b: Inter-hemispheric transport of air in the lower troposphere over the western Indian Ocean. *Quarterly Journal Royal Meteorological Society* **95**, 400.

FINDLATER, J. 1971: Mean monthly air-flow at low levels over the western Indian Ocean. *Geophysical Memoirs* **115**. London: HMSO.

FINDLATER, J. 1972: Aerial explorations of the low-level cross-equatorial current over eastern Africa. *Quarterly Journal Royal Meteorological Society* **98**, 274.

FLOHN, H. 1955: Tropical circulation patterns. *WMO Technical Note No.* **9**, *Technical Paper* **13**, Geneva.

FLOHN, H. 1958: Monsoon winds and general circulation. In Basu, S. *et al.* (editors), *Monsoons of the world*. New Delhi: Hindi Union Press, p. 65.

FLOHN, H. 1964a: Investigations on the tropical easterly jet. Bonn. *Bonner Meteorologische Abhandlungen* **4**.

FLOHN, H. 1964b: The tropical easterly jet and other regional anomalies of the tropical circulation. *Proceedings of WMO/IUGG Symposium Tropical Meteorology*, Rotorua, New Zealand, p. 160.

FROST, R. and STEPHENSON, P. M. 1965: Mean streamlines and isotachs at standard pressure levels over the India and west Pacific Oceans and adjacent land areas. *Geophysical Memoirs* **109**. London: HMSO.

GHOSH, A. 1952: The Rajputana Desert—its archaeological aspect. *Proceedings of Symposium on the Rajputana Desert. Bulletin National Institute Science India* **1**, 37.

GODBOLE, R. V. and KELKAR, R. R. 1969: Radiation over the Himalayas. *Japan Meteorological Agency Technical Report* **67**.

HALLEY, E. 1686: *An historical account of trade-winds and monsoons with an attempt to assign the physical causes of the said winds. Philosophical Transactions Royal Society*, London, p. 153.

LOCKWOOD, J. G. 1965a: The Indian monsoon—a review. *Weather* **20**, 2.

LOCKWOOD, J. G. 1965b: Synoptic disturbances near the equator. *Weather* **20**, 279.

LOCKWOOD, J. G. 1968: Some uses of profiles and charts of total energy content. *Meteorological Magazine* **97**, 145.

MILLER, F. R. and KESHAVAMURTHY, R. N. 1968: *Structure of an Arabian Sea summer monsoon system.* Honolulu: East–West Center Press.

NAQVI, S. N. 1960: Storm waves in the Bay of Bengal. *Proceedings of the Fourth Pan–Indian Ocean Science Congress*, Karachi.

RAGHAVAN, K. 1964: Influence of the Western Ghats on the monsoon rainfall at the coastal boundary of Peninsula India. *Indian Journal Meteorology Geophysics* **15**, 617.

RAGHAVAN, K. 1967: Influence of tropical storms on monsoon rainfall in India. *Weather* **22**, 250.

RAMAGE, C. S. 1965: The summer atmospheric circulation over the Arabian Sea. *Proceedings of Symposium on meteorological results of International Indian Ocean Expedition*, Bombay, July 1965, p. 197.

RAMAN, C. R. V. and RAMANATHAN, Y. 1964: Interactions between lower and upper tropical tropospheres. *Nature* **204**, 31.

RAO BHASKARA and MAZUMDAR, S. 1965: A technique for storm-wave forecasting. *Proceedings of Symposium on Meteorological results of International Indian Ocean Expedition*, Bombay, July 1965, p. 250.

RAO, K. N. and RAMAMOORTHY, K. S. 1958: Seasonal (monsoon) rainfall forecasting in India. In Basu, S. *et al.* (editors), *Monsoons of the world*. New Delhi: Hindi Union Press, p. 237.

*Report of the Indian Irrigation Commission*, 1901–03, Part **1**.

SARKER, R. P. 1966: A dynamical model of orographic rainfall. *Monthly Weather Review* **94**, 24.

STAFF MEMBERS OF THE ACADEMIA SINICA, PEKING. 1958: On the general circulation over eastern Asia (11). *Tellus* **10**, 58.

SUBRAHMANYAM, V. P. and MURTY, P. S. N. 1968: Ecoclimatology of the tropics with special reference to India. In Misra, R. and Gopal, B. (editors), *Proceedings of Symposium on Recent Advances in Tropical Ecology, International Society for Tropical Ecology*, Varanasi, p. 67.

SUBRAHMANYAM, V. P. and SUBRAMANIAM, A. R. 1964: Application of water balance concepts for a climatic study of droughts in South India. *Indian Journal Meteorology Geophysics* **15**, 393.

SUTCLIFFE, R. C. and BANNON, J. K. 1954: Seasonal changes in the upper air conditions in the Mediterranean–Middle East area. *Scientific Proceedings International Association Meteorology*, Rome, p. 322.

WORLD METEOROLOGICAL ORGANIZATION. 1965: *Proceedings of the Symposium on Meteorological Results of the International Indian Ocean Expedition*, Bombay, 22–26 July 1965. Geneva.

YEH TU-CHENG, DAO SHIH-YEN, LI MEI- TS'UN, 1959: The abrupt change of circulation over the northern hemisphere during June and October. In Bolin, B. (editor), *Rossby Memorial Volume: The atmosphere and sea in motion*. New York: Rockefeller Institute Press, p. 249.

YIN, M. T. 1949: A synoptic–aerologic study of the onset of the summer monsoon over India and Burma. *Journal Meteorology* **6**, 393.

# 7
# Equatorial climates

Vertical air motion and rainfall are closely correlated, and in middle latitudes there is a further correlation with the horizontal air flow. This follows from the simplified version of the vorticity equation, which was introduced in chapter 3:

$$\frac{d}{dt}(\zeta + f) = -(\zeta + f)\, \text{div}_p\, V$$

This equation states that in middle latitudes, convergence leads not only to ascent but also to cyclonic rotation. The Coriolis parameter (f) is small near the equator, changes sign at the equator itself, and so the nature of closed circulation is uncertain near or over the equator where, in fact, they are but rarely observed. Since the large-scale rotating systems of the subtropics do not come within 5 degrees of the equator, low-pressure systems observed within this 5-degree band on either side of the equator are normally orographically induced. The small or zero value of the Coriolis parameter found near or at the equator also leads to the breakdown of the geostrophic wind equation; the relationship between the pressure and wind fields is often uncertain with strong flow across the isobars. The band extending for about 10 degrees on either side of the equator can be considered to have atmospheric dynamics which differ significantly from those found in higher latitudes and this difference explains its treatment as a separate climatic zone.

Trade winds from the northern and southern hemispheres meet in the equatorial trough. Figure 7.1 portrays idealized surface wind conditions for the months of January and August. Two types of equatorial trough may be distinguished (Gray, 1968): the doldrum equatorial trough and the trade wind equatorial trough. The former is an area along which calm surface conditions prevail, for example, in the northwest Pacific in August. When the equatorial trough is not displaced far from the equator, the northeast and southeast trade winds tend to meet along a line of lowest pressure and calm conditions are not observed. This type is known as the trade wind equatorial trough. It can be seen from figure 7.1 that the equatorial trough is sometimes displaced to outside the 10°N to 10°S equatorial zone; this is particularly so in the Indian Ocean in August. The diagrams also suggest that at times general convergence into the 10°N to 10°S equatorial zone may exist while other sections of the zone have general cross-equatorial flow, implying that weather and climate in the equatorial zone will be far from uniform.

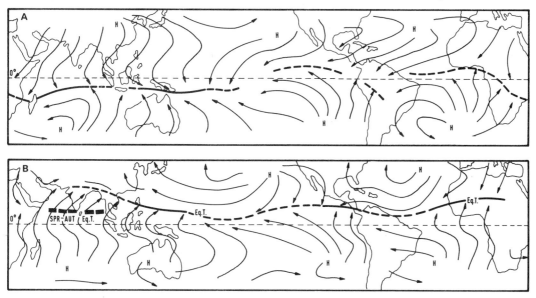

**Figure 7.1**  Idealized surface wind conditions. A : January. B : August. (*After Gray, 1968.*)

Several distinctive atmospheric flow patterns can be recognized in the equatorial zone, and their properties explored using the equations for the three components of atmospheric motion. According to Lettau (1956), these equations can be written as follows:

$$\frac{du}{dt} = -\frac{1}{\rho}\frac{\partial p}{\partial x} + fv - lw,$$

$$\frac{dv}{dt} = -\frac{1}{\rho}\frac{\partial p}{\partial y} - fu,$$

$$\frac{dw}{dt} = -\frac{1}{\rho}\frac{\partial p}{\partial z} - g + lu,$$

where $\frac{d}{dt}$ means the change with time of the three wind components, u, v, and w are the zonal, meridional, and vertical wind components respectively, $\frac{\partial p}{\partial x}$, $\frac{\partial p}{\partial y}$ and $\frac{\partial p}{\partial z}$ are the three local components of the pressure gradient, and l is the vertical Coriolis parameter ($2\,\Omega\cos\phi$).

Large-scale changes of wind speed in the atmosphere normally take place only slowly, and therefore the acceleration terms in the above equations can often be neglected. Flow near the equator frequently has large easterly or westerly components, and so the last two equations may be particularly relevant. If the acceleration term in the second equation is nearly zero, the possibility of quasi-geostrophic zonal flow arises.

### 1  Contour models
Broad-scale easterly flow can arise in a contour model called the Equatorial Duct by Johnson and Mörth (1960). In the commonest case, illustrated in figure 7.2, there are large

subtropical anticyclones in each hemisphere with the lowest pressure in any given longitude at the equator itself. Confluence and diffluence occur in the eastern and western ends of the Duct respectively, but in the central regions the flow is nearly zonal. Johnson and Mörth suggest that for steady zonal flow approximately parallel to the contours in a steady pressure field, the pressure gradient and Coriolis forces must be of the same order of magnitude down to very low latitudes. Under these conditions and assuming uniform zonal flow, the meridional pressure profile will be described by the second of Lettau's equations $\left(\text{for } \dfrac{dv}{dt}\right)$, that is to say, it will be bowl-shaped with the meridional component of the pressure gradient vanishing at the equator.

General westerly flow occurs in a contour model called the Equatorial Bridge, and in this case the meridional pressure profile takes the form of a broad ridge which is flat at the equator. In some ways this model is the opposite of the Equatorial Duct, in that the flow is

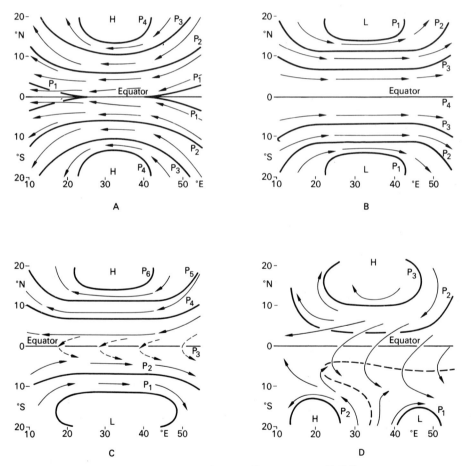

**Figure 7.2** Equatorial contour and streamline patterns. Full lines are contours and arrows represent segments of streamlines. **A**: The Equatorial Duct. **B**: The Equatorial Bridge. **C**: The Equatorial Step. **D**: The Simple Cross-Equatorial Drift. (*After Johnson and Mörth, 1960.*)

180

still quasi-geostrophic to low latitudes but the pressure and wind fields are inverted. Another possibility for low-latitude quasi-geostrophic flow exists if the meridional pressure profile has an inflexion at the equator, where the pressure gradient vanishes and the flow is zonal. This model is called an Equatorial Step and is illustrated together with the two previous models in figure 7.2.

Often there is an anticyclone on one side of the equator and a low on the other with marked cross-equatorial flow, as shown in figure 7.2; this particular model is known as the Cross-Equatorial Drift. Clearly, the flow near the equator is far from geostrophic, and Johnson and Mörth have suggested that, for a linear variation of pressure with latitude, zonally-orientated isobars, and initial zonal flow, there is for a given pressure gradient a critical latitude equatorwards of which any small geostrophic departure is sufficient to cause air particles to stream across the equator.

## 2 Zonal flow

It has been suggested that zonal flow is not infrequent near the equator. The last of Lettau's equations $\left(\text{for } \dfrac{dw}{dt}\right)$ contains a relationship between zonal components (u) and vertical motion (w). It indicates two possible consequences of zonal flow near the equator. First, the presence of the term lu, which is a maximum at the equator, could mean that air flowing from the east is slightly denser than air flowing from the west, if temperature and pressure are equal. Alternatively, a second consequence of the term lu could be a tendency for lifting to be associated with westerly winds and subsidence with easterly winds. Flohn (1964) has found a strong correlation between westerly surface winds and lifting and between easterly surface winds and subsidence. He has discovered, for instance, that the average frequency of rain associated with tropical easterlies is 8·1 per cent, but with tropical westerlies this increases to 25·1 per cent.

Vertical motion and meridional flow may be correlated according to the first equation for $\dfrac{du}{dt}$, but Flohn suggests that this is less significant than the correlation between zonal flow and vertical motion. He finds that tropical winds with an equatorward component are accompanied by a rainfall frequency of 8·1 per cent, and the winds with an opposite component with a rainfall frequency of 12·0 per cent.

Lettau (1956) has investigated the fluctuation of the vertical wind component (w') with the zonal wind, and he derives the following equation:

$$lu = \frac{\partial\,(w'^2)}{\partial z}$$

which demonstrates that the energy of the fluctuations of the vertical wind component increases with height in large-scale westerly winds but decreases with height in easterly winds. This suggests that in low latitudes westerly winds are positively correlated with increasing convective activity.

## 3 Factors determining weather

Though the direction of air flow in the equatorial trough is probably important in determining the weather, the presence of externally imposed convergence or divergence is of paramount importance. Developments within the subtropics generate convergence or divergence within equatorial trough and thus control the broad-scale weather. A rainfall maximum exists over Indonesia during the northern winter because of the massive convergence between the northern and southern trades, but in the northern summer there is slight divergence and consequently a rainfall minimum. Equatorial dry zones exist between latitudes 0° and 10°s over the eastern and central Pacific and these seem to be related to general divergence in the trades.

A glance at the map of world mean annual rainfall (figure 2.16) shows that precipitation within the equatorial trough is extremely variable. Rainfall maximums exist over the equatorial regions of South America and Africa, and also over the equatorial islands from Indonesia to the Carolines; but in contrast rainfall minimums tend to exist over the oceans. Variations within the trough become clearer if the annual percentage of days with thunder are considered (figure 7.3). Thunderstorm peaks exist over South America, Africa, and Indonesia, while they are comparatively rare over the oceans. The bulk of the rainfall over the continents and Indonesia falls from thunderstorms; but this is not true of the oceans. For example, according to Ramage (1968), Fanning Island (Line Islands) with an annual average rainfall of 1992 mm, has only one thunderstorm-day a year.

Zonal flow along the equator was described in chapter 4, together with the dry zone in the equatorial eastern Pacific. The meridional circulation in the central and eastern Pacific may be rather complex, since several authors have reported a closed equatorial cell between the two Hadley cells. Bunker (1971) found that the meridional circulation over the

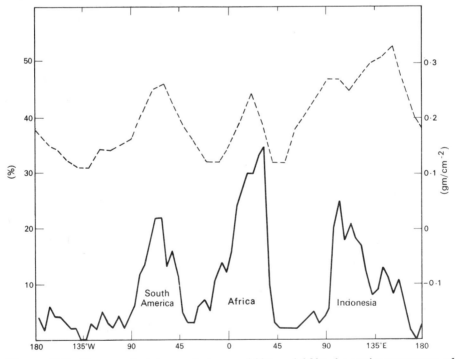

**Figure 7.3** The equatorial zone between 10°N and 10°S. Annual percentage of days with thunder by 5° longitude intervals (solid line). Mass of water vapour in gm cm$^{-2}$ above 500 mb for 10° longitude intervals (dashed line); averages of February, April, and June 1962. (*After Ramage, 1968.*)

equatorial central Pacific Ocean in April 1967 consisted of a shallow northern Hadley cell, an extensive equatorial cell centred south of the equator with maximum subsidence over the equator, and a strong southern Hadley cell. Part of the high-level outflow of the southern Hadley cell subsided across the equator and decreased the activity of the northern convergence zone.

The drier parts of the equatorial trough are really an extension of the subtropical deserts, and the typical climatic type is best developed over South America, Africa, and southeast Asia. The climates of these three regions are not exactly the same, but they are similar in most important aspects. For this reason it is proposed to discuss the climate of a typical region in detail and assume that the basic principles will apply to the two remaining areas. The region chosen is southeast Asia because it is the most densely populated part of the equatorial world and the climate of the area is reasonably well documented.

## The climate of equatorial southeast Asia

The area under consideration extends from 95°E to about 150°E, and is bounded by the parallels 10°N and 10°S. It includes nearly the whole of Indonesia, all Malaysia, Australian New Guinea (Papua), Mindanao Island in the Philippines, and part of southern Thailand. The total population of the area is at least 120 million, and includes several cities with populations exceeding one million. The mean annual rainfall is mostly above 2,000 mm, the sea-level temperature above 25°C, and the natural vegetation consists of tropical evergreen forest. Because of its nearness to the equator, only the extreme northern and southern regions are crossed by major tropical storms, and this gives some unity to the climate. Further unity exists in that massive convergence takes place in the northern winter and therefore most of the region has a rainfall maximum between October and March.

Climatically, the region is very similar to equatorial Africa or South America, but about two-thirds of the area is water and the rest consists mostly of mountainous islands or peninsulas. Because climatically the many mountainous islands act as if they were a much larger land-mass, the whole region has been called by Ramage (1968) a 'Maritime Continent'.

*1 Circulation systems*
The surface air is always warm and humid, and therefore the annual variation of climate is controlled by the movement of convergence zones. During the northern winter a convergence zone lies near to the equator, and heavy rains occur. In the northern summer this convergence zone has moved outside the region to the northern subtropics and a dry season is created.

*a Wet season* December, January, and February cover the wet season in much of Indonesia, and figure 7.4 illustrates daily 700 mb flow patterns which tend to occur frequently during these months. A belt of dynamic anticyclones occurs along 15°N. Normally, at this season, a high will be present on nearly every 700 mb chart, and indeed on most charts a similar dynamic anticyclone will be found over central or southern Australia. Weak stationary lows frequently exist in the South China Sea and cyclonic activity occurs during this season over northern Australia and southern Indonesia. A narrow belt of weak westerly winds existing along the equator over Sumatra and Borneo are caused by the lows forming

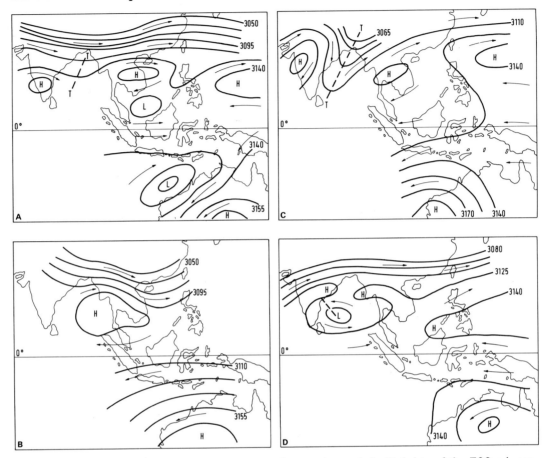

**Figure 7.4** Selected 700-mb contour patterns for southeast Asia. Heights of the 700 mb surface are in metres and contours are drawn at 15-metre intervals. Arrows indicate wind directions. H, high; L, low; T, tropical storm, centre probably deeper than indicated; T with broken line trough. Broken arrows indicate direction of movement of lows. **A:** January and February—an equatorial bridge. **B:** January and February—an equatorial duct. **C:** April—a trough in Bay of Bengal. **D:** May—a low developing in Bay of Bengal.

an equatorial bridge. In contrast equatorial easterlies occur over New Guinea and easterly winds are also found in the subtropics over northern Australia and southern Thailand. The rather simple flow patterns at 700 mb are reflected by the surface circulations. Convergence areas are caused by the equatorial westerlies over Borneo and the Pacific equatorial easterlies, and also by the flow from the subtropics into the equatorial westerlies over Sumatra. The main intertropical convergence zone occurs mostly over the Java Sea during January, from where it runs south–east to the Arafura Sea and the Torres Strait. The intensity of convergence increases from the surface up to about 2,000 m, and from there upward it decreases gradually until above about 6,000 m the converging currents cannot be easily identified.

March, April, and May represent the transition period from the wet to the dry season

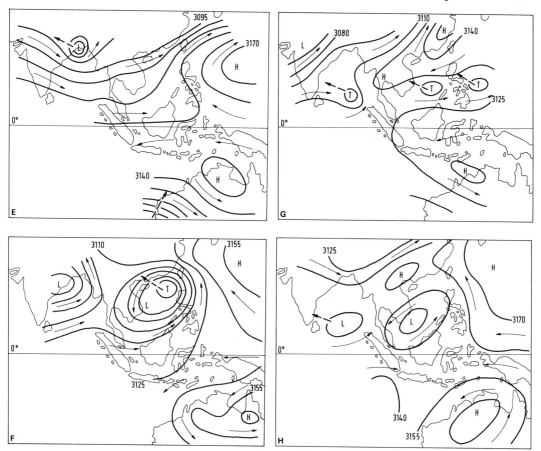

**Figure 7.4** **E:** June—the start of the southwest monsoon season. **F:** July, August, and September—tropical storms and strong westerlies. **G:** July, August, and September—a ridge situation. **H:** October and November—a South China Sea low. (*After Lockwood, 1966.*)

over much of Indonesia. A typical 700 mb chart from this period is shown in figure 7.4. During April, surface temperatures reach their annual maximums at many places in southern Asia and the 700 mb meridional temperature gradients reverse. A marked trough develops in the Bay of Bengal, and in May low-pressure systems start to move from the Bay into India. As the trough develops, the winter highs over India and mainland southeast Asia decline allowing the mid-latitude westerly winds to advance south towards the equator. As the season advances, the high over Australia drifts north, bringing dry easterly winds to within about 5 degrees of the equator. During these three months the main convergence zones travel slowly north and by May lie over the Sulu Sea and Mindanao.

*b Dry season* June, July, and August form the dry season over most of the region, and some typical 700 mb charts for this season are shown in figure 7.4. Over most of southern Asia, June marks the beginning of the southwest monsoon season. The dynamic highs have retreated from India and mainland southeast Asia and been replaced by a broad belt of westerly winds. During the southwest monsoon season, deep tropical storms move westwards across the Philippines towards China and similar storms develop over the Bay of Bengal and move westwards towards India. The dynamic high over Australia is now near to its most northerly position, and easterly winds exist over Indonesia. The easterly winds just to the south of the equator are normally very dry, indicating subsidence, in contrast to the moist southwesterlies to the north. Steep humidity-gradients across the equator are not uncommon; the 700 mb dew point at Djakarta may reach $-16°$c while at Singapore it may be only $2°$c.

September, October, and November represent the transition back to the wet season. In October, the 700 mb meridional temperature-gradients are changing sign, and for some weeks near-uniform temperatures may be found over wide areas. During this period dynamic anticyclones start to extend across China, and later across India. There is often a well-developed low in the South China Sea, as illustrated in figure 7.4; it has northeasterly winds on the poleward side, the first sign of the northeast monsoon in West Malaysia. During October the Australian high is still near to its extreme northern position, but it moves south in November as cyclonic activity increases over northern Australia. The South China Sea lows are probably most intense during November and bring heavy rain to the east coast of West Malaysia.

## 2   Radiation and temperature

Charts of global and net radiation which are reproduced in figure 3.2, indicate the broad radiation conditions of the area. The annual heat balance at a site over the Pacific Ocean near New Guinea was discussed in chapter 1 (figure 1.21), where it was noted that net radiation changes only slightly during the year. Indeed, radiation changes near the equator are always small, the duration of daylight varying only slightly from 12 hours. Since the high rainfall supplies abundant moisture for evaporation, which uses up large amounts of net radiation, the temperatures tend to be not only nearly uniform throughout the year but also rather low when compared with the arid tropics. There are few measurements of radiation within the region, but there has been a station in existence at Singapore for some years; mean values are given in table 7.1.

The thermal regime of the region is characterized by general uniformity, the sea-level temperatures being almost constant throughout the whole area and throughout the year. At any given site the largest variations in temperature are between day and night, rather than between seasons. The coldest parts of the region are on the mountain peaks.

Diurnal temperature variations follow the same pattern in most parts of Malaysia and Indonesia; the daily minimum is reached between 5 and 7 a.m. local time, and the maximum between 12 a.m. and 3 p.m. A fairly rapid rise in temperature takes place during the first 6 or 7 hours of daylight, and after reaching the maximum the temperature falls again, fairly rapidly in the late afternoon and slowly through the night. Annual and diurnal temperature curves for Singapore are illustrated in figure 7.5. The annual temperature range at Singapore is about $2.2°$c while the diurnal range reaches $6.2°$c, but annual ranges increase with distance from the equator.

Mean annual sea-level temperatures throughout the region are close to $26°$c. In Java, Braak (1929) states that between sea level and 1500 m the decrease in mean temperature is $0.6°$c/100 m, but from this point upward the lapse rate becomes smaller and decreases to

**Table 7.1** Radiation at Singapore:
**A**: Global short-wave radiation at the University of Singapore and Paya Lebar (1952–1967 inclusive)

| Month | Mean (cal cm$^{-2}$) |
|-------|----------------------|
| Jan   | 391·4 |
| Feb   | 431·3 |
| Mar   | 453·4 |
| Apr   | 430·1 |
| May   | 403·9 |
| Jun   | 399·7 |
| Jul   | 398·7 |
| Aug   | 417·0 |
| Sept  | 419·7 |
| Oct   | 393·9 |
| Nov   | 374·8 |
| Dec   | 336·2 |

**B**: Mean hourly values of global short-wave radiation (cal cm$^{-2}$) at Paya Lebar (April 1965–March 1968)

| Local time | 06–07 | 07–08 | 08–09 | 09–10 | 10–11 | 11–12 | 12–13 |
|------------|-------|-------|-------|-------|-------|-------|-------|
| Jan        | 2·4   | 14·6  | 29·3  | 41·6  | 50·3  | 55·3  | 50·8  |
| July       | 4·1   | 18·4  | 32·3  | 44·7  | 51·7  | 55·5  | 55·6  |

| Local time | 13–14 | 14–15 | 15–16 | 16–17 | 17–18 | 18–19 | 19–20 |
|------------|-------|-------|-------|-------|-------|-------|-------|
| Jan        | 47·2  | 37·8  | 29·1  | 17·8  | 6·7   | 0·2   | 0·0   |
| July       | 48·6  | 39·1  | 31·9  | 19·6  | 6·9   | 0·1   | 0·0   |

*After Chia (1969).*

0·55°C/100 m. According to Braak the average temperature in Java can be computed from:

$$T = 26·3 - 0·61h,$$

in which the temperature is in °C, and h is the altitude above sea level in hectometers. Above 2,000 m the formula becomes

$$T = 14·1 - 0·52h$$

Formulas of this type can probably be applied to other parts of the region with a moderate degree of confidence.

Radiosonde data from Singapore (Frost and Stephenson, 1965) indicate that the mean height of the 0°C isotherm is 579 mb (4,694 m), being highest in May at 569 mb (4,846 m) and lowest in September at 588 mb (4,572 m). The snow-line, that is the line of permanent snow, is found therefore only on the most lofty peaks, which are in New Guinea. The snow-line on Sukarno Peak (5,030 m), has been located at about 4,300 m.

*a Frost* The areas with near freezing average temperatures are extremely limited, but frost can be important over rather wider areas. Within the tropics, frost is usually due to intense radiational cooling rather than to the movement of cold air-masses, and because of this its incidence will be related to the air temperature and humidity. Braak has given some details (table 7.2) on the number of frost nights on the tea estate Kertosarie (west Java)

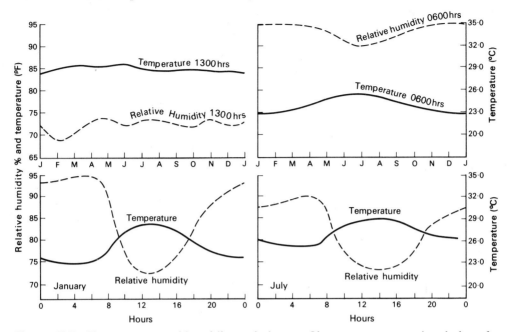

**Figure 7.5** Temperature and humidity variations at Singapore : seasonal variations for 0600 and 1300 local time (*top*), and diurnal variations for January and July (*bottom*). (*After Dale, 1963.*)

situated at 1,550 m. From this table it appears that the dry years of 1914, 1915, 1918, and 1922 were also 'frost years'; similarly frost occurs most frequently in the dry months of July and August.

Height itself, below the 0°c isotherm, is not sufficient to cause frost within the tropics. Clear dry calm nights are required, but the influence of topography is normally decisive, because it is only found in hollows into which cold air drains at night. In Java, according to Van Steenis (1968), frost hollows are frequently found above 2,000 m, but their number at about 1,500 m is restricted; nevertheless one even appears to exist at 900 m at Blawan, on the north side of the Idjen Caldera in east Java.

The radiating surface is also important in frost formation, and in particular the type of vegetation cover can determine whether frost will form in the tropics. At 2,000 m altitude

**Table 7.2** Number of frost nights on tea estate Kertosarie, west Java (1550 m) in different years

| Month | Years | | | | | | | | | | | | | |
|-------|------|------|------|------|------|------|------|------|------|------|------|------|------|------|
| | 1909 | 1910 | 1911 | 1912 | 1913 | 1914 | 1915 | 1916 | 1917 | 1918 | 1919 | 1920 | 1921 | 1922 |
| Jun | | | | | | | | | | 2 | | | | 3 |
| Jul | | | 1 | 1 | 5 | 7 | 8 | 2 | | 8 | 5 | 4 | | 8 |
| Aug | | 2 | | 2 | 16 | 13 | 3 | 7 | | 8 | 4 | 1 | 10 | 9 |
| Sept | | | | | | 4 | | | | 10 | 2 | | 3 | |
| Oct | | | | | | 4 | | | | 10 | 2 | | 3 | |
| Nov | | | | | | | 3 | | | | | | 2 | |

*After Braak (1929).*

188

in Java over short grass, Van Steenis has measured temperatures at about 5.30 a.m. of −5° to −10°C. He reports that on such mornings the grass, especially the dry, withered blades, carry hoar-frost and on shallow puddles a thin ice-crust can be found. The situation is entirely different if the same site is covered by evergreen forest. The radiating leaf surfaces are now high above the ground, and the cold air sinks from these surfaces and mixes with warmer air below, and so a cold surface layer is not generated. It is interesting to note that Kajewski (1932) wrote concerning the Atherton Tableland in north Queensland (800–1,500 m altitude), that 'it is noticed that after the decidedly tropical rain-forest is felled, the climate is altered, and the soil becomes subject to severe frosts'.

The mountains of West Malaysia are somewhat lower than those of Java, and Dale (1963) states that as far as is known, the surface temperature has never fallen to freezing point. The lowest minimum temperature observed in a thermometer screen at the Cameron Highlands, West Malaysia (altitude 1,448 m) is 2·2°C, which would suggest that on this occasion the temperature just above bare soil or short grass would be below freezing and a ground frost would occur. Since in West Malaysia the mountainous terrain reaches 2,100 m, it would appear that ground frost probably does occur in high enclosed hollows which are not covered by forest, and similar comments probably apply to other mountainous islands in the region.

*b Soil temperatures* Braak (1929; 1931) has investigated soil temperatures under grass at depths of 3, 5, 10, 15, 30, 60, 90 and 110 cm near Djakarta over a period of 2 years and his results are summarized in table 7.3. The soil temperatures show the usual lag of the time of maximum daily temperature with depth, but more interesting is the difference between soil and air temperature. Table 7.3 indicates that the mean temperature of the soil from the surface to a depth of at least 110 cm exceeds that of the atmosphere by about 3·4°C. This finding appears to apply to other areas of bare soil, the difference decreasing with the density of vegetation. In the full shade of a large tree in Djakarta, at a depth of 30 cm,

**Table 7.3**   Soil temperature at Djakarta

| 1 | 2 | 3 | 4 | 5 | 6 | 7 | 8 | 9 | 10 | 11 | 12 |
|---|---|---|---|---|---|---|---|---|---|---|---|
| Depth in cm | Mean daily maximum temperature | Time of the day | Retardation | Mean daily minimum temperature | Time of the day | Retardation | Yearly mean of temperature 24 hours | Yearly mean of daily amplitude | Absolute maxima of temperature | Absolute minima of temperature | Absolute Amplitude of temperature |
| (Air) | 29·97 | 1 pm | — | 23·06 | 6 am | — | 26·11 | 6·91 | 35·8 | 18·3 | 17·5 |
| 3 * | 32·12 | 2½ pm | 1½ h | 26·92 | 7 am | 1 h | 29·20 | 5·20 | — | — | — |
| 5 | 32·11 | 3 pm | 2 h | 27·06 | 7 am | 1 h | 29·30 | 5·05 | 37·9 | 23·1 | 14·8 |
| 10 | 31·09 | 4½ pm | 3½ h | 27·99 | 8 am | 2 h | 29·48 | 3·10 | 35·6 | 23·6 | 12·0 |
| 15 | 30·19 | 7½ pm | 6½ h | 28·70 | 9½ am | 3½ h | 29·46 | 1·49 | 33·3 | 23·9 | 9·4 |
| 30 | 29·56 | 1 am | 12 h | 29·28 | 2 pm | 8 h | 29·44 | 0·28 | 31·5 | 23·4 | 8·1 |
| 60 | 29·55 | 4 pm | 27 h | 29·50 | 5 pm | 23 h | 29·53 | 0·05 | 30·6 | 23·8 | 6·8 |
| 90 | 29·51 | 5 pm | 52 h | 29·47 | 4 pm | 46 h | 29·49 | 0·05 | 30·1 | 26·9 | 3·2 |
| 110 | 29·49 | 2 pm | 73 h | 29·45 | 4 am | 68 h | 29·47 | 0·04 | 29·9 | 26·3 | 3·6 |

* This depth may be a little more, near to 4 cm (Braak).

*After Braak (1929).*

Braak observed a mean temperature almost equivalent to that of the air. Similar results have been registered at other locations in Java, e.g., under a virginal forest at Tjibodas (1,450 m), and near the summit of the volcano Panggeranggo (1,450 m) (Möhr and Van Baren, 1959). In Java, soil temperatures at 1 m depth appear to decrease with altitude at a rate of 0·6°c/100 m.

### 3  Rainfall

Over the equatorial continents, the bulk of the rainfall comes from showers or thunderstorms, and in this respect the scattered islands of Indonesia act as a continent. Figure 7.3 illustrates the distribution of thunderstorms in the equatorial zone and clearly shows a secondary maximum over the Malaysian/Indonesian region. In general, annual rainfall totals exceed 2,000 mm, and stations near the equator tend to have it almost evenly distributed throughout the year, but further away from the equator the rainfall becomes more seasonal with one pronounced maximum. Throughout most of Malaysia and Indonesia the maximum rainfall occurs during the months from November to February. These are the months when active convergence is taking place over the region and satellite pictures show extensive cloud cover.

Rainfall distribution patterns tend to reflect the major features of the topography. For example, in West Malaysia there are two particularly wet belts: one running the full length of the South China Sea side of the eastern mountains, and the other confined to the northern part of the peninsula, on the Malacca Straits' side of the western mountain ranges. As is illustrated in figure 7.6, the axes of the high rainfall belts do not, in general, coincide with the highest land, but rather with the foothills of the mountains. The axis of the western wet belt in West Malaysia appears to approximate with an altitude of between 150 and 350 m, beyond which the rainfall decreases with increasing height. The eastern wet belt covers the mountain slopes and foothills facing the South China Sea and extends across the lowlands to the coast. Indeed one of the wettest places in the eastern part of the peninsula is situated on the lowlands at less than 130 m above sea level and only 19 km from the coast. Sukanto (1969) comments that a similar phenomenon is observed in Indonesia but that it is difficult to determine the altitude of maximum precipitation due to a shortage of observations. The maximum annual rainfall recorded in Indonesia is 7,069 mm at Baturaden, central Java, which has an altitude of 700 m and the minimum annual rainfall observed occurs in the Palu Valley of Celebes. Similarly in West Malaysia dry areas are found in the interior valleys.

Within the region much of the rainfall comes from convective storms, for according to Sukanto, showers with an intensity of 0·5 mm/min are common in Indonesia. Thunder is very frequent in Sumatra and west Java; Buitenzorg records it on 322 days a year, which is almost a world record, but many stations have much lower means—Djakarta 136 days, Surabaya 73 days, and Singapore 41 days. A very good idea of the intensity of rainfall can be obtained by the study of cloudbursts (showers having an intensity of at least 1 mm per minute for not less than 5 minutes). Möhr and Van Baren (1959) have calculated the total rainfall resulting from cloudbursts at a number of Indonesian stations, and some of these results are summarized in table 7.4. This table indicates that, on average, about 22 per cent of the annual rainfall is produced by cloudbursts; a similar estimate by Geiger (1927) for Bavaria, Germany is 1·5 per cent. If an allowance is made for the higher total rainfalls found in the tropics as compared with the temperate latitudes, then up to 40 times as much

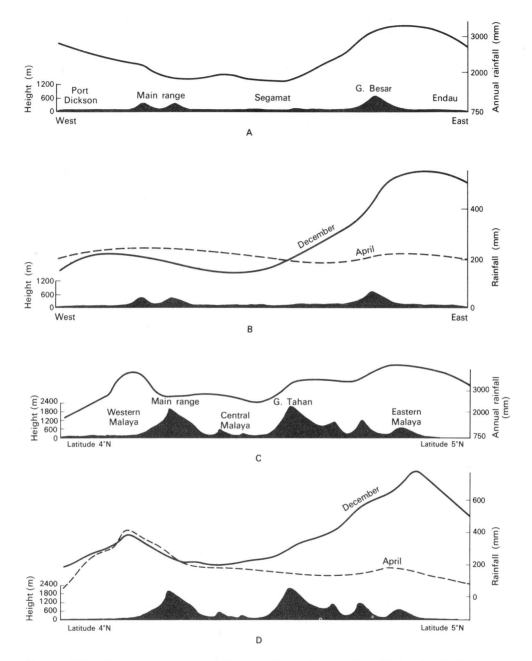

**Figure 7.6** The relation of rainfall to configuration in West Malaysia. **A**: Average annual rainfall from Port Dickson to Endau (southern Malaya). **B**: Average April and December rainfalls from Port Dickson to Endau (southern Malaya) **C**: Average annual rainfall from 4°N on the west coast to 5°N on the east coast (northern Malaya). **D**: Average April and December rainfalls from 4°N on the west coast to 5°N on the east coast (northern Malaya). (*After Dale, 1959.*)

**Table 7.4**  Cloudburst rainfall as a percentage of the annual total for selected Indonesian stations

| | |
|---|---|
| Tjinjinian | 8 |
| Ambon | 13 |
| Menado | 14 |
| Bandung | 17 |
| Asembagus | 19 |
| Padang | 19 |
| Djakarta | 22 |
| Medan | 24 |
| Djember | 27 |
| Bogor | 29 |
| Pasuruan | 30 |
| Bangelan | 32 |
| Sawaham | 37 |

*After Möhr and Van Baren (1959).*

rainfall is precipitated in Indonesia by cloudbursts as in temperate latitudes. An analysis by Möhr and Van Baren of rainfall at Bogor further illustrates the importance of heavy showers. Bogor has an annual rainfall of about 4,230 mm, of which about 900 mm is due to some 60 showers of 10 to 20 mm, 3,500 mm to some 65 showers of 20 to 100 mm, and showers of less than 10 mm contribute only about 700 mm of the yearly total.

*a  Diurnal variations*  Shower formation is controlled, in the equatorial trough, by local relief and the diurnal cycle of radiation. For most inland stations, the daily rainfall cycle retains the same general pattern throughout the year. The morning hours are relatively dry, especially between 8 and 11 a.m., while the hours from noon until shortly before midnight are much wetter. Ramage (1964) has studied the diurnal variation of August rainfall in West Malaysia, and 3-hourly running means of rainfall at 11 stations are given in figure 7.7. All the stations show marked diurnal variations, though the time of maximum rainfall varies greatly. As the diurnal march of temperature is practically constant over the whole peninsula, it cannot explain all the widely varying diurnal rainfall regimes, some of which must be due to convergence resulting from local land and sea breezes. Because of land and sea breezes it appears that general low-level convergence and lifting occurs over West Malaysia during the day and low-level divergence and sinking at night. According to Ramage the diurnal rainfall regimes can be categorized as shown in figure 7.8.

Qualitative explanations for the regimes presented in figure 7.7 have been suggested by Ramage. The southwest coast–Malacca Strait regime is caused by the interaction of land and sea breezes from Sumatra and West Malaysia over the Malacca Strait. At night the convergence of the land breezes causes upward motion over the Malacca Strait; and since the Sumatra land breeze reinforces and the Malaysian land breeze opposes the westerly synoptic wind, the convergence zone in which showers develop probably lies close to the Malaysian coast. Showers drift over the Malaysian coast and cause the heaviest rain to fall between midnight and dawn; nevertheless the showers decrease during the day when the sea breezes set in generating divergence and sinking in the Strait and along the coast.

Further north the Malacca Strait is much wider and consequently the Malaysian and Sumatran land and sea breezes do not significantly interact. Therefore, in the west coast regime the diurnal rainfall variation is slight.

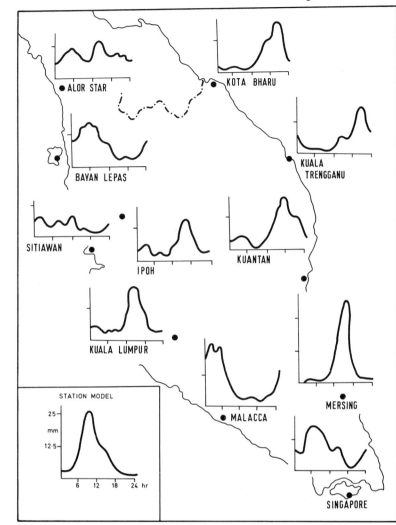

**Figure 7.7** Diurnal variation of August rainfall in West Malaysia. Three-hourly running means for selected stations. (*After Ramage, 1964.*)

Over much of the interior there is the simple inland-mountain regime. At night the air is dried by katabatic winds, but during the day solar heating gives rise to an afternoon rainfall maximum. This particular type of regime, which is common over many of the inland parts of the islands and peninsulas of the region, shows little change throughout the year. The diurnal variations have been graphically described by Ashton (1964) during a day and night spent on the summit of Bukit Retak, Brunei State (1,750 m). Ashton made the following notes during his stay.

6:15 pm–5:30 am    Clouds mainly confined to the valley bottoms, forming a dense unbroken even carpet throughout the valleys in the hilly regions, becoming more broken and scattered near the coast and in flat areas. Mountain tops almost cloudless.
5:30 am–8:15 am    As the sun rises, the valley clouds completely dissipate. Clouds on mountains above 1,200 m tend to accumulate, but to be strictly confined to the peaks.

193

8:15 am–11:30 am    Continued accumulation of clouds over the hills, particularly over the steep narrow ridges above 1,000 m, but by 11:30 am beginning to form over narrow ridges down to 500 m. View becoming increasingly obscured.

11:30 am–4:30 pm    View obscured by cloud.

4:30 pm–6:15 pm    Clouds above about 1,300 m becoming patchy, exposing densely cloud-filled valleys and widespread rain which lessened later in the afternoon. The clouds in the valleys, initially very uneven, settling to below about 500 m as an even blanket.

Ashton has also noted marked differences in vegetation on some high mountains in Brunei State. While tree flora of these exposed rocky ridges is physiognomically and in part floristically similar at all altitudes, the striking difference between the vegetation above and below

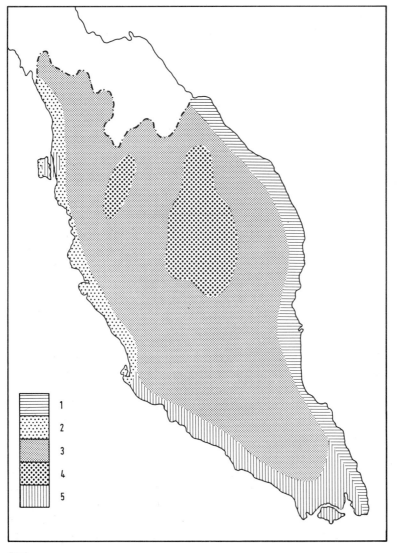

**Figure 7.8** Boundaries of August diurnal rainfall regions in West Malaysia.
1 East coast;
2 West coast;
3 Inland–mountain;
4 Mountain–valley;
5 Southwest coast–Malacca Straits.
(*After Ramage, 1964.*)

about 1,400 m is that below this altitude the formation of deep moss tussocks and blankets, on the ground and up the bases of the tree, is absent; absent too is the rich flora of epiphytic and terrestrial herbs, orchids and ferns, and other cryptogams, but all these features are found above about 1,400 m. This is probably because the higher summits are cloaked in cloud during the hottest part of the day, whereas the structurally similar forest at lower altitudes is exposed to greater extremes of temperature.

In some inland valleys of the peninsula a night rainfall maximum may exist, and this is caused by katabatic winds producing low-level convergence. During the day, anabatic winds cause low-level divergence in the valleys and a rainfall minimum. According to Ramage this particular phenomenon is found in the areas marked mountain-valley regimes on figure 7.8. Braak (1940) reports a similar diurnal variation in some of the inland valleys of Indonesia.

Schematic representation of the development of the east coast evening rainfall maximum is presented in figure 7.9, where in August the sea breeze blows against the prevailing westerly wind. At night the katabatic wind reinforces the westerly synoptic wind and very little rain falls. As the sea breeze strengthens during the day, the sea breeze convergence moves inland, intensifies, and by 3 p.m. the sea breeze is at its strongest and the convergence may have penetrated up to 35 km inland. As the sea breeze convergence stops and then retreats toward the coast, it draws in relatively moist air from the seaward side; thus it gives more rain when it retreats than when it advances and a late afternoon maximum occurs as the convergence crosses the coast.

During the wet season over West Malaysia, the low-level flow is predominantly from the northeast, and therefore many of the dry season diurnal rainfall regimes will be altered. The major changes will be along the coast, with the inland-mountain and mountain-valley regimes remaining basically the same. Because of the reversal of the low-level flow, the dry season regime along the east coast will be found along the west coast during the wet season; similarly during the wet season, the east coast will experience the dry season regime of the northern part of the west coast. As the diurnal rainfall regimes of West Malaysia have now been discussed in detail, it will not be necessary to examine also the similar rainfall regimes with similar causes which are found in the Indonesian Islands.

*b   Daily maximum rainfalls*   Absolute daily maximum rainfalls can be surprisingly high in both Indonesia and Malaysia. According to Möhr and Van Baren (1959), 12 stations in Indonesia have recorded daily values between 500 and 600 mm, 6 between 600 and 700 mm, and one (Ambon) has recorded 702 mm, while in West Malaysia the highest recorded daily rainfall is about 470 mm. Daily rainfall totals likely to be equalled or exceeded once every 10 years in West Malaysia are presented in figure 7.10. The distribution of maximum rainfall in West Malaysia is clearly partly determined by the general relief. There is a general tendency for daily maximum rainfalls to be highest on the east coast and to decrease westwards, but even in the west, daily maximum rainfalls tend to be highest along the coast and to decrease inland, before rising again in the mountainous interior. It is often stated, with some justification, that extremely high daily rainfalls occur over tropical mountains and plateaux. Analysis of data from seven rain gauges in the Cameron Highlands, West Malaysia, suggests that daily rainfall totals for a given return period are either equal to or below those noted for the nearby lowlands. An analysis of maximum observed rainfall intensities over durations from 0·1 to 100 hours by the Malayan Drainage and

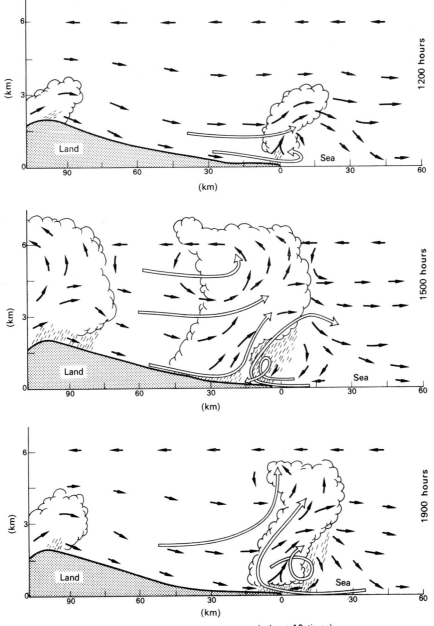

**——** Instantaneous vectors (vertical component exaggerated about 10 times)

**⟹** Trajectories

**Figure 7.9** Schematic representation of the development of the east coast evening rainfall maximum in West Malaysia during August. Local time is used throughout. (*After Ramage, 1964.*)

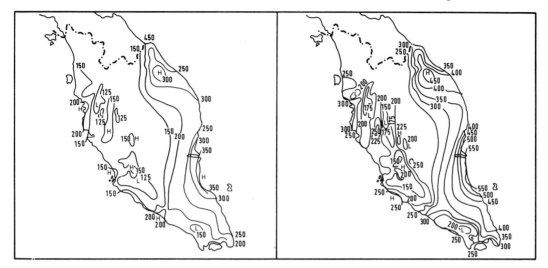

**Figure 7.10** Extreme rainfalls in West Malaysia (mm). *Left:* daily precipitation likely to be equalled or exceeded once every 10 years. *Right:* daily precipitation likely to be equalled or exceeded once every 100 years. (*After Lockwood, 1967.*)

Irrigation Department reveals that the values recorded in the Cameron Highlands are slightly below those recorded elsewhere in the country. Results from other mountainous regions in Malaysia and Indonesia are inconclusive, but it appears that heavy daily rain-falls are more likely to occur over steep mountain slopes or in narrow valleys rather than over interior plateaux. The rainfall values suggested in figure 7.10 are for points, but areal rainfalls are often of greater interest. Much of the normal rainfall of the region comes from convective showers and it is a well-known property of these storms that rainfall decreases rapidly with distance from the centre of the storm. This is illustrated in figure 7.11, which shows the variation of rainfall intensity with area in four storms.

There are reasons for expecting extreme rainfall to be higher in the maritime tropics than near the equator, and extreme rainfalls recorded in the subtropics tend to be higher than those observed near the equator. Extreme rainfalls in the subtropics normally result from highly organized storms of the hurricane type, but because of the low value of the Coriolis parameter, highly organized large-scale storms do not exist near the equator and therefore extreme rainfalls are probably lower.

The nature of rainfall can have important consequences for the development of land-scapes. There is an interesting difference, for example, in the nature of the river channels between Queensland, which is in the subtropics and subject to tropical cyclones, and West Malaysia. According to Douglas (1967), Queensland rivers often have channels which have cut through to the bedrock, while West Malaysian streams are not so deeply entrenched. The Queensland streams require much larger channels than West Malaysian streams which carry the same mean discharge, because so much of the Queensland discharge is concentrated into a few hydrologic events of great magnitude. For instance, the daily rainfall likely to be equalled or exceeded once every 100 years at stations around the Gombak catchment, near Kuala Lumpur, is less than 250 mm; in northeast Queensland, Babinda, Behana, and Campbell Creeks regularly experience daily rainfalls over 250 mm and the

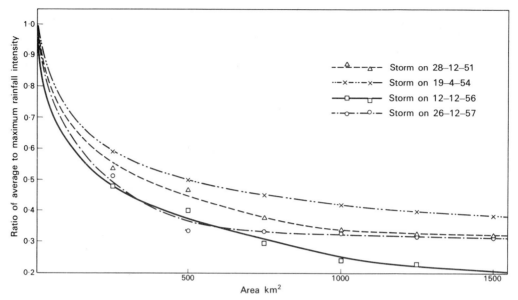

**Figure 7.11**  Variation of rainfall intensity with area for four storms in West Malaysia. (*After Malaysian Drainage and Irrigation Department, 1962.*)

daily rainfall likely to be equalled or exceeded once every 100 years is of the order of 700 mm.

### 4  Evapotranspiration and water balance

Nieuwolt (1965) has studied the problem of evaporation in West Malaysia. He took the available evaporation pan readings and correlated them with various climatic factors, concluding that there is a close correlation between evaporation pan readings, saturation deficit, and actual sunshine hours. He found that wind speed did not correlate well with measured evaporation values. This approach seems promising until the nature of evaporation pan readings are considered. It is well-known that they do not correspond to the evaporation from a large water surface or indeed to evapotranspiration from a vegetation cover; this is partly because of the oasis effect.

The alternative method to the actual measurement of evapotranspiration is to attempt to calculate evapotranspiration values using climatological data. Nieuwolt has calculated potential evapotranspiration values using the method of Thornthwaite and Mather (1957). Figure 7.12 illustrates some of Nieuwolt's results for West Malaysia together with two similar diagrams for Java. Penman's (1948) approach to the problem of the estimation of potential evapotranspiration is probably superior to that suggested by Thornthwaite and Mather in that it takes into account solar radiation. Calculated potential evapotranspiration values by this method for a number of Malaysian stations are shown in table 7.5.

### a  Tropical rain-forest

Theoretical potential evapotranspiration estimates can only be applied to a particular site after some modification to allow for the nature of the surface, which can vary in albedo and roughness. Little is known about the influence of tropical

**Table 7.5**   Potential evapotranspiration values at selected West Malaysian sites (mm/day)

|  | Jan | Feb | Mar | Apr | May | Jun | Jul | Aug | Sept | Oct | Nov | Dec |
|---|---|---|---|---|---|---|---|---|---|---|---|---|
| Alor Star | 4·7 | 5·1 | 5·7 | 5·3 | 4·7 | 4·6 | 4·7 | 4·6 | 4·4 | 4·1 | 3·9 | 4·2 |
| Kota Bharu | 3·8 | 4·7 | 5·3 | 5·5 | 5·2 | 4·9 | 4·9 | 4·8 | 4·8 | 4·3 | 3·8 | 3·6 |
| Penang | 5·2 | 5·3 | 5·4 | 5·2 | 4·9 | 4·8 | 4·7 | 4·8 | 4·9 | 4·5 | 4·0 | 4·6 |
| Kuala Trengganu | 4·1 | 4·6 | 5·4 | 5·6 | 5·2 | 5·0 | 5·0 | 4·8 | 5·1 | 4·5 | 4·0 | 3·8 |
| Ipoh | 4·4 | 4·9 | 5·1 | 4·9 | 4·9 | 4·8 | 4·5 | 4·5 | 4·3 | 4·8 | 4·1 | 4·1 |
| Sitiawan | 3·9 | 4·5 | 4·7 | 4·7 | 4·6 | 4·7 | 4·6 | 4·6 | 4·5 | 4·4 | 4·0 | 3·8 |
| Kuantan | 3·4 | 4·2 | 4·8 | 4·8 | 4·6 | 4·3 | 4·3 | 4·2 | 4·4 | 4·0 | 3·6 | 3·3 |
| Mersing | 4·1 | 5·0 | 5·7 | 5·4 | 5·1 | 4·7 | 4·3 | 4·4 | 4·7 | 4·7 | 4·2 | 3·9 |

rain-forest on potential evapotranspiration, but the evidence that exists indicates that the evapotranspiration rate may approach the evaporation rate from an open water surface as long as the water supply is ample.

Brunig (1970) finds that according to Thornthwaite's method, the actual and potential annual evapotranspiration values at Kuching are equally 1,728 mm, with monthly amounts between 118 (February) and 165 mm (June), and no mean soil moisture deficit. He considers that these values may be on the low side for evapotranspiration from high mixed forests, because tall crops with irregular surfaces may transpire at more than the calculated potential evapotranspiration rates. The range of roughness length estimates for various forest types is shown in table 7.6. Diffusion resistance to water vapour will increase

**Table 7.6**   Calculated relationships between estimated aerodynamic roughness length $Z_O$ and forest types (cms)

| Forest type | $Z_O$ estimated from | |
|---|---|---|
|  | $\log h - 0·98$ | $\log (h_T - d) -0·98$ |
| Mixed dipterocarp forest | 555 | 209 |
| Kerangas forest, wet | 471 | 125 |
| Kerangas forest, dry | 366–293 | 66–31 |
| Kerangas forest, peat | 346–272 | 73–42 |
| Peatswamp forest, peripheral | 408 | 241 |
| Shorea albida association, tall | 513 | 178 |
| Shorea albida consociation | 502 | 83 |
| Interior closed Shorea albida | 399 | 26 |

$h_T$ is the top canopy height
d is the estimated level of the zero-plane

*After Brunig (1970).*

when the aerodynamic roughness and consequently the turbulence decrease; therefore the potential evapotranspiration rate, for a given net radiation, will be lower in forest stands with smoother canopy surfaces. Brunig considers that the mean annual evapotranspiration from tall mixed dipterocarp forest (7 layers, height 45 m) with irregular surface, near Kuching, is at least 1,700 mm, but more likely to be nearer 2,000 mm. Forests on less fertile sites are lower, simpler structured, and have smoother canopies; they will therefore transpire at lower rates. Estimates by Brunig of average annual evapotranspiration rates for simple types of forest near Kuching are as follows:

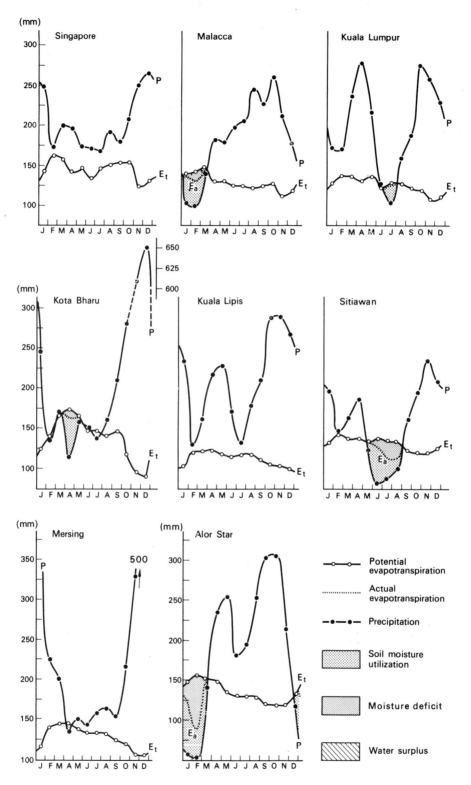

**Figure 7.12** *See above opposite.*

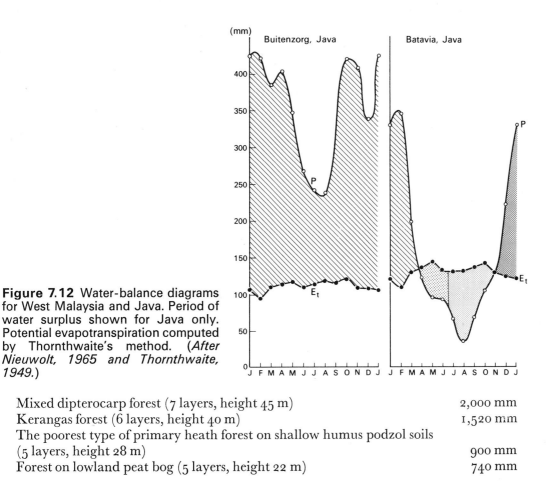

**Figure 7.12** Water-balance diagrams for West Malaysia and Java. Period of water surplus shown for Java only. Potential evapotranspiration computed by Thornthwaite's method. (*After Nieuwolt, 1965 and Thornthwaite, 1949.*)

| | |
|---|---|
| Mixed dipterocarp forest (7 layers, height 45 m) | 2,000 mm |
| Kerangas forest (6 layers, height 40 m) | 1,520 mm |
| The poorest type of primary heath forest on shallow humus podzol soils (5 layers, height 28 m) | 900 mm |
| Forest on lowland peat bog (5 layers, height 22 m) | 740 mm |

Another approach to the problem of evapotranspiration is to study the water balance of selected catchments, but unfortunately this has barely been attempted in southeast Asia. Kenworthy (1970) studied the water balance of a very small forest catchment on the north side of the Sungei Gombak, Central Range, West Malaysia. This particular catchment drained an area with an elevation between about 580 m and 250 m, and on the higher slopes the vegetation was primary dipterocarp forest, but on the lower slopes it had been disturbed by logging operations. Over a relatively short period of time, Kenworthy estimated that with an annual rainfall of 2,500 mm, the actual evapotranspiration was about 1,750 mm per year. He further estimated that 1,300 mm per year was used in transpiration, and 450 mm was intercepted by the canopy and lost again as direct evaporation; 100 mm was lost as direct runoff over the soil surface, the remaining 650 mm being lost by flow through the soil and the deeper rocks.

*b Drought* Regular droughts are unusual in the region, but dry spells and even droughts do occur at irregular intervals. Periods of 30 days with low rainfalls are unusual if not particularly rare, but periods of up to 6 weeks without effective rainfall are rare. In tropical rain-forest a fair amount of water is stored in the vegetation as well as in the soil. Assuming a mixed dipterocarp forest and transpiration rates of between 3 and 5 mm per day, the water stored in the plant tissue would last for 2–5 days. Allowing similar rates of evapo-transpiration, the minimum times to permanent water stress for various forest types on a selection of soils are shown in table 7.7. In practice the time to permanent water stress will

**Table 7.7**   Depletion of soil water during a bright dry period

| Forest type | Soil type | Water available in | | Minimum time to permanent water stress |
|---|---|---|---|---|
| | | (a) Soil | (b) Crop | |
| | | mm | mm | days |
| Mixed dipterocarp forest | Red–Yellow podzols | 180 | 20 | 33 |
| ,, | Lateritic | 160 | 25 | 23 |
| Kerangas forest | Shallow humus podzols | 60 | 5 | 16 |
| ,, | Shallow humus podzols | 144 | 10 | 38 |
| ,, | Medium humus podzols | 264 | 20 | 47 |
| ,, | Deep humus podzols | 552 | 25 | 115 |
| ,, | Podzolic gley soil | 66 | 7 | 18 |

*After Brunig (1970).*

be longer because evapotranspiration rates fall with decreasing soil moisture. Dewfall can also have a survival value by resaturating leaf tissue in critical situations. Drought will occur most frequently in poor vegetation on thin soils, but it may occasionally occur even on good soils. For example, Guha (1969) estimates that at Kuala Lumpur, loamy soils under rubber trees will have a water content as low as 0·25 per cent of the field capacity on 18–100 days a year. He concludes that soil moisture deficit is likely to be a limiting factor for productivity.

# References

ARAKAWA, H. 1969: *Climates of northern and eastern Asia. World survey of climatology*. Vol. **8**. Amsterdam: Elsevier Publishing Co.

ASHTON, P. S. 1964: Ecological studies in the mixed dipterocarp forest of Brunei State. *Oxford Forest Memoir* **26**, Oxford.

BRAAK, C. 1929: Het Klimaat van Nederlands—Indice, I. *Verhandel. Kon. Magnetisch. Meteorol. Obs. Batavia*. **8**.

BRAAK, C. 1931: *Klimakunde von Hinterindien und Insulinde. Handbuch der Klimatologie*. Vol. **4**, Part R. Berlin: Borntraeger.

BRAAK, C. 1940: Over de Oorzaken van te Tijdelijke en Plaatselijke Vershillen in den Neerslag. Koninklijk Nederlands Meteorologisch Instituut, *Mededeelingen en Verhandelingen* **45**, 57.

BRUNIG, E. F. 1970: On the ecological significance of drought in the equatorial wet evergreen forest of Sarawak. In Flenley, J. R. (editor), *The water relations of Malesian forests*. University of Hull, *Department of Geography Miscellaneous Series* **11**, 66.

BUNKER, A. F. 1971: Energy transfer and tropical cell structure over the central Pacific. *Journal Atmospheric Science* **28**, 1101.

CHIA, LIN SIEN. 1969: Sunshine and solar radiation in Singapore. *Meteorological Magazine* **98**, 265.

COCHEMÉ, J. 1960: Some streamlines and contours over the Equator. *Proceedings of Munitalp/WMO Joint Symposium Tropical Meteorology in Africa*, Nairobi, p. 181.

DALE, W. L. 1956: Wind and drift currents in the South China Sea. *Journal Tropical Geography* **8**, 1.

DALE, W. L. 1959: The rainfall of Malaya, Part I. *Journal Tropical Geography* **13**, 23.

DALE, W. L. 1960: The rainfall of Malaya, Part II. *Journal Tropical Geography* **14**, 11.

DALE, W. L. 1963: Surface temperatures in Malaya. *Journal Tropical Geography* **17**, 57.

DALE, W. L. 1964: Sunshine in Malaya. *Journal Tropical Geography* **19**, 20.

DOUGLAS, I. 1967: Erosion of granite terrains under tropical rain-forest in Australia, Malaysia and Singapore. *Symposium on River Morphology*, General Assembly of Bern, Sept–Oct 1967, p. 31.

FLOHN, H. 1964: Intertropical convergence zone and meteorological equator. In *WMO Technical Note* **64**, Geneva, 21.

FROST, R. and STEPHENSON, P. M. 1965: Mean streamlines and isotachs at standard pressure levels over the Indian and west Pacific Oceans and adjacent land areas. *Geophysical Memoirs* **109**. London: HMSO.

GEIGER, R. 1927: *Das Klima der bodennahen Luftschicht*. Braunschweig: Friedr. Vieweg und Sohn.

GRAY, W. M. 1968: Global view of the origin of tropical disturbances and storms. *Monthly Weather Review* **10**, 669.

GRIMES, A. 1937: *Malayan Meteorological Service Memoirs No. 2*.

GUHA, M. M. 1969: A preliminary assessment of moisture and nutrients in soil as factors for production of vegetation in W. Malaysia. *Malayan Forester* **32**, 423.

JOHNSON, D. H. and MÖRTH, H. T. 1960: Forecasting research in east Africa. *Proceedings of Munitalp/ WMO Joint Symposium Tropical Meteorology in Africa*, Nairobi, p. 56.

JOHNSON, D. H. 1963: Tropical meteorology. *Science Progress* **51**, 587.

KAJEWSKI, S. F. 1932: In C. T. White, *Ligneous plants collected for the Arnold Arboretum in north Queensland by S. F. Kajewski in 1929*. Contribution Arnold Arboretum **4**: 6.

KENWORTHY, J. B. 1970: Water and nutrient cycling in a tropical rain-forest. In Flenley, J. R. (editor), *The water relations of Malesian forests*. University of Hull, *Department of Geography Miscellaneous Series*, **11**, 49.

LETTAU, H. 1956: Theoretical notes on the dynamics of the equatorial atmosphere. *Beiträge zur Physik der Atmosphäre* **29**, 107.

LOCKWOOD, J. G. 1966: 700-mb contour charts for south-east Asia and neighbouring areas. *Weather* **21**, 325.

LOCKWOOD, J. G. 1967: Probable maximum 24-hour precipitation over Malaya by statistical methods. *Meteorological Magazine* **96**, 11.

MALAYSIAN DRAINAGE AND IRRIGATION DEPARTMENT. 1962: *Hydrological Data–Rainfall Records 1879–1958*. Kuala Lumpur.

MISRA, R. and GOPAL, B. 1968: *Proceedings of the Symposium on Recent Advances in Tropical Ecology*. Banaras: International Society Tropical Ecology, Varanasi, India.

MÖHR, E. C. J. and VAN BAREN, F. A. 1959: *Tropical soils*. The Hague: N. V. Uitgeverÿ W. Van Hoeve.

NIEUWOLT, S. 1965: Evaporation and water balances in Malaya. *Journal Tropical Geography* **20**, 34.

PALMER, C. E., WISE, C. W., STEMPSON, L. J. and DUNCAN, G. H. 1955: The practical aspect of tropical meteorology. *Air Weather Service Manual* **105–48**, Washington.

PENMAN, H. L. 1948: Natural evaporation from open water, base soil and grass. *Proceedings of Royal Society*, London, **A193**, 120.

RAMAGE, C. S. 1964: Diurnal variation of summer rainfall of Malaya. *Journal Tropical Geography* **19**, 62.

RAMAGE, C. S. 1968: Role of a tropical 'maritime continent' in the atmospheric circulation. *Monthly Weather Review* **96**, 365.

RASCHKE, E. and BANDEEN, W. R. 1967: A quasi-global analysis of tropospheric water vapour content from TIROS IV radiation data. *Journal Applied Meteorology* **6**, 468.

RIEHL, H. 1954: *Tropical meteorology*. New York: McGraw-Hill Book Co.

SCOTT, J. R. 1958: An essay on low-latitude wind analysis. *Quarterly Journal Royal Meteorological Society* **84**, 166.

SUKANTO, M. 1969: Climate of Indonesia. In Arakawa, H. (editor), Climates of northern and eastern Asia. Amsterdam: Elsevier Publishing Co., p. 215.

THORNTHWAITE, C. W. 1949: Review of 'Climate and soil moisture in the tropics' by Möhr. *Geographical Review* **39**, 98.

THORNTHWAITE, C. W. and MATHER, J. R. 1957: *Instructions and tables for computing potential evaporation and water balance*. Centerton, N.J.: John Hopkins University, Laboratory of Climatology.

VAN STEENIS, C. G. G. J. 1968: Frost in the tropics. In Misra, R. and Gopal, B. (editors), *Proceedings of Symposium on Recent Advances in Tropical Ecology, International Society Tropical Ecology*, Varanasi, p. 154.

WATTS, I. E. M. 1955a: *Equatorial weather*. London: University of London Press.

WATTS, I. E. M. 1955b: Rainfall of Singapore Island. *Journal Tropical Geography* **7**, 71.

WORLD METEOROLOGICAL ORGANIZATION 1953: World distribution of thunderstorm days, Part I. Tables. *WMO No.* **21**, *Technical Paper* **6**, Geneva.

WORLD METEOROLOGICAL ORGANIZATION. 1956: World distribution of thunderstorm days, Part II. Tables of marine data and world maps. *WMO No.* **21**, *Technical Paper* **21**, Geneva.

WORLD METEOROLOGICAL ORGANIZATION 1964: High-level forecasting for turbine-engined aircraft operations over Africa and the Middle East. *WMO Technical Note* **64**, Geneva.

WYCHERLEY, P. R. 1968: Climatological and phenological phenomena in Malaysia. In Misra, R. and Gopal, B. (editors), *Proceedings of Symposium on Recent Advances in Tropical Ecology, International Society Tropical Ecology*, Varanasi, p. 138.

# Part III
# The Middle and High Latitude Atmosphere

Plate 11    *Above right:* Severe icing on the trawler *Notts County* during the winter of 1967/68. *Reproduced by permission of Thomson Newspapers Ltd. Above left:* Tornado funnel. *Reproduced by permission of the United States Department of Commerce, National Oceanic and Atmospheric Administration.  Below:* Beaufort force 9 wind and snow showers in the North Atlantic at Ocean Weather Ship Station *Alfa. Reproduced by permission of the Controller of Her Majesty's Stationery Office; Crown Copyright reserved.*

# 8
# The middle
# latitude atmosphere

The greatest atmospheric variability occurs in middle latitudes, from approximately 40° to 70°N and s, where large areas of the earth's surface are affected by a succession of cyclones (depressions) and anticyclones or ridges. This is a region of strong thermal gradients with vigorous westerlies in the upper air, culminating in the polar-front jet stream near the base of the stratosphere. The zone of westerlies is permanently unstable and gives rise to a continuous stream of eddies near the surface, the cyclonic eddies being thrown poleward and the anticyclonic ones equatorward.

Recent work with simple laboratory apparatus (Hide, 1969) suggests that while condensation of water vapour may play an essential role in the tropics, it appears to be no more than a modifying influence in temperate latitudes, because hydrodynamical phenomena found in the atmosphere, including cyclones, jet streams, and fronts, also exist in the laboratory apparatus where there is no analogue of the condensation process. Sutcliffe (1966) has further commented that in recent years it has been proved by reliable calculation that the essential major properties of depressions can be derived without introducing the concept of a front at all, and that not only depressions but also anticyclones emerge from the same basic dynamical theory. To understand the climate of the temperate latitudes, it is therefore necessary to look at some basic atmospheric dynamics.

## Basic dynamics

Daily weather charts for any large region in middle and high latitudes reveal well-defined dynamic systems that normally move from west to east with a speed that is considerably smaller than the speed of the air currents. The structure of these systems varies considerably with altitude, for near the earth's surface they are of relatively small dimensions (1,000 to 3,000 km) and show a maximum of complexity, while in the middle and upper troposphere the systems are relatively large and have a great simplicity. At the surface the predominant features are closed cyclonic and anticyclonic systems of irregular shape, while higher up smooth wave-shaped patterns are the general rule. The dimensions of these upper waves are much larger than those of the surface cyclones and anticyclones, and only rarely is there a one-to-one correspondence. In typical cases there are four or five major waves around the hemisphere, and superimposed upon these are smaller waves which travel through the slowly moving train of larger waves. The major waves are called long waves or Rossby waves, after Rossby (1939, 1940, 1945) who first investigated their principal properties.

### 1 Long waves

In a simple mathematical model long waves could arise anywhere in the middle latitude atmosphere, but in the real atmosphere waves seem to have preferred locations which are shown, for the northern hemisphere, in figure 8.1. These preferred locations may arise

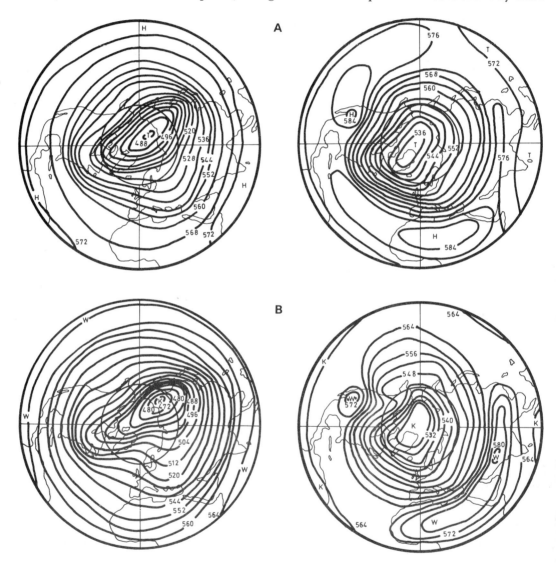

**Figure 8 1** Mean topography of the 500 mb surface and mean 1,000–500 mb thickness pattern over the northern hemisphere. **A**: mean topography of the 500 mb surface. *Left:* January. *Right:* July. **B**: mean 1,000–500 mb thickness pattern. *Left:* January. *Right:* July. (Units: dynamic decameters) (*After Scherhag, 1948 and Sutcliffe, 1951.*)

because the atmospheric circulation is influenced not only by the thermal properties of land and sea masses, but also by high mountain ranges and highlands in general. When large-scale currents approach a mountain range, there is on the windward side vertical shrinking and isentropic divergence, and it follows therefore from the isentropic vorticity equation

$$\frac{d}{dt}\left(\zeta_\theta + f\right) = -\left(\zeta_\theta + f\right)\operatorname{div}_\theta V$$

that the vorticity change is negative. Isentropic surfaces are surfaces of constant potential temperature. Similarly, on the leeward side of the range, vertical stretching and isentropic convergence exist and the vorticity increases. Consequently, anticyclonic deformation of the flow is found over the mountain range and cyclonic deformation to leeward. By assuming an average zonal flow of about 10 m/sec and applying general lee-wave theory (Queney *et al.*, 1960) to a large-scale mountain (100–1,000 km), Flohn (1969b) has calculated that there is an anticylonic displacement of the flow immediately before reaching the obstacle, followed by a first cyclonic lee-wave trough at a distance of about 2,000 km downwind, while the crest of the next anticyclonic wave occurs after some 6,000 km and a second cyclonic trough after some 9,000 km. In short, a whole series of stationary long waves develop downwind of the obstacle. Rossby has established for a barotropic atmosphere that in the case of horizontal non-divergent flow, the speed C of propagation of a long wave is given by

$$C = u - \beta\left(\frac{L}{2\pi}\right)^2$$

where L is the wave length, $\beta$ the Rossby parameter ($\beta = \partial f/\partial y$), and the stationary case is given by C = 0. With a zonal flow of 10 m/sec at about 60°N, the waves will become stationary with a wave length of approximately 5,800 km, and this is in moderate agreement with Flohn's suggested wavelength, but the agreement is poor with the same flow at lower latitudes.

Bolin (1950) has argued that the main troughs and ridges which appear in the mean contour charts for the middle and upper troposphere are not the net result of complex thermal and synoptic processes but are caused by wave disturbances in the westerlies created by mountain barriers, and in particular, the Rockies. In support of this theory he points to an in-phase relation between summer and winter trough–ridge patterns when land–sea thermal differences might more naturally indicate antiphase. Sutcliffe (1951) among others does not agree with this theory, and suggests that a thermal-synoptic explanation, resting mainly on the direct thermal effects of land and sea modified by baroclinic synoptic processes, is a satisfactory explanation of the mean positions of the long waves.

## 2 Thermal patterns

The basic thermal pattern of the lower half of the atmosphere in the temperate latitudes is partly controlled by the prevailing mean surface temperatures. This mean thermal pattern may be usefully investigated by using the concept of thickness lines. The thickness (or the depth) of an isobaric layer is given by

$$z - z_0 = \frac{RT_m}{g}\ln\frac{p_0}{p}$$

where z and $z_0$ are the heights above sea level of the top and bottom of the isobaric layer; p and $p_0$ are the pressures at the top and bottom of the isobaric layer; R is the gas constant for dry air; and $T_m$ is the mean temperature of the layer. Clearly, the thickness of an atmospheric layer bounded by two fixed pressure surfaces, 1,000 and 500 mb for example, is directly proportional to the mean temperature of the layer. Charts showing the thickness of the 1,000–500 mb layer are produced on a routine basis by many meteorological services and mean charts for January and July are contained in figure 8.1.

A parallel exists between the mean topography of the 500 mb level and the mean thickness patterns which, according to Sutcliffe (1951), is a consequence of the application of the hydrostatic equation upwards from a boundary with relatively weak pressure differences. The mean January thickness pattern for the northern hemisphere shows two dominant troughs near the eastern extremities of the two continental land-masses, while ridges lie over the eastern parts of the oceans. A third weak trough extends from north Siberia to the eastern Mediterranean. Climatologically, the positions of the main troughs may be associated with cold air over the winter land-masses, and the ridges with relatively warm sea surfaces.

In July, the mean ridge in the thickness pattern found in January over the Pacific has moved about 25° west, and now lies over the warm North American continent, while there is a definite trough over the east Pacific. Patterns elsewhere are less marked, but a weak trough does appear over Europe, and may perhaps be connected with the coolness of the North Sea, the Baltic, the Mediterranean, and the Black Sea.

Because of the obvious associations between thickness patterns and surface thermal conditions, Sutcliffe has argued in a qualitative manner that surface thermal features are the dominant control in producing the observed mean upper wave patterns. Smagorinsky (1953) has calculated that the influence of large-scale asymmetries of non-adiabatic heating and cooling on mid–tropospheric flow is of the same magnitude as the influence of broad mountain areas.

Laboratory experiments of the type described in chapter 2 (see Hide, 1969) and general experience with mathematical models of the atmosphere suggest that the large-scale upper waves observed in the atmosphere are not strictly Rossby waves and that the tendency for the relative vorticity to decrease in northward flow and increase in southward flow is of lesser importance than was once assumed. It appears that the large-scale waves observed in the upper troposphere are of a baroclinic nature and that baroclinic instability is responsible for at least part of the development of wave disturbances within the strong westerly wind flow. The growth of these wave disturbances is characterized by the ascent of the warmer, and the descent of the colder, air-masses, thus causing a decrease of potential energy and an associated release of kinetic energy. This ascent and descent of air-masses takes place in a manner described by the term slantwise convection. In the normal atmosphere, equal potential temperature (isentropic) surfaces slope upwards from lower to higher latitudes, and in slantwise convection the trajectories of individual air parcels are tilted at an angle to the horizontal that is comparable with, but less than, the slope of the isentropic surfaces. Thus while an air parcel may be prevented from rising vertically because of a subadiabatic lapse rate, poleward travel brings it to an environment more dense than itself, thus enabling it to rise. Air at higher levels and higher latitudes may similarly descend by equatorward movement. The process of slantwise convection has been fully described by Eady (1949, 1950) and the recognition of the baroclinic nature of atmospheric disturbances throws further doubt on suggestions that mountain ranges restrict waves to preferred locations. The discussion above must be considered with this fact in mind.

Comment was made earlier that a marked parallel exists between mean contour and mean thickness patterns and it may further be noted that in temperate latitudes wind speed increased rapidly with altitude in the lower troposphere. The change of the geostrophic wind with altitude is given by:

$$V' = -\frac{g}{f} \nabla p \, (\Delta z) \times k$$

where $\Delta z = z - z_o$ is the vertical thickness between the pressure surfaces $p_2$ and $p_1$, which have geopotential heights $z_2$ and $z_1$, respectively and k is a unit vector. Inasmuch as the thickness is but another expression for the mean temperature of the isobaric layer, the foregoing equation states that the wind shear will be along the mean isotherms with lower temperature to the left in the northern hemisphere. Consequently the shear of the geostrophic wind has become known as the thermal wind (V'), which is usually measured over the layer 1,000–500 mb. Figure 8.1 suggests that in temperate latitudes the thermal wind will be westerly, and also since temperature gradients are steep, it will be relatively strong. The 500 mb wind field is the vector sum of the rather weak 1,000 mb wind field and a vigorous thermal wind field. This implies that in the lower troposphere, winds will become increasingly westerly with altitude and also that the 500 mb wind field is strongly controlled by the thermal wind.

## 3   Development

Storm models have been discussed in earlier chapters, and it is clear that vertical motion accompanies cyclonic and anticyclonic development. This vertical motion must be accompanied by surface and upper level convergence or divergence. It is difficult to measure atmospheric vertical motions directly, but they can be estimated if the net difference between the divergence and convergence at lower and upper levels of the storm are known.

In synoptic meteorology the term 'development' is applied to the intensification of both cyclonic and anticyclonic circulations. Sutcliffe (1947) in his 'development theorem' demonstrated, for the flow of the atmosphere at any instant, that areas favourable to cyclogenesis and anticylogenesis can be determined from a knowledge of the vorticity and the vertical shear of the horizontal wind. He found that the horizontal divergence ($\text{div}_p V_1$) at the top of an atmospheric layer relative to the bottom ($\text{div}_p V_o$), is, with various approximations, given by

$$\text{div}_p V_1 - \text{div}_p V_o = -\frac{1}{f} \left( V' \frac{\partial f}{\partial s} + 2V' \frac{\partial \zeta_o}{\partial s} + V' \frac{\partial \zeta'}{\partial s} \right)$$

where $\text{div}_p$ is the divergence in an isobaric surface, $V_1$ and $V_o$ the wind vectors at the upper and lower pressure levels, f the Coriolis parameter, V' the thermal wind vector between the selected levels, and $\zeta'$ its vorticity, $\zeta_o$ the vorticity at the lower level, and where $\partial/\partial s$ denotes differentiation in the direction of the thermal wind.

Sutcliffe has also established that the left-hand side of this equation may be interpreted in terms of vertical motion, and in particular in terms of cyclonic or anticyclonic development, since it is positive in cyclonic development areas (where there is net divergence and therefore a lowering of surface pressure), and negative in anticyclonic development areas (where there is net convergence and therefore an increase in surface pressure).

Each of the three terms which are contained on the right-hand side of the above equation, has some influence on the development of surface pressure systems. The equation is normally applied to the 1,000–500 mb layer. The first term represents the latitude effect and can usually be neglected. The second term is the thermal-steering term, and controls the direction of movement of individual depressions and anticyclones. A surface cyclone is a region of maximum relative vorticity, and at the maximum the rate of change of $\zeta_0$ along the thickness lines is zero, which indicates that the second term vanishes. However, ahead of a cyclone, the thermal wind is directed toward decreasing values of $\zeta_0$, and there is relative divergence, but behind the cyclone the thermal-steering term leads to relative convergence. Since similar arguments apply to anticyclones, pressure systems tend to be propagated along the thickness lines at a speed proportional to the thermal wind; this is particularly true in the early stages when the systems are shallow.

The third term is known as the thermal-vorticity effect and is determined entirely by the topography of the thickness chart and the distribution of the thermal winds. It signifies broadly, cyclonic development where the thermal vorticity $\zeta'$ decreases in the direction of

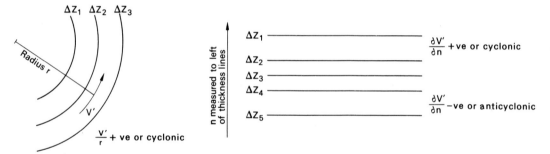

**Figure 8.2** Definition of curvature (*left*) and shear (*right*) terms in thermal vorticity.

the thermal wind, and anticyclonic development where it increases. Thermal vorticity may be considered as composed of curvature and shear terms:

$$\zeta' = \frac{V'}{r} - \frac{\partial V'}{\partial n}$$

where the terms are as defined in figure 8.2. This equation implies that cyclonic development will be associated with marked ridges and troughs, and also with the entrances and exits to thermal jets. A selection of possible cases are shown in figure 8.3. Given a strong thermal wind, cyclones are likely to form downwind of thermal troughs and anticyclones downwind of thermal ridges. Similarly, cyclones tend to form on the equatorward side of the entrances to thermal jets, and on the poleward side of the exists.

The climatological relationships between thickness patterns and surface features now becomes clear. Cyclones will tend to form downwind of the mean thermal troughs in winter over the east coasts of North America and Asia, move northeastwards with the thermal wind, and decay in the thermal ridges over the western regions of the oceans.

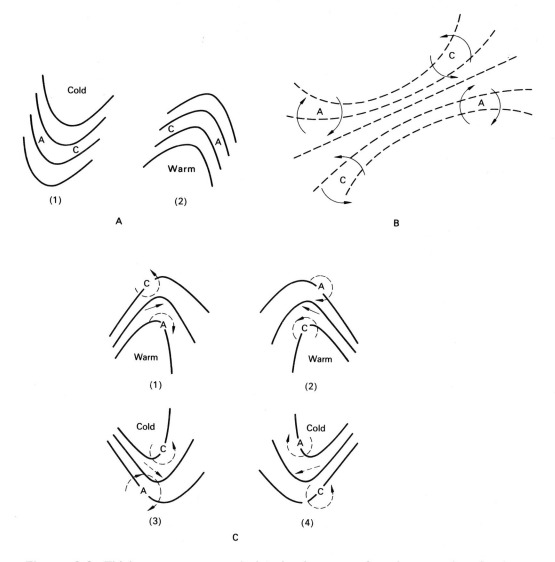

**Figure 8.3** Thickness patterns and the development of cyclones and anticyclones. **A:** (1) cold thermal trough; (2) warm thermal ridge. Cyclonic development, C, occurs forward of the thermal trough, behind the thermal ridge; anticyclonic development, A, on the opposite sides. **B:** A thermal jet. The development regions favour the existence of depressions in the C regions and anticyclones in the A regions, which tend to distort the thermal pattern by advection and maintain or develop the jet structure. **C:** Combinations of thermal ridge and trough with difluence and confluence: (1) difluent thermal ridge; (2) confluent thermal ridge; (3) difluent thermal trough; (4) confluent thermal trough. (*After Sutcliffe and Forsdyke, 1950.*)

## 4 Index cycles

Long-term mean circulation charts for the middle latitudes show only a faint resemblance to individual synoptic charts. The long-term mean charts are composed of averages of circulation patterns for a great variety of scales and types and do not reveal the traits of individual synoptic situations. A study of daily charts has disclosed that middle latitude circulation patterns undergo a series of irregular quasi-cyclic changes, which are best analysed in terms of the zonal index and the index cycle. The strength of the zonal circulation is conveniently measured by the mean pressure difference between two latitude circles, which is called the zonal index; for sea-level conditions 35° and 55°N are often used.

Willett (1949) found, from a study based on 7 years of northern hemisphere charts, that a typical index cycle can be divided into four stages with the following characteristics:

1   an initial high zonal index with strong sea-level westerlies and a long-wavelength pattern in the upper atmosphere;
2   an initial lowering of the sea-level zonal index with an associated shortening of the wavelength pattern aloft;
3   the lowest sea-level zonal index characterized by a complete breakup of the sea-level zonal westerlies into closed cellular centres, and with a corresponding breakdown of the wave pattern aloft;
4   finally, an initial increase of the sea-level zonal index with a gradual increase of the westerlies and the development of an open wave pattern aloft.

*a Blocking*   Since the high index situation is characterized by strong latitudinal temperature gradients in middle latitudes, and by little air-mass exchange, the cyclones and anti-cyclones of the mid-latitude belt drift eastward, often with considerable speed. Under conditions of low zonal index, a warm cutoff high forms in the middle and upper troposphere, and sea-level cyclones have a tendency to be steered around these highs, either to the north of the cutoff high or to the south along the base of the pattern. Because the normal westerly current is blocked and reversed, this last arrangement is often called a 'blocking pattern' or simply a 'block'. Blocking patterns can have a variety of shapes, but a common one has the shape of the Greek capital $\Omega$ and the pattern has been called an Omega-block. Anticyclonic development is associated with the warm ridge and sometimes cyclonic development with the cold trough of the blocking pattern, and so an extensive blocking anticyclone often exists in middle and high latitudes, while a non-frontal low exists to the south.

Clearly, very different weather types will be allied to high index and blocking situations. In order to isolate typical blocking cases, Rex (1950) has introduced the following criteria which apply to the 500 mb level:

1   the basic westerly current must split into two branches;
2   each branch current must transport an appreciable mass;
3   the double-jet system must extend over at least 45° of longitude;
4   a sharp transition from zonal type flow upstream to meridional type downstream must be observed across the current split; and
5   the pattern must persist with recognizable continuity for at least 10 days.

Blocking patterns once formed are relatively stable and usually persist for a period of

12–16 days. In the northern hemisphere they are most frequently initiated in two relatively narrow longitudinal zones, one (Atlantic) centred aloft at 10°w and the other (Pacific) at 150°w. The number of cases occurring over the Atlantic exceeds those over the Pacific, probably by a factor of two to one.

*b   Climatic characteristics*   Rex has studied the regional climatic characteristics of periods dominated by strong zonal flow and blocked periods over Europe. His results (figure 8.4) may not apply to other regions, but they do indicate the control on climate exerted by the zonal index. Under conditions of strong zonal flow, Rex found the following climatic characteristics over Europe.

1   Precipitation amounts over Scotland, Scandinavia, and adjacent regions reach values several times normal. A secondary area of weak positive anomalies is produced in the mountains of southern Europe and along the North African coast.

2   Abnormally high mean temperatures (2° to 4°c above normal) are observed over the

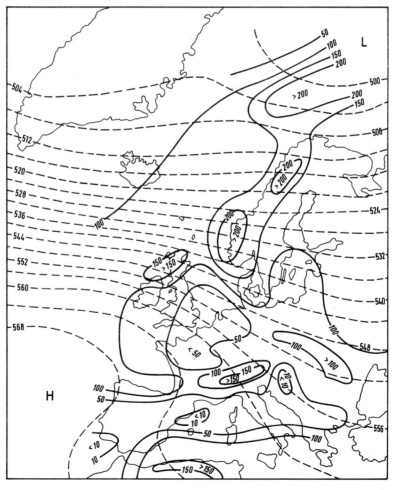

**Figure 8.4** Isanomals of precipitation over the Atlantic–European area during a winter example of strong zonal flow aloft (1–31 January 1949). Isanomal lines are given as a percentage of normal precipitation. The mean contour pattern at the 500 mb level is shown by long-dashed lines; contour heights are in dynamic decameters. (*After Rex, 1950.*)

Continent, Scandinavia, and the British Isles during winter zonal periods. The summer temperature patterns are more complex but indicate slightly below normal temperatures over the coastal areas of Norway and western Europe.

3    The major portion of all precipitation recorded is produced by frontal action in combination with orography.

4    Strong westerly surface winds with a stormy weather regime, connected with the passage of intense cyclonic systems prevail over northern Europe and Scandinavia, with light winds and settled weather generally observed over southern Europe.

In contrast, Rex discovered (figure 8.5) the following regional climatic characteristics associated with blocked periods.

1    Moderate, mostly northeasterly, surface winds prevail over the Continent generally. Storminess, as associated with the passage of intense cyclonic systems, is more or less absent over the entire area.

2    Convective and orographic rain and snow showers account for a major portion of the total precipitation recorded over the European area. Precipitation amounts over the Continent, Scandinavia, and the British Isles are generally much below normal. Normal or slightly above-normal values are recorded only in scattered small areas along windward coasts and on windward mountain slopes. Precipitation amounts over the Scandinavian and alpine mountain complexes are usually much below normal.

3    Winter surface temperatures are 2° to 6°c above normal over central and northern

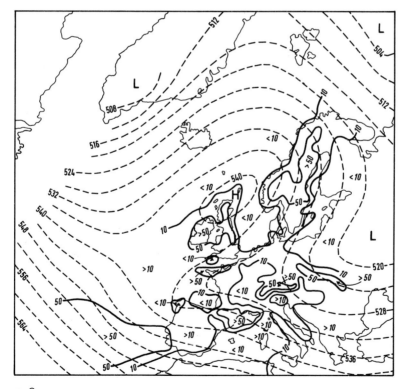

**Figure 8.5** Isanomals of precipitation over the Atlantic–European area during a winter example of strong Atlantic blocking (16–31 January 1947). Key as figure 8.4. (*After Rex, 1950.*)

Scandinavia, but are below normal over the entire Continent and the British Isles, reaching a minimum of 8° to 10°C below normal over the Balkan area. Summer blocking produces even higher surface temperatures over Scandinavia and positive anomalies over the British Isles and west–central Europe.

## 5   Southern hemisphere

So far, only atmospheric dynamics have been discussed mostly with reference to the northern hemisphere. The basic dynamics apply equally to both hemispheres, but there are some interesting differences between the northern and southern temperate latitudes. For instance, it has long been known that the southern hemisphere westerlies are stronger than their northern hemisphere counterparts (Lamb, 1959).

Mean zonal winds for the extreme seasons are shown in figure 8.6, where it is seen that that seasonal changes in the southern hemisphere troposphere are much smaller than those in the northern hemisphere. Also in the southern hemisphere, the stratospheric westerly jet stream extends down into the troposphere, so that there are two jet streams in the troposphere in three seasons out of four.

Profiles illustrating the mean height of the 500 mb surface are contained in figure 8.7. They show a ridge over the Andes to about 50°s, but from 50°–70°s a ridge in the western Pacific–Australasian sector (150°E–150°W) is a more noticeable feature; there is also a secondary ridge just east of southern Africa. There is a marked trough downstream from the Andes, which south of 45°s is less pronounced than the broad trough in the Indian Ocean. Rather scanty evidence also exists for a trough in the Pacific. Seasonal changes in longitude of the troughs or ridges appear rather ill-defined, apart from the possible westward displacement of the broad Indian Ocean trough in summer to about 60°E.

The greater total zonal circulation of the southern hemisphere implies that the overall range of mean 500 mb heights between the tropical and polar zones will be larger than in the northern hemisphere. The greatest mean annual 500 mb heights are believed to be about 5,900 m over the Bolivian Altiplane and 5,870 to 5,880 m over the central African highlands (Flohn, 1955). The nearest approach to these in the northern hemisphere appear to be values between 5,824 m and 5,883 m over Mexico and the southwestern United States (Lamb, 1959). The lowest mean–annual 500 mb heights of 4,920–4,990 m occur over Antarctica and are some 300 m below the lowest in the Arctic. The steepest thermal gradient is found over the Indian Ocean, between the trough in the higher latitudes and the particularly warm tropical zone of this ocean. This thermal gradient is well illustrated by the fact that the islands exhibit climates ranging from fully tropical with tropical vegetation at Mauritius (20°s) to ice age at Kerguelen (49°s) and Heard (53°s). Very strong upper westerlies are found in this sector, as would be expected.

*Quasi-stationary anticyclones*   In the general discussion on atmospheric dynamics, it was considered that blocking patterns were important in determining the weather in the northern temperate latitudes. A major feature of a blocking situation is the quasi-stationary anticyclone, and Lamb states that south of 40°s these are limited to positions over Antarctica and three sectors of the surrounding ocean, namely

1   from close to the south coast of Australia to 140°w;
2   near the tip of South America and southeast over the Scotia Sea;
3   the region southeast of South Africa near the Marion and Crozet Islands.

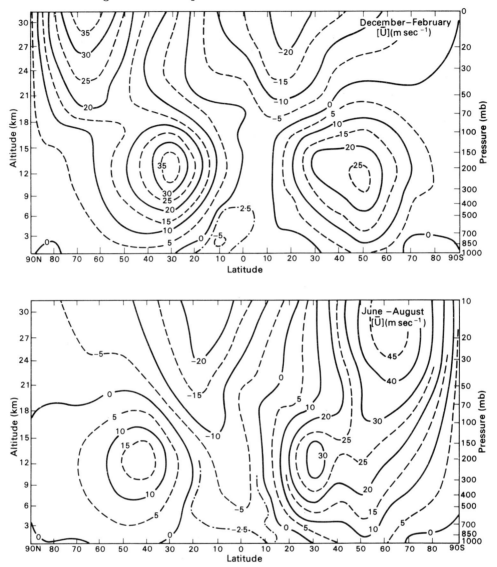

**Figure 8.6** Mean zonal wind for December–February and June–August (m sec$^{-1}$). Positive values denote eastward flow. (*After Newell* et al., *1969.*)

Even in these regions the frequencies of stationary anticyclones do not compare with those reported from the corresponding zones in the northern hemisphere, and vast regions exist in the southern hemisphere where there are virtually no anticyclones, neither stationary nor mobile—a state of affairs unmatched in the north. Blocking patterns do occur fairly frequently over certain parts of the Southern Ocean but stationary situations seldom result, for the anticyclones always tend to drift east. Lamb thinks that in the northern hemisphere the large continents in the temperate zone play an important part in permitting stationary patterns to develop, mainly because of the slight thermal gradients often associated with

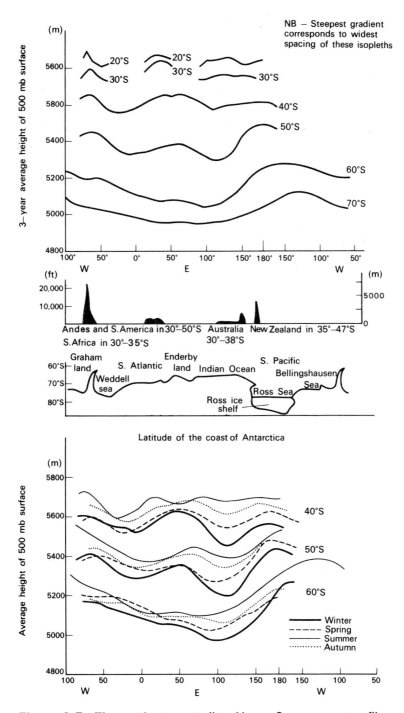

**Figure 8.7** The southern westerlies. *Upper:* 3-year mean profiles of the 500 mb surface along each 10° parallel of latitude (December 1951–November 1954). *Lower:* seasonal profiles of 500 mb height along 40°S, 50°S, and 60°S (1952–1954). (*After Lamb, 1959.*)

the main winter snow cover and also because of the development of distorted thermal patterns between oceans and continents in the same latitude zone. In the southern hemisphere the geography does not generally permit large depressions or anticyclones to produce a thermal pattern capable of maintaining the system in a stationary position.

## Radiation and temperature

The general features of the radiation balance of the earth's surface were discussed in chapter 2. Table 2.1 reproduces details of the seasonal changes in mean meridional net radiation, and the swing in middle latitudes from a marked deficit in winter to a clear surplus in summer is obvious. In winter in the northern hemisphere the net radiation deficit extends to about 15°N, while in summer there is a surplus to 75°N, the corresponding values for the southern hemisphere being 15°s and 65°s. These marked changes in radiation balance impose distinct seasonal temperature changes on the middle latitude areas, whereas similar changes are not normally found in the tropics. The exact nature of the seasonal temperature changes depends partly on the nature of the interaction of the radiation with the surface. Since the nature of the surface can vary widely, a large variety of annual temperature curves might be expected.

*1    Meridional temperature gradients*

Simpson (1939) has investigated the influence of the distribution of land and sea upon temperature. A circle of latitude in the northern hemisphere receives almost exactly the same amount of solar radiation in the course of a year as the corresponding latitude in the southern hemisphere. The amount of land along a specified circle of latitude is, however, often quite different in the two hemispheres. Hence, a comparison of the mean annual temperatures of a given latitude in the northern and southern hemispheres should indicate what effect the amount of land has on temperature. Simpson found that all latitudes in the northern hemisphere are warmer, but the mean difference is only about 2°C, and since there is twice as much land in the northern as in the southern hemisphere (34 and 19 per cent respectively), he concluded that the effect of land is only small, even if the whole temperature difference can be ascribed to that cause. This conclusion was reinforced when Simpson examined the individual latitudes. No circle in the northern hemisphere is more than 3°C warmer than the corresponding latitude in the southern hemisphere, yet the amount of land along some of the norther circles of latitude is many times greater than along the corresponding latitudes.

The mean temperature of the zones of latitude decreases poleward from the equator, at nearly the same rate in each hemisphere. This shows that the mean meridional temperature distribution is almost independant of surface conditions. Simpson considers that the general temperature distribution is determined by two main factors: the spherical shape of the earth which determines the distribution of solar radiation; and certain physical qualities of the atmosphere.

Within any given latitudinal zone the temperatures are far from uniform, for within 40° of the equator the continents are warmer than the oceans, while over the remainder of the earth the mean temperature of the continents is below that of the oceans. For example, the North Atlantic Ocean in latitude 70°N off the coast of Norway is throughout the year more than 12°C warmer than its latitude warrants, but it is compensated by cold regions over the continents so that the mean temperature of the latitude as a whole falls into its right position in the normal decrease of temperature from equator to the pole.

## 2 Continentality

The degree of influence of land on climate is expressed in the concept of continentality, which is usually applied to the temperature characteristics of a given place. Annual temperature curves can conveniently be represented by three characteristics, which are the mean, the amplitude, and the phase; it is sometimes convenient to include a fourth parameter which is indicative of the degree of symmetry of the temperature curve. All these parameters are influenced by changes in the distribution of land and sea, but not always in the same manner, for example the amplitude of the curve may be increased while the mean remains unchanged.

Most climatologists take the mean annual temperature range as the best measure of continentality, and this is used in the index suggested by Conrad (1946):

$$K = 1 \cdot 7 \, A/\sin (\phi + 10) - 14$$

where K is the index of continentality; A is the average annual temperature range; and $\phi$ is the latitude. K should be zero for a completely oceanic earth and 100 for the corresponding continental planet. In practice the formula yields values of nearly zero for Thorshavn (62° 2′N, 6° 44′W) and nearly 100 at Verkhoyansk (67° 33′N, 133° 24′E) in eastern Siberia. Both D'ooge (1955) and Fobes (1954) have used this equation to investigate continentality within the United States. The lowest values of K are found in the west where over the coastline it varies between 0 and 10, whereas by comparison coastal New England is relatively continental with values between 30 and 40. Major highs are recorded in southwest Colorado, eastern Montana, and North Dakota, but the values of between 60 and 65 are considerably lower than those reported from Siberia.

An examination of the lag of temperature behind solar radiation has been made by Prescott and Collins (1951) using the technique of harmonic analysis. This lag is an index of continentality in that coastal stations are late in phase and continental stations relatively early. They discovered the largest lags along the western margins of the continents, with substantial differences between west and east coasts. Outside of the tropics considerable areas with lags of 25 days or less occur in Argentina, South Africa, and central Asia.

An extremely continental climate is therefore characterized by very large changes in mean temperature between winter and summer, with the temperature following very closely the annual march of solar radiation. Thus the temperature changes in spring and autumn are relatively rapid. Oceanic climates have small annual ranges with a large lag of temperature behind solar radiation, and there is also some evidence that the mean annual temperatures of land areas in high latitudes are below those in oceanic locations at a similar latitude. Land or sea ice to the west appears to influence the temperature regime far more than a similar amount of land or ice to the east, and this probably reflects the mean westerly winds in middle latitudes.

## 3 Interannual variability

Mean sea-level temperatures form one basis for the demarcation of the world into climatic regions, for they indicate the general suitability for human habitation and various forms of agriculture. However, the variability of individual years about the long-term average is also important and this has been studied by Craddock (1964), who charted the standard deviation of monthly mean temperatures. He considers that the interannual variability is greater in winter than in summer, greater in the interior of the continents than it is in the oceanic and maritime areas, and very low in all months in the tropics. The regions of highest variability are generally grouped round the Arctic Ocean, and southwards there is a general tendency for variability to decrease with latitude. Craddock's findings confirm some of the comments made earlier about the relationship between temperature regimes and land distribution.

Charts showing the standard deviation of monthly mean temperatures for January and July are reproduced in figure 8.8. In January, areas of greatest variability exist over the west coast of Hudson Bay, from Greenland to Novaja Zemlja in central Siberia, and over the northern Rocky Mountains, Alaska, and northeast Siberia. The variability decreases during the summer and by July weak maximums are found over northern Siberia and parts of northern Canada.

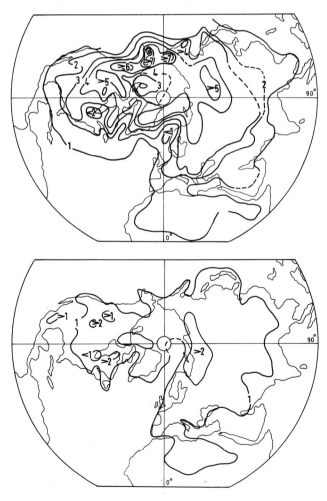

**Figure 8.8** The interannual variability of monthly mean air temperature over the northern hemisphere. *Upper:* January. *Lower:* July. (*After Craddock, 1964.*)

Craddock further suggests that the ideal climate for agriculture would have no inter-annual variation at all because crops sown on a suitable calendar date would be always ready for harvest a certain number of days later. He comments that while this ideal does not occur in reality, it is nearly approached in the ancient cradles of civilization, especially around the eastern Mediterranean and in the Euphrates Valley. Fortunately, the areas of greatest variability are not very significant agriculturally, because they are mostly too cold.

## 4 Topography

Regional temperature distributions can be modified by local topography. For example water bodies tend to moderate temperature regimes by leading to warmth in winter and coolness in summer, and mountain ranges can do likewise by the action of foehn winds. This name originated in the Alps and in general terms applies to a warm, dry wind descending in the lee of a mountain range. Winds of the foehn type are found in most of the world's mountainous areas, and are known by a variety of local names; the chinook in the Rocky Mountains, the berg wind of South Africa, the Santa Ana of southern California, the zonda and puelche of the Andes, and the northwester of New Zealand.

The chief characteristics of foehn winds, which are caused by air descending to leeward of mountain ranges and undergoing dynamic warming, are their high temperatures, low relative humidities, and often extreme gustiness. Temperature and humidity changes at the beginning and end of a spell of foehn winds are sometimes quite rapid, rises of up to 25°c in one hour being reported in extreme cases. The rise of temperature is most marked in winter and the foehn can very effectively melt snow in mountain valleys, whereas the low relative humidities can cause wood to become extremely dry and create a fire risk in buildings and forests.

## Moisture advection and water balance

The atmospheric water balance of a subcontinental area has been considered by Drozdov and Grigor'eva (1963), and a simplified model was discussed in chapter 4 and illustrated in figure 4.1. Suppose that through the windward vertical boundary an amount of moisture arrives per unit time which is equivalent to an amount A per unit area. Within the bounded area an amount of precipitation r falls during the same time, which is made up of water of an amount $r_A$ formed of advected moisture, plus an amount $r_E$ due to evaporation within the area. Assuming, therefore, that the precipitation comes from water vapour of mixed origin, which is formed from local and advective vapour, it is possible to write

$$r = r_A + r_E$$

Part of the moisture evaporating from the territory goes to form the local precipitation $r_E$, while an amount, C, is carried beyond the boundaries of the region. The quantity C is often called the atmospheric runoff:

$$E_t = r_E + C$$

Consequently moisture is carried beyond the boundaries of the territory either by advection plus atmospheric runoff $(A - r_A + C)$, or by stream–flow $f_R$. The long-term average moisture balance for a region yields the following relationships:

$$r = E_t + f_R; \quad r_A = C + f_R$$

The coefficient of the hydrologic cycle, k, is the number of times on average that each molecule of external water vapour recirculates over the given territory before it is carried beyond its boundaries, and it is given by

$$k = \frac{r}{r_A} = \frac{A + \beta E_t}{A}$$

where $\beta$ is a constant, partly controlled by the distribution of evaporation within the region. For a territory 500 to 2,000 km across, in which there is complete mixing, $\beta$ probably equals about 0·5.

Similarly, the intensity of the hydrologic cycle, I, is defined as that fraction of the water vapour located over a given territory that participates in the hydrological cycle:

$$I = \frac{r_A}{A} = \frac{r}{kA} = \frac{r_E}{\beta E_t}$$

The various components of this model should be studied in more detail, starting with the inflow A. The total flux of an atmospheric property is made up of two major components known as the advective flux and the eddy flux (see chapter 2). Bannon, Matthewman, and Murray (1961) estimate that the advective flux of water vapour in middle latitudes is a fairly good approximation to the total flux, with the eddy flux often an order of magnitude smaller. These findings are confirmed by Hutchings (1957), who concluded that the eddy flux made only a small contribution to the total flux over southern England during the summer of 1954. Similarly, Flohn and Oeckel (1956), while investigating the water-vapour flux during the summer rains of Japan and Korea, judged the magnitude of the eddy flux to be about one-sixth of the total flux. The ratio of the advective to the eddy flux appears to vary with the direction of the total flux; Benton and Estoque (1954) have estimated for the USA that the eddy flux contributes 42 per cent of the total meridional flux of water vapour, but only 6 per cent of the total zonal flux. Barry (1967) discovered for northeast North America that the eddy contribution varies with different synoptic situations, and that in individual winter months it can range from 5 to 20 per cent of the zonal transport and from 40 to 100 per cent of the meridional transport.

Bannon, Matthewman, and Murray conclude (figure 8.9) that in winter the advective flux reaches a maximum over the Atlantic and Pacific Oceans at around latitudes 32° to 35°N, that is to say, in the region of strong geostropic winds at the 850 mb level. Further south, towards the light winds of the anticyclonic belt, the advective flux decreases to very small values. In summer, it shows great complexity, with minimums around latitude 35°N over the Atlantic and over the eastern Pacific, corresponding to areas of light geostrophic winds in the anticyclonic belt and also, in the eastern Pacific, to relatively low water-vapour content. Summer maximums do exist, however, over the western Pacific and in the central Atlantic.

Continental advection of moisture in North America for the year 1949 has been studied by Benton and Estoque (1954). They have found that the average annual moisture influx to North America (figure 8.10) is caused by two flow patterns—a more intensive flux from the Gulf of Mexico and a less intensive one from the Pacific Ocean. The maximum flux from the Gulf of Mexico is observed at approximately the 925 mb level, since above this it decreases rapidly. The air-masses from the Pacific lose a great deal of moisture on the western slopes of the western cordilleras, but still a considerable part moves on to the east;

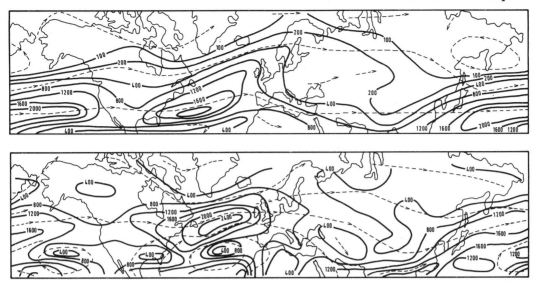

**Figure 8.9**  The flux of water vapour due to the mean winds: magnitudes shown by full lines with values in g cm⁻¹ sec⁻¹; directions shown by pecked lines with arrows. *Upper:* January (1951–1955). *Lower:* July (1951–1955). (*After Bannon* et al., *1961*.)

indeed, in 1949, 50 per cent of the moisture arrived over the North American continent through the Pacific-coast region between 35° and 63°N, while 38 per cent arrived from the Gulf of Mexico through the region between 80° and 101°W. In winter a more intensive flux from the Gulf of Mexico is observed than in summer, and the maximum flux occurs at more southerly latitudes. Zonal moisture flux increases in intensity in the summer and the maximum moves northward and to a higher level, approximately at 700 mb.

It is very difficult to distinguish the various components of the water-balance model presented earlier, but this has been managed by Begermann and Libby (1957). They used the tritium produced during thermonuclear weapons' tests in the spring of 1954 to study the circulatory rates for waters in the northern hemisphere, and particularly in the northern Mississippi Valley, where they concluded that one-third of the average rainfall is re-evaporated water and two-thirds ocean water. Since the average annual precipitation is 770 mm, this means that about 520 mm is ocean water and about 250 mm re-evaporated rain water. The complete water balance of the catchment is shown in figure 8.11, which demonstrates that on average about 8,000 mm of ground water or equivalent (including bound hydroxyl groups in clays) are available for mixing with the rainfall; about 280 mm of water runoff annually in rivers, and 240 mm per year are carried back to the oceans by the winds. Their calculations also establish that the average residence time of water in the upper Mississippi Valley is about 15 years.

From calculations based on the data of Begermann and Libby, the hydrologic cycle coefficient, $k$, has a value of 1·48, which can be compared with an estimated value of 1·30 for the whole continent. Similarly, the hydrologic cycle intensity, $I$, over the upper Mississippi is nearly 0·53, while for the continent as a whole, it appears to reach 0·65.

It is suggested that much of the rainfall over the continents consists of water which has been directly evaporated from the oceans, and this conclusion appears to hold even in the

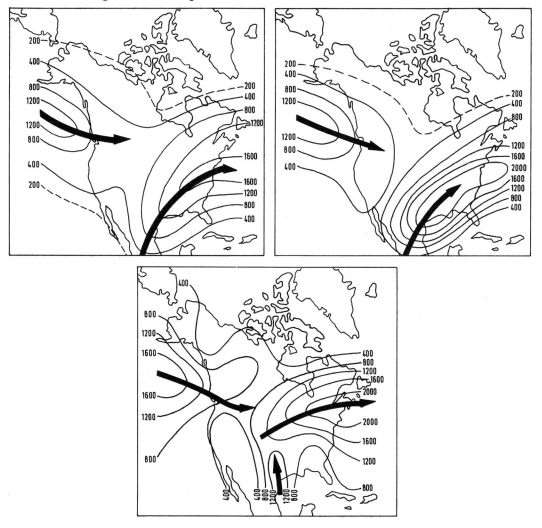

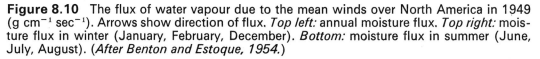

**Figure 8.10** The flux of water vapour due to the mean winds over North America in 1949 (g cm$^{-1}$ sec$^{-1}$). Arrows show direction of flux. *Top left:* annual moisture flux. *Top right:* moisture flux in winter (January, February, December). *Bottom:* moisture flux in summer (June, July, August). (*After Benton and Estoque, 1954.*)

arid regions of central Asia. According to Drozdov and Grigo'eva the proportion of precipitation in central Asia derived from locally evaporated water vapour is only 4 per cent, and even in spring it does not rise above 8 per cent. Further details are given in table 8.1; the annual average value of k being 1·04 and the corresponding value of I, 0·104.

The importance of moisture advection in creating middle latitude precipitation patterns has now become clear. Topographical barriers and great distance from the oceanic source areas will decrease the atmospheric moisture content and therefore lower the precipitation. This is the explanation of the low rainfall areas in the hearts of the temperate latitude continents. It also implies that local surface conditions only have a small influence on rain-

**Table 8.1**  Hydrologic-cycle components for the plains and foothill regions of central Asia

|  | Jan | Feb | Mar | Apr | May | Jun | Jul | Aug | Sept | Oct | Nov | Dec | Year |
|---|---|---|---|---|---|---|---|---|---|---|---|---|---|
| r cm | 1·3 | 1·2 | 1·6 | 1·8 | 1·3 | 0·9 | 0·7 | 0·5 | 0·6 | 0·9 | 1·2 | 1·4 | 13·4 |
| $E_t$ cm | 0·3 | 0·5 | 1·5 | 2·9 | 3·9 | 2·9 | 1·9 | 1·5 | 1·2 | 1·0 | 0·7 | 0·4 | 18·7 |
| W gm/cm² | 0·8 | 0·8 | 1·1 | 1·6 | 2·1 | 2·4 | 2·7 | 2·8 | 2·1 | 1·8 | 1·3 | 0·8 | 1·7 |
| $r_A$ cm | 1·3 | 1·2 | 1·5 | 1·7 | 1·2 | 0·9 | 0·7 | 0·5 | 0·6 | 0·9 | 1·2 | 1·4 | 13·1 |
| $r_E$ cm | 0·0 | 0·0 | 0·1 | 0·1 | 0·1 | 0·0 | 0·0 | 0·0 | 0·0 | 0·0 | 0·0 | 0·0 | 0·3 |
| k |  | 1·013 | 1·025 | 1·053 | 1·079 | 1·075 | 1·056 | 1·030 | 1·023 | 1·026 | 1·023 | 1·022 | 1·029 | 1·0398 |

See text for definition of symbols.
*After Drozdov and Grigor'eva (1963).*

fall totals and that in temperate latitudes the extensive clearing or planting of forests, for instance, will not lead to great changes in long-term precipitation. The origin of droughts and wet periods also should be sought in atmospheric flow patterns and possibly subtropical sea-surface temperatures.

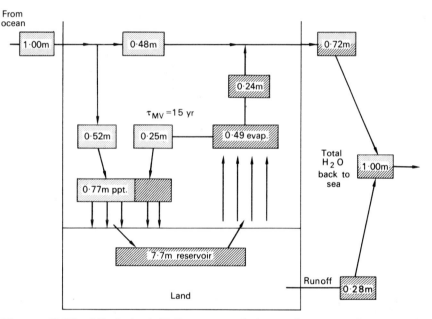

**Figure 8.11**  Mississippi Valley water balance from cosmic-ray and bomb tritium. Water from the oceans stays in the northern Mississippi Valley 15 years on average. Units: rainfall equivalent depth in m. (*After Begermann and Libby, 1957.*)

## Air-masses and middle lattitude disturbances

So far, the development of cyclones and anticyclones has been considered in terms of upper flow patterns and thermal zones, for fronts and air-masses no longer appear to be essential parts of the explanation of the development of synoptic systems. Even so, in terms of every-day weather and the environment, air-masses and fronts are of some importance.

## I  Air-masses

The lower troposphere can be divided into moving masses of air of varying geographical dimensions up to continental size, which are known as air-masses. An air-mass, separated from another by a boundary called a front, may be defined as a large body of air whose physical properties, especially temperature, moisture content and lapse rate, are more or less uniform for hundreds of kilometres, and are closely related to the characteristics of the source region, of which there are a large variety in temperate latitudes. This is in complete contrast to the tropical world where surfaces tend to be uniform and contrasting air-masses are rare. Areas with extensive uniform surfaces which are overlain by quasi-stationary pressure systems form the chief source regions of air-masses. These requirements are most nearly fulfilled by slow divergent flow from a major high-pressure cell, for low-pressure regions are normally zones of convergence towards which air-masses tend to move. Air-masses may be classified as arctic, polar, and tropical depending on temperature, and as maritime or continental according to the humidity of the source region.

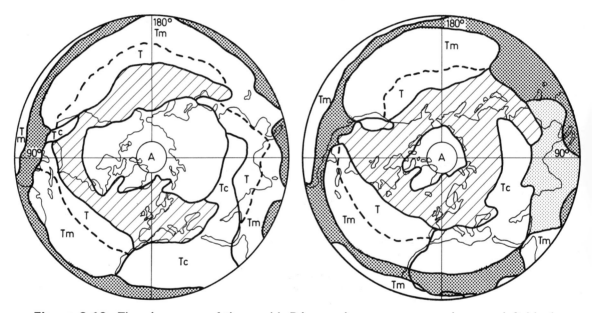

**Figure 8.12**  The air-masses of the world. Primary air-mass source regions are left blank Secondary air-masses, which owe their nature as much to past weather processes as to the intrusive qualities of a source region, are shaded. The shading distinguishes mid-latitude air, equatorial air, and southern monsoon air. A, arctic; T, maritime sectors with weak circulation; T$_m$, trade-wind systems; T$_c$, tropical continental air. *Left:* January. *Right:* July. (*After Crowe, 1971.*)

Principal sources (figure 8.12) of cold air are the winter continental anticyclones of Siberia and northern Canada, and the two polar regions. Extensive snow cover in the source regions usually leads to intense surface cooling with subsidence and the formation of a low-level inversion. In their source regions, the air-masses are characterized by extreme cold, extreme dryness, and small cloud amounts with only occasional light snowfalls. Such air-masses are called continental-polar or continental-arctic, depending on their exact point of origin. Maritime polar air-masses form over cool ocean surfaces either directly or as a result of the extensive modification of continental polar air. They are relatively warm compared with continental polar air, and much more humid.

Warm air-masses (figure 8.12) have their origins in the subtropical high-pressure cells and over the large land areas in summer. Both tropical maritime and tropical continental air-masses were described in chapter 3.

Given the juxtaposition of contrasting surfaces and air-mass source regions in middle latitudes, wind direction will be an important factor determining the local weather, since certain air-masses will always approach from the same direction and become associated with certain weather types. For example, the coldest weather in western Europe is often associated with the outflow of continental-polar air from the continental interior, and this appears as an easterly or northeasterly wind on the western margins of the continent. Similarly, westerly winds in winter are accompanied by relatively mild weather. Indeed, typical temperatures and humidities occurring with various air-mass types can be compiled for numerous localities in the middle latitudes.

## 2   Extratropical cyclones

In 1863, after only a few years of experimentation with the synoptic analysis of the conditions at sea level, Fitz-Roy (1863) proposed a model of the essential features of an extratropical cyclone. He observed that cyclones were normally composed of two air-masses of different temperatures, moisture content, and motion, with the warmer and moister of these originating in subtropical latitudes, while the colder and drier mass came from the polar regions. Bjerknes (1919) analysed a large number of extratropical cyclones and proposed a model which both outlined the structure of a typical mobile cyclone and also went far to indicate the underlying dynamical processes. The essential features of Bjerknes' model are reproduced in figure 8.13. He visualized the cyclonic disturbances propagating along the polar front, more or less like a wave disturbance, and explained the typical distribution of clouds and precipitation as the result of adiabatic cooling of the warm air which was ascending over the warm and cold fronts.

Bjerknes and Solberg (1921, 1922) discovered that cyclones normally pass through a series of typical stages which are represented diagrammatically in figure 8.14. They found that the initial disturbance could be identified with a wave of small amplitude which had formed on a more or less straight quasi-stationary front, with a marked cyclonic wind shear at the front. The initial wave disturbance grows in a period of about a day into the young cyclone shown in figure 8.14. As the warm air ascends and escapes to higher levels over the warm-front surface and as the cold front undercuts the warm air, the warm sector narrows, with the result that the cold front tends to overtake the warm front, and the cyclone is said to be occluded. The front resulting from the combination of the warm and cold front is called an occlusion. In the early stages of cyclone development the lowest pressure is found at the tip of the wave formed on the polar front, but as the occlusion process continues the fronts gradually become separated from the region of lowest pressure. In the later stages of the cycle, the occlusion process becomes almost complete, and the warm air is lifted completely off the surface. A frequently-occurring feature of the partly-occluded and occluded stages is the development in the rear of the cyclone centre of a trough, which is often accompanied by bad weather due to general convergence in the lower troposphere.

The formation and movement of extratropical cyclones are closely linked to the upper

231

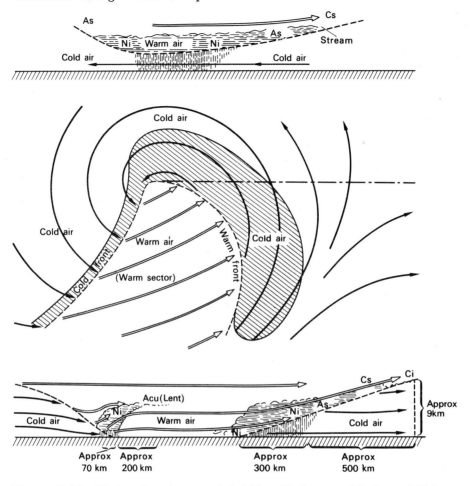

**Figure 8.13** Bjerknes' cyclone model. (*After Bjerknes and Solberg, 1921 and Petterssen, 1956.*)

flow patterns discussed earlier. In the northern hemisphere they typically form downward of the large thermal troughs in the thickness patterns over the eastern coasts of Asia and North America. During the early stages the cyclones move rapidly northeast with the thermal wind, but after about 24–30 hours they start to occlude and slow down. They frequently become stationary over the eastern parts of the oceans, just upwind of the major thermal troughs over the western regions of North America and Europe. Once stationary and fully occluded, the cyclones decay slowly over several days. It is the frequency with which they become stationary near Iceland or the Aleutian Islands which gives rise on mean pressure charts to the Iceland and Aleutian lows. From this discussion it will have become apparent that extratropical cyclones, or frontal depressions as they are sometimes called, are primarily maritime storms which are best developed over the oceans. In the interior of continents, particularly to the east of the Rocky Mountains and in central and eastern Eurasia, the areas of continuous cloud cover and precipitation may be small, and in

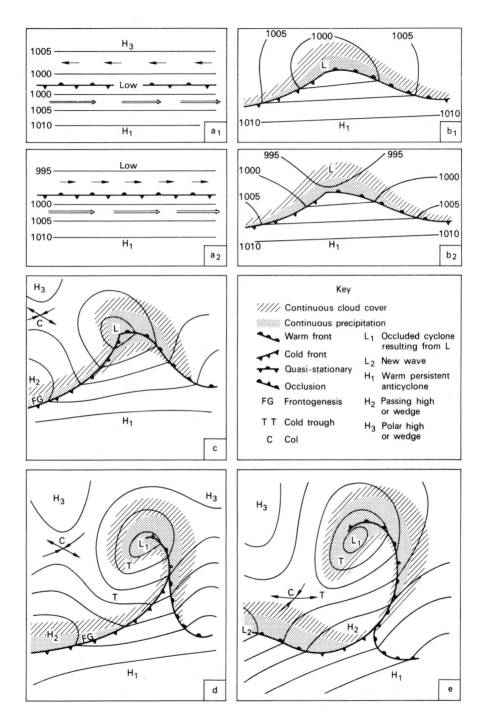

**Figure 8.14** The life-cycle of extratropical cyclones. (*After Bjerknes and Solberg, 1922 and Petterssen, 1956.*)

*The middle and high latitude atmosphere*

many cases the precipitation area may be broken or absent. Indeed, the weather distribution within a cyclone depends very much on the location and the time of year and it is not possible to generalize with safety. For example, cold fronts are often active with severe thunderstorms over the eastern USA, but they rarely give any significant weather in western Europe.

Fitz-Roy, and later Bjerknes and Solberg observed that an extratropical cyclone rarely appeared alone, for normally there would be two, three, or more in a series, each in the wake of the other with a general tendency to move northeast. These cyclone families may be identified with the high index situation discussed earlier.

### 3 Anticyclones

The other major building block in temperate latitude synoptic systems is the anticyclone. These often appear as sluggish and passive systems which fill the spaces between the far more active cyclonic systems. Considered as circulation systems, they never acquire intensities comparable with those of mature cyclones, since it can be seen from the vorticity equation that, when the anticyclonic vorticity approaches that of the Coriolis parameter, no further increase is possible. Anticyclones may be divided very broadly into cold or polar continental highs and warm or dynamic highs.

Polar continental highs develop over the northern land-masses in winter. They are caused by the intense cooling of the snow-covered surface giving rise to a shallow layer of very cold air. The relatively high density of the cold air increases the surface pressure above normal and so an anticyclone appears on the surface pressure chart. These cold anticyclones are clearly shallow formations and it is rare that they can be traced more than 2 km above sea level. Indeed, it is possible for disturbances in the upper westerlies to pass

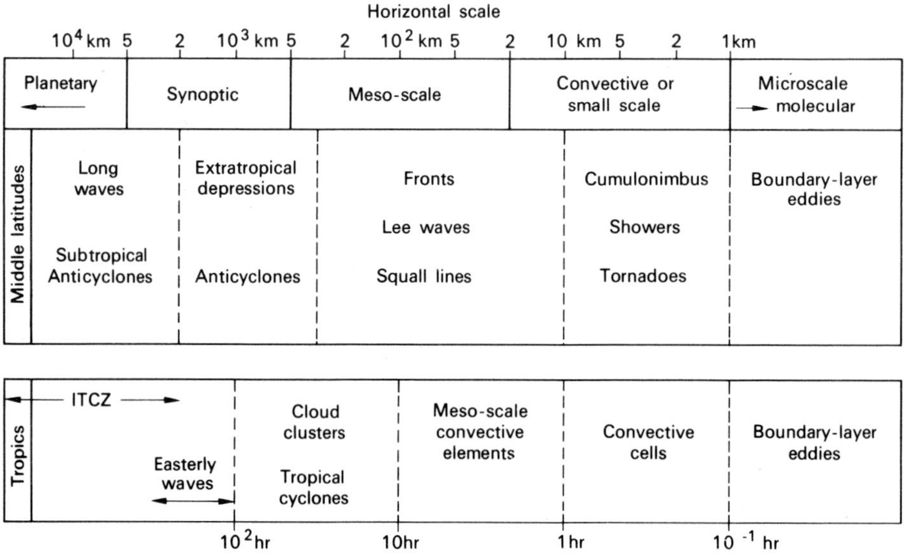

**Figure 8.15**  Scales of atmospheric motion. (*After Mason, 1970.*)

234

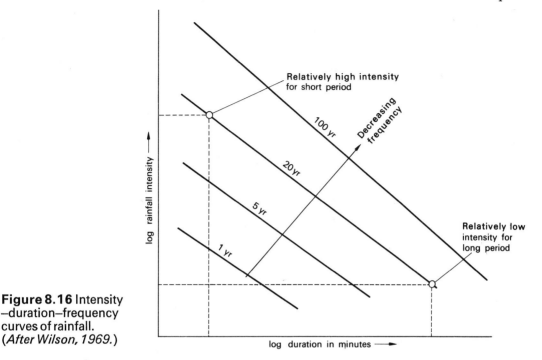

**Figure 8.16** Intensity
–duration–frequency
curves of rainfall.
(*After Wilson, 1969.*)

over them almost completely unmodified. The most pronounced and persistent example is found over Siberia, the corresponding high over North America being far less regular. These highs are, of course, the main sources of continental polar air.

The warm or dynamic anticyclone is caused by large-scale subsidence throughout the depth of the lower atmosphere. Good examples are the highs associated with blocking systems and the descending limb of the Hadley cell. The subtropical highs were discussed in detail in chapter 3. An essential feature of all dynamic anticyclones is a low-level temperature inversion. The general subsidence usually ceases at the top of this inversion and the weather associated with the high often depends on the relatively cool air near the surface. If the surface air is moist, cloud may form below the inversion and light rain may even fall. This is particularly likely where the surface air flows from a warm sea across a cold land surface. In contrast, when the surface air is relatively dry, the skies may be almost cloudless, leading to great heat in summer and intense cold in winter. As with depressions, it is not possible to generalize about the weather distribution within these systems. Dynamic anticyclones can occur anywhere in middle latitudes at any season of the year, since the surface inversion makes them almost independent of the temperature of the surface.

### 4  Weather systems

The main aspects of temperate latitude synoptic meteorology have been discussed in general. In practice, there is a large spectrum of weather systems and these are illustrated in figure 8.15. They range from long waves down to boundary layer eddies. An assessment of typical amounts of divergence, together with the time required for the absolute vorticity to increase (decrease) to twice (one-half) its original value, is also given in figure 8.15. It is apparent that the largest divergence values are associated with the smallest systems, neglecting the boundary layer, and this implies that vertical motion will increase with decreasing scale of system. Consequently the most intense rainfalls are likely to be found in cumulonimbus showers and related systems. As the scale of system increases, the precipitation will become less intense and more widespread (figure 8.16). The scale of the significant rain-forming disturbances varies geographically, for cyclones are most important over the sea, while cumulonimbus and shower clouds are dominant in the continental interiors.

## References

BANNON, J. K., MATTHEWMAN, A. G. and MURRAY, R. 1961: The flux of water vapour due to the mean winds and the convergence of this flux over the northern hemisphere in January and July. *Quarterly Journal Royal Meteorological Society* **87**, 502.

BARRY, R. G. 1967: Variations in the seasonal flux of water-vapour over north-eastern North America during two winter seasons. *Quarterly Journal Royal Meteorological Society* **93**, 535.

BEGERMANN, F. and LIBBY, W. F. 1957: Continental water balance, ground water inventory and storage times, surface mixing rates and world-wide water circulation patterns from cosmic-ray and bomb tritium. *Geochimica et Cosmochimica Acta* **12**, 277.

BENTON, G. S., and ESTOQUE, M. A. 1954: Water-vapour transfer over the North American continent. *Journal Meteorology* **11**, 462.

BJERKNES, V. 1919: On the structure of moving cyclones. *Geofisiske Publikasjoner*, Oslo **1**.

BJERKNES, V. and SOLBERG, H. 1921: Meteorological conditions for the formation of rain. *Geofisiske Publikasjoner*, Oslo **2**.

BJERKNES, V. and SOLBERG, H. 1922: Life cycle of cyclones and the polar front theory of atmospheric circulation. *Geofisiske Publikasjoner*, Oslo **3**.

BOLIN, B. 1950: On the influence of the earth's orography on the westerlies. *Tellus* **2**, 184.

CONRAD, V. 1946: Usual formulas of continentality and their limits of validity. *Transactions American Geophysical Union* **27**, 663.

CRADDOCK, J. M. 1964: Interannual variability of monthly mean air temperatures over the northern hemisphere. *Meteorological Office Scientific Paper* **20**, London.

CROWE, P. R. 1971: *Concepts in climatology*. London: Longman.

D'OOGE, C. L. 1955: Continentality in the western United States. *Bulletin American Meteorological Society* **36**, 175.

DROZDOV, O. A. and GRIGOR'EVA, A. S. 1963: *The hydrologic cycle in the atmosphere. (Vlagooborot v atmosfere.)* Leningrad: Gidrometeorologicheskoe Izdatel'stvo.

EADY, E. T. 1949: Long waves and cyclone waves. *Tellus* **1**, 33.

EADY, E. T. 1950: The cause of the general circulation of the atmosphere. *Centenary Proceedings*, Royal Meteorological Society, p. 156.

FITZ-ROY, R. 1863: *Weather book. A manual of practical meteorology*. London: Longman.

FLOHN, H. 1955: Zur Vergleichenden Meteorologie der Hochgebirge. *Archiv Meteorologie Geophysik und Bioklimatologie Ser. B.* **6**, p. 193.

FLOHN, H. (editor), 1969a: *General climatology*, **2**. *World survey of climatology*. Vol. **2**. Amsterdam: Elsevier Publishing Co.

FLOHN, H. 1969b: *Local wind systems*. In Flohn, H. (editor), *General climatology*, **2**. Amsterdam: Elsevier Publishing Co., p. 139.

FLOHN, H. and OECKEL, H. 1956: Water vapour flux during the summer rains over Japan and Korea. *Geophysical Magazine* **27**, 527.

FOBES, C. B. 1954: Continentality in New England. *Bulletin American Meteorological Society* **35**, 197.

HIDE, R. 1969: Some laboratory experiments on free thermal convection in a rotating fluid subject to a horizontal temperature gradient and their relation to the theory of the global atmospheric circulation. In Corby (editor), *The global circulation of the atmosphere*. London: Royal Meteorological Society, p. 196.

HUTCHINGS, J. W. 1957: Water-vapour flux and flux divergence over southern England: Summer 1954. *Quarterly Journal Royal Meteorological Society* **83**, 30.

LAMB, H. H. 1959: The southern westerlies: a preliminary survey; main characteristics and apparent associations. *Quarterly Journal Royal Meteorological Society* **85**, 1.

MASON, B. J. 1970: Future developments in meteorology; an outlook to the year 2,000. *Quarterly Journal Royal Meteorological Society* **96**, 349.

NEWELL, R. E., VINCENT, O. G., DOPPLICK, T. G., FERRUZZA, D. and KIDSON, J. W. 1969: The energy balance of the global atmosphere. In Corby, G. A. (editor), *The global circulation of the atmosphere*. London: Royal Meteorological Society, p. 42.

PETTERSSEN, S. 1956: *Weather analysis and forecasting*. New York: McGraw-Hill.

PRESCOTT, J. A. and COLLINS, J. A. 1951: The lag of temperature behind solar radiation. *Quarterly Journal Royal Meteorological Society* **77**, 121.

QUENEY, P., CORBY, G. A., GERBIER, N., KOSCHMIEDER, H. and ZIEREP, J. 1960: The airflow over mountains. *WMO Technical Note* **34**, Geneva.

REX, D. F. 1950: Blocking action in the middle troposphere and its effect upon regional climate. *Tellus* **2**, 196, 275.

ROSSBY, C. G. 1939: Relation between variations in the intensity of the zonal circulation of the atmosphere and the displacements of the semipermanent centres of action. *Journal of Marine Research* **2**, 38.

ROSSBY, C. G. 1940: Planetary flow patterns in the atmosphere. *Quarterly Journal Royal Meteorological Society* **66**, 68.

ROSSBY, C. G. 1941: The scientific basis of modern meteorology. In Hambridge (editor), *Climate and man; US yearbook of agriculture*. Washington: US Department of Agriculture, p. 599.

ROSSBY, C. G. 1945: On the propagation and energy in certain types of oceanic and atmospheric waves. *Journal Meteorology* **2**, 187.

SCHERHAG, R. 1948: Neue Methoden der Wetteranalyse und Wettervorhersage. Berlin: Springer.

SIMPSON, G. 1939: Possible causes of change in climate and their limitations. *Proceedings of the Linnaean Society of London*, Part 2, **152**, 190.

SMAGORINSKY, J. 1953: The dynamical influence of large-scale heat sources and sinks on the quasi-stationary mean motions of the atmosphere. *Quarterly Journal Royal Meteorological Society* **79**, 342.

SUTCLIFFE, R. C. 1947: A contribution to the problem of development. *Quarterly Journal Royal Meteorological Society* **73**, 370.

SUTCLIFFE, R. C. 1951: Mean upper contour patterns of the northern hemisphere—the thermal synoptic view-point. *Quarterly Journal Royal Meteorological Society* **77**, 435.

SUTCLIFFE, R. C. 1966: *Weather and climate*. London: Weidenfeld and Nicolson.

SUTCLIFFE, R. C. and FORSDYKE, A. G. 1950: The theory and use of upper air thickness patterns in forecasting. *Quarterly Journal Royal Meteorological Society* **76**, 189.

TUCKER, G. B. 1957: Evidence of a mean meridional circulation in the atmosphere from surface wind observations. *Quarterly Journal Royal Meteorological Society* **83**, 290.

WILLETT, H. C. 1949: Long-period fluctuations of the general circulation of the atmosphere. *Journal Meteorology* **6**, 34.

WILSON, E. M. 1969: *Engineering hydrology*. London: Macmillan.

# 9
# Temperate maritime climates

This chapter is concerned with the climates of the temperate oceans and with those of the neighbouring land areas which are substantially influenced by the North Atlantic, the North Pacific and the Southern Ocean around Antarctica. Because of the generally westerly flow of the atmosph'ere, maritime influences are most marked on the eastern sides of the oceans, and are particularly strong in western Europe, western North America, the southern parts of western South America, New Zealand, and the extreme south of Australia. High mountain ranges limit oceanic influences to the coast in North and South America, and the other land areas in the southern hemisphere are extremely limited, so the only extensive land area with a temperate maritime climate is found in western Europe.

## General Climate

These climates are characterized by oceanic or suboceanic temperature regimes with mild to moderately cold winters and moderately warm to warm summers. The more oceanic climates have an autumn and winter maximum of precipitation, the less oceanic climates have a summer to autumn maximum. These climates are not generally given to extremes of heat or cold and the rainfall is usually adequate.

The climatic type is both well developed and well documented over the North Atlantic and western Europe, and since the land area in other parts of the world experiencing this type of climate is limited, the discussion will be restricted to these two regions, which may be considered as being typical. As most of Europe is open to invasions of the maritime zonal westerlies, the change eastward from marine to continental climates is very gradual, and marine effects are still obvious in eastern Europe. It is therefore difficult to fix an eastern boundary for this climatic type, but it can probably be taken as the 50 per cent line in figure 9.3, for east of this line the climate is dominated by continental air-masses. The southern limit in Europe can conveniently be taken as the Mediterranean Sea and the northern limit as the Arctic Circle.

### *1  Seasonal variation*
The temperate maritime climates are markedly seasonal but the seasonal changes do not take place gradually, but irregularly and often suddenly. These sudden changes are present in both temperature and rainfall records and are of some environmental importance. Some

appear to be completely random, while others occur at nearly the same date each year and have been described as singularities. In chapter 8, it was suggested that the upper flow patterns in middle latitudes have an irregular cyclic rhythm, and that this is reflected in surface pressure distribution and weather. A useful tool in this connection is the concept of 'Grosswetterlagen' or 'weather types', which helps in the understanding of the cause of singularities and of weather variations throughout the year.

## 2 Weather types

Weather types have been extensively studied by the German meteorologists, in particular Baur (1947a and b) who concentrated upon the years 1881–1943 over central Europe. Lamb (1950) has classified 50 years of surface weather charts of the British Isles and surrounding regions. His classification, described in table 9.1, has the advantages of being

**Table 9.1**   Classification of weather types over the British Isles according to Lamb (1950)

*Anticyclonic type:* Anticylcones centred over, near, or extending over the British Isles ; therewith also cols situated over the country, between two anticyclones.
*Cyclonic type:* Depressions stagnating over, or frequently passing across, the British Isles.
*Westerly type:* High pressure to the S (also sometimes SW and SE) and low pressure to the N of the British Isles. Sequences of depressions and ridges travelling E across the Atlantic.
*Northwesterly type:* Azores anticyclone displaced NE towards the British Isles or N over the Atlantic west of the UK, or with extensions in these directions.  Depressions (often forming near Iceland) travel SE or ESE into the North Sea and reach their greatest intensity over Scandinavia or the Baltic.
*Northerly type:* High pressure to the W and NW of the British Isles, particularly over Greenland, and sometimes extending as a continuous belt S over the Atlantic Ocean towards the Azores.  Low pressure over the Baltic Scandinavia, and the North Sea. Depressions move S or SE from the Norwegian Sea.
*Easterly type:* Anticyclones over, or extending over, Scandinavia and towards Iceland.  Depressions circulating over the western North Atlantic and in the Azores–Spain–Biscay region.
*Southerly type:*  High pressure covering central and N Europe. Atlantic depression blocked west of British Isles or travelling N and NE off UK western coasts.

relatively simple, easy to use, and also reasonably typical of other classifications in use. In theory, each category should produce a characteristic type of weather depending on the season.

Long spells of weather, marked by the persistence of one or another easily recognized and definable type, are well-known features of middle latitude climates. Lamb defines a long spell as one lasting over 25 days with interruptions not exceeding 3 days. He found that spells show certain groupings according to the calendar, and that some dates come within periods of a persistent weather type much more often than others. This is indicated by the frequency curves found in figure 9.1, which suggest the division of the year for the British Isles into five natural seasons, defined by Lamb as follows:

1   high summer (18 June–9 September);
2   autumn (12 September–19 November);
3   early winter (20 November–19 January);
4   late winter or early spring (20 January–29 March);
5   spring or early summer (30 March–17 June).

## 3 Singularities

Certain weather types have a tendency to occur around fixed dates each year and give rise to climatic occurrences which have become known as singularities. Schmauss (1938) and

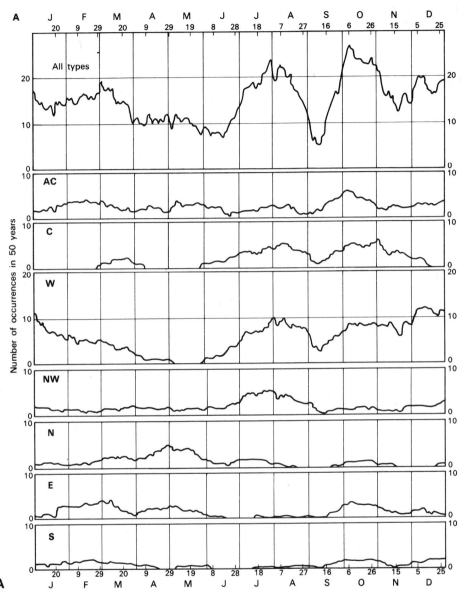

**Figure 9.1A**

Flohn (1954) have listed European singularities, which they define as recurring in a well-marked form in about 50 per cent of the years and marked in various degrees by other effects in a number of other years. For these singularities the mean variation in the date of occurrence is only about 3 days and the extreme variation in individual years amounts to no more than 6–10 days. Singularities which are regular in their recurrence from year to year will be reflected by appropriate features in the trend curves for daily and weekly mean values of the meteorological elements. Lamb has listed the important singularities affecting the British Isles and they are shown in table 9.2.

The periods 5–11 January and 27 January–3 February may be identified with general

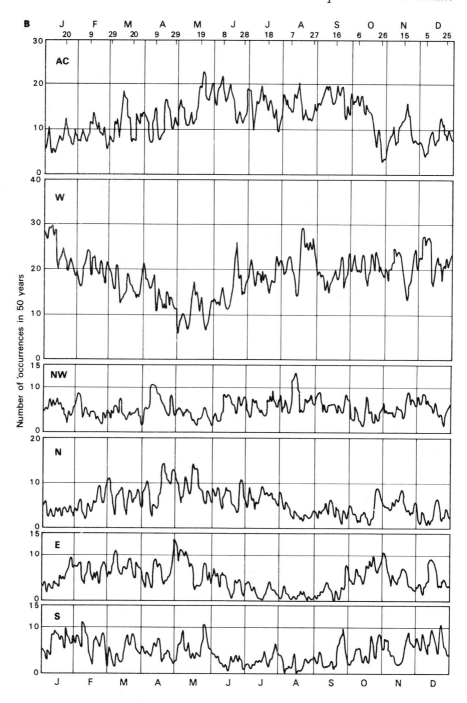

**Figure 9.1** **A**: Frequency curves for long spells of over 25 days of persistent weather in the British Isles, 1898–1947. **B**: Frequency curves for daily occurrences in the British Isles, 1898–1947. (*After Lamb, 1950.*)

**Table 9.2**   Important singularities affecting the British Isles

| | |
|---|---|
| 5–11 January | Stormy period of early January. |
| 20–23 January | Anticyclonic in Europe and southern and eastern Britain. |
| 27 January–3 February | Renewed storminess, gales, and rain or snow. |
| 8–13 February | February anticyclones. |
| 26 February–9 March | Cold, stormy period. |
| 12–22 March | Early spring anticyclones over Britain and Europe. |
| 28 March–1 April | Cold, stormy period. |
| 12–19 April | Cold, stormy period. |
| 29 April–16 May | Northerly weather in some anticyclonic intervals. |
| 21–31 May | Fore-monsoon fine weather periods. |
| 1–4 June and 12–14 June | First waves of European summer monsoon; stormy, cool episodes. |
| 5–11 June | June anticylones over the British Isles and W Europe. |
| 18–22 June and following two weeks | Return of the westerlies. |
| 23–30 July and following week | Thundery, cyclonic weather over Europe and the British Isles. |
| 16–30 August | First storms of autumn. |
| 5–30 September | Old-wives summer anticyclones. |
| 24 October–13 November | Late autumn rains. |
| 15–24 November | Quiet, foggy, anticyclonic interlude. |
| 24 November–10 December | Early winter storms and rains. |
| 19–23 December | Continental and N European anticyclones of the winter solstice. |
| 25–31 December | Christmastide thaw and storms of the end of the year. |

*After Lamb (1950).*

storminess in the British Isles. Lowest pressures of the year are observed at Arctic and northern oceanic stations during the first period, which also corresponds to the maximum intensity of zonal westerlies in the northern hemisphere. Precipitation frequency exceeds 70 per cent in north Germany during the second period. March 12–22 is often a period of anticyclonic weather over both Britain and continental Europe, giving dry weather with large diurnal temperature variations. Peaks occur on the mean pressure curves in Britain, Germany, and southern Scandinavia. April 19–16 May is strongly associated with northerly weather over the British Isles, and also coincides with the minimum strength of the zonal westerlies in the northern hemisphere. Cyclonic, often stormy weather, is associated with the period 24 October–13 November, making 22–29 October one of the three wettest weeks of the year in most districts of England and Ireland. Peaks of raininess also occur in continental Europe, most notably in Italy.

The European/Atlantic singularities are of further interest in that they suggest a maximum zonal circulation in early January and a minimum in late April–early May. This is confirmed by studies of the zonal index (Putnins, 1963) for latitudes 50–60°N between longitudes 0° to 40°W which show (table 9.3) a maximum in December/January and a minimum in May. These circulation changes are paralleled by annual precipitation curves over many parts of northwest Europe, which denote a minimum in spring.

**Table 9.3**   500 mb zonal indices for zone 50°–60° N and 0°–40° W (gpm)

| | | | |
|---|---|---|---|
| Jan | 197·08 | Jul | 149·68 |
| Feb | 156·02 | Aug | 148·58 |
| Mar | 107·52 | Sept | 158·30 |
| Apr | 138·04 | Oct | 181·78 |
| May | 100·44 | Nov | 166·04 |
| Jun | 125·40 | Dec | 203·36 |

*After Putnins (1963).*

*4 Southern Europe*

During the summer months the prevailing westerly winds retreat northwards from southern Europe leaving it under the influence of the subsiding air of the subtropical highs. Under these conditions the weather is similar to that in North Africa and only occasional precipitation falls from convective showers. In winter the zonal westerlies return to southern Europe bringing with them frequent cyclones which cause a precipitation maximum in the winter half of the year. Escardo (1970) considers that cyclonic disturbances from the Atlantic, which are most frequent in the autumn, winter, and spring, are responsible for 54 per cent of the annual precipitation in Iberia. Further east, Atlantic depressions are not so important and much of the precipitation is caused by upper cold pools which are connected with low index situations further north.

## Radiation and temperature

The discussion of the broad nature of the radiation balance of the earth in chapters 2 and 8 made it clear that in winter the middle and high latitudes experience a net radiation loss, while in summer there is a large net radiation gain. Although the winter net radiation loss is partly counteracted by the heat advection of the atmosphere, nevertheless temperatures can fall to low levels. As in other temperate and high latitude areas, it is the annual march of net radiation and temperature which determines the important seasons.

*1 Radiation*

The southernmost rim of Europe lies at about latitude 35°N, almost within the subtropics, while the northern extensions reach 70°N, almost in the Arctic. One important consequence of this large latitudinal spread is that the length of daylight varies significantly over the region, as illustrated in table 9.4 from which it can be seen that this is a variation in time as well as in space. At the winter solstice day-length varies from zero at 70°N to 9 hr 48 min at 35°N; at the corresponding summer solstice the day-length is 24 hours at 70°N and only 14 hr 31 min at 35°N. The long summer days in northern Europe are important for vegetation growth and partly compensate for the net radiation deficit in winter. Near 35°N the total

**Table 9.4**  Variation of day-length with latitude

| Latitude | 21 December | 21 June |
|---|---|---|
| 70°N | nil | 24 hrs |
| 60°N | 5 hrs 52 mins | 18 hrs 52 mins |
| 50°N | 9 hrs 4 mins | 16 hrs 23 mins |
| 40°N | 9 hrs 30 mins | 15 hrs 1 min |
| 35°N | 9 hrs 48 mins | 14 hrs 31 mins |

incoming radiation over the year from sun and sky is about 150,000 cal cm$^{-2}$, and of this about 65 per cent is due to direct solar radiation and 35 per cent to diffuse radiation from the sky. At 70°N the total incoming radiation has fallen to about 70,000 cal cm$^{-2}$ and of this about 60 per cent results from diffuse sky radiation. This increase in the relative amount of diffuse sky radiation is normal in high latitudes and makes the influence of the angle of slope of the land on radiation totals relatively less important than nearer the equator, since diffuse radiation is received nearly equally on all slopes of moderate inclination.

In figure 2.4 which indicates estimates of net radiation at the surface over Europe, very wide variations are immediately apparent. The annual net radiation in the south is almost four times that in the extreme north. The net radiation may be considered as the basic background energy which is available at the soil surface for plant growth and other natural energy-consuming activities. From figure 2.4, a large variation in vegetation type should be expected along lines of longitude in Europe, and this is indeed the case. In northern Europe the net radiation and associated temperature changes are the main controls on vegetation growth, but further south the lack of summer rainfall is also a very important factor. In the extreme south of Europe, net radiation hardly limits the growth of vegetation, which occurs throughout the winter, but in the Mediterranean lands it is rather the lack of water which limits plant growth.

## 2   Mean temperature

Mean temperature charts for Europe show that the isotherms have a zonal trend in summer but are more meridional in winter (figure 9.2). Wallén (1970) considers that this is because zonal flow is more dominant at the surface in summer, and meridional flow in winter, but it could also reflect the seasonal variations in the heat balance of the land-mass. In winter, the net radiation balance of both the continental surface and the earth-atmosphere system over the continent is negative, and the radiation loss is made up by the advection of heat from the west and south. Therefore, the progressive cooling of the air-masses as they pass over the cold continent is probably the explanation for the trend of the January isotherms. In summer, the net radiation balances over the continent are positive and the advection of warm air is not of great importance, except perhaps in the south. The summer net radiation values fall towards the north and this is reflected by the zonally decreasing temperatures.

*a   Winter*   The coldest European winters are found in the northeast of Scandinavia with average temperatures of $-12°$C, and the warmest in the Mediterranean peninsulas with temperatures of $10°$C, an isotherm which migrates to the extreme north of the continent in July. In most of west and northwest Europe, and on the Mediterranean coasts, no month has a mean temperature below $0°$C, but inland winter temperatures fall rapidly and well below freezing-point in the interior of the continent. The steepest temperature gradients are found at latitudes $65$–$70°$N, when the mean January temperature drops from $0°$C on the coast of Norway to $-12°$C in the interior of Scandinavia and Finland, a distance of about 200 km.

*b   Summer*   July mean temperatures are wholly above $10°$C, except in the very extreme north, and reach $20°$C over large parts of southern Europe. Because of the differences in the distributions of the January and July mean temperatures, the annual temperature range increases rapidly inland, being about $10°$C along the coast but reaching $20°$C in the continental interior and above $30°$C in the northeast. From the agricultural viewpoint, the large annual range and associated high summer temperatures in the continental interior compensate for the low winter temperatures.

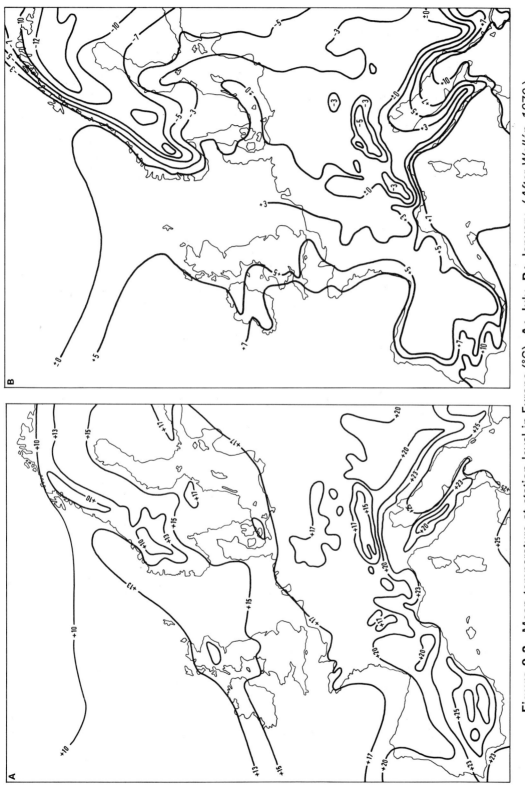

**Figure 9.2** Mean temperature at station level in Europe (°C). **A:** July. **B:** January. (*After Wallén, 1970.*)

### 3   Air-masses

Actual day to day surface temperatures in Europe are partly correlated with the prevailing air-mass. Winter temperatures fall inland as the frequency of continental air increases, and this has been used as an index of continentality, K, by Berg (1940):

$$K = \frac{C}{N} \times 100$$

where C is the frequency of continental air-masses, and N is the frequency of all other air-masses.

As illustrated by the results in figure 9.3, the ratio of continental air-masses rises to 50 per cent in the east, but is below 25 per cent in the west. According to Belasco (1952), maritime polar air accounts for nearly 59 per cent of all air-masses observed at Stornoway, and 42 per cent of those at Kew and Scilly. It is the most frequent air-mass appearing over northwest Europe.

### 4   Low temperatures

Very low temperatures in western Europe are often connected with continental polar air derived from northern Russia. This is air which has cooled over snow surfaces in a stagnant cold high. It generally crosses the Russian frontier with a temperature which is sometimes below −18°C, and although its temperatures rises slowly as it crosses Germany largely as a result of mixing with warmer air ahead and above, even in Holland and eastern France it may still be below −12°C. The North Sea warms the air a little during its passage across,

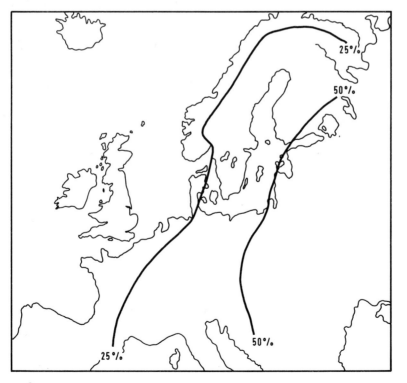

**Figure 9.3**
Continentality (K) in Europe according to Berg. K = C/N × 100 where C is the frequency of continental air-masses and N is the frequency of all other air-masses. (*After Berg, 1940 and Barry and Chorley, 1968.*)

but it can still arrive over the British Isles with midday temperatures below 0°C. Continental polar air is rare in the extreme west and is observed on only a very few days each year at Kew, but cold air can penetrate through mountain gaps into the Mediterranean, where it often arrives as a violent wind with temperatures sometimes below 0°C.

Cold, often dry, northerly winds are frequent along the Mediterranean coasts of Europe in winter, and are given a variety of local names. The mistral is a northwesterly or northerly wind which blows offshore along the north coast of the Mediterranean from the Ebro to Genoa, but it is most intense on the coasts of Languedoc and Provence, especially in and off the Rhone delta, where speeds of about 20 m sec$^{-1}$ are recorded. It is an antitriptic wind in that surface friction causes it to blow directly from high-pressure areas over Europe to low-pressure areas over the sea, along the valleys through the mountains of southern Europe. A related wind is the bora, which is a cold, often very dry, northeasterly wind blowing, sometimes in violent gusts, down the mountains on the eastern coast of the Adriatic. Winds of this type blow on average up to 100 days a year at Marseilles and 39 days at Trieste.

## 5 High temperatures

Continental tropical air comes mainly from North Africa during the summer, but it is usually much modified by its passage across the Mediterranean. The sea cools and moistens the current so that it arrives in southern Europe as a hot damp wind. It is rare for continental tropical air to travel much further north than southern Europe though it does occur occasionally in Britain, giving high summer temperatures.

It should be noted that unusual warmth can be caused by foehn winds, which are warm, dry winds blowing to leeward of a ridge of mountains. They are caused by the passage of the air currents over the mountain range, where the release of latent heat by precipitation warms the air. There is also evidence that foehn winds can be caused by subsidence to leeward of the mountains when precipitation may be absent under conditions suitable for lee waves. They are best developed in the valleys of the Alps, but can even be observed along the south shore of the Moray Firth in Scotland.

## 6 Seasonal variation

Temperatures show marked seasonal variations, but neither actual nor average annual curves are simple and smooth in shape. Day-to-day temperatures vary widely, depending on weather types and air-masses, so that unseasonable temperatures can occur at any time. Even mean curves are not smooth, since there are many breaks in the seasonal warming and cooling limbs, and some of these can be identified with the singularities mentioned earlier in this chapter.

## 7 Upland temperatures

Western European uplands are interesting because they are abnormally cool as compared with those further east, and in particular, landscapes resembling those of the tundra may be found at elevations as low as 600 m–800 m in Britain. This is partly a result of the dominance of maritime polar air in northwest Europe, with its associated strong winds, frequent low cloud, and steep lapse rate. Further east, with clearer skies and lighter winds, the positive radiation balance in summer gives rise to relatively high temperatures in the uplands and causes the cultivation limit to increase in height inland. The summer temperatures experienced on temperate latitude mountains depends largely on how effective the mountains are in creating their own local climates. If the winds are light and the uplands are very large and massive, they will be able, under the influence of the positive net radiation in summer, to create a local climate which is largely independent of that in the free atmosphere. Therefore, temperatures in upland areas are strongly controlled by the exposure of local topography to solar radiation and the atmospheric circulation, both of which were discussed in some detail in chapter 1.

## 8   *Plant growth*

Within the tropics, short-wave radiation and temperature values are always sufficient for plant growth, but in middle latitudes they both place limits on vegetation development. According to Budyko (1968) two relationships hold, provided that there is always adequate soil moisture for plant growth. First, an increase in the radiation flux leads at all temperatures to an advance in the vegetation productivity. Second, with a rise in temperature the productivity first increases and then, after reaching some maximum value depending on the short-wave radiation flux, decreases. These relations are illustrated in figure 9.4, where productivity is expressed as the mass of carbon dioxide assimilated per unit of area assuming optimum agrotechnique. At low temperatures and low values of incoming radiation (such as are found over Europe in winter), plant productivity will be extremely small, but

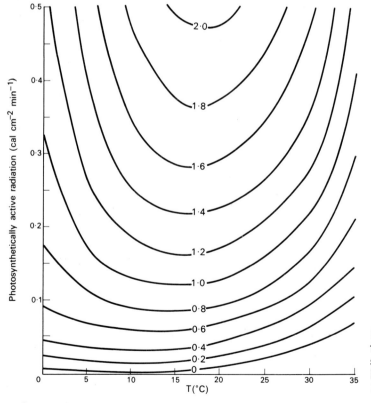

**Figure 9.4** Radiation and temperature effect on photosynthesis (moisture non-limiting). (*After Budyko, 1968.*)

it increases rapidly with increasing values of both factors. Incoming radiation and surface temperature are closely correlated, so it is unlikely, for instance, that low temperatures will be normally associated with a large radiation flux. Unfortunately the individual effects of these two parameters are often confused in plant growth studies because it is not fully understood that both radiation and temperature are important. Further complications arise because while temperature measurements are plentiful, radiation measurements are comparatively scarce. It has therefore become usual to express the start and finish of the growing season simply in terms of temperature, and the vegetation growth period is usually considered to cover the period of the year when the mean daily air temperature is 6°c or higher.

Taking 6°c as a rough threshold value for plant growth, it is possible to calculate the length of the growing season at selected sites in northern Europe. In England and Wales, this varies from about 200 days over the upland areas of Wales, the Pennines, and the Lake District, to 365 days along parts of the west coast of Wales and in lowland areas of Devon and Cornwall (Hogg, 1967b). According to Johannessen (1970), the vegetation growth period in south Denmark begins on average about 10 April and lasts for between 195 and 212 days; but in southern Sweden, in the coastal districts of eastern and southern Norway, and at the inner fjord districts of western Norway, it starts between 20 and 25 April and lasts about 206 days. It begins during the first days of May in southern Finland, and at the end of May or the beginning of June in the mountains of Norway and Sweden, in interior Finnmark and in the northernmost part of Finnish Lapland. In interior Finnmark, the vegetation growth period lasts for about 125 days, but because of the long days, plant development is rapid. Calculations for southern Europe are not particularly meaningful because of the long summer dry period. The maritime areas of Devon and Cornwall might appear to have a great advantage agriculturally over areas further east, but in practice the amount of plant growth is partly proportional to the average number of accumulate degree-days above 6°c, and therefore not only to the length of the season but also to the average summer temperatures. The latter are normally higher in the continental east than nearer the Atlantic and this can more than compensate for the shorter growing season. Manley (1970) calculates totals of 1,600 degree-days in the favourable areas of Britain, diminishing to 1,200 in marginal arable areas on the flanks of the uplands, and to less than 800 around the highest parts. Corresponding values for Denmark range from 1,590 to 1,251, showing that in this respect Denmark is comparable to southern Britain. Values decline to the north, falling to 415 degree-days at Utsjok in northern Finland (Johannessen).

## 9 Frost

Frost and low temperatures are important to farmers or growers because they can kill some plants and damage others. Sometimes, on clear nights, the soil surface temperature falls below 0°c giving a ground frost, while the screen temperature remains above freezing-point; but most damage is caused by air frost when the screen temperature is 0°c or below. Air frost can be caused in two ways: first, by the advection of very cold air into the area, and second, by intense radiational cooling under clear skies during winter nights. Advection of very cold air, normally from the east or northeast in winter, can cause air temperatures to fall below 0°c even during the day over vast areas of Europe, and cold winds can reduce temperatures to below 0°c even on the Mediterranean coasts. Marseilles, for instance, has recorded temperatures below 0°c in every month from October to April. The lowest temperatures, however, are caused not by advection, but by the nocturnal cooling of already cold air.

In winter most of Europe suffers a net radiation deficit. Under a dry clear polar air-mass, the infra-red radiation loss from the surface becomes large, and intense cooling takes place at night. The lowest temperatures are normally reached in open valley-bottoms, on clear calm-nights when the ground is snow-covered. This is because snow radiates very effectively in the far infra-red and the cold air tends to accumulate in hollows. Norway is particularly suited for the study of this phenomenon, because the relief is well developed, and so in winter cold mountain air drains down-valley and down-fjord towards the sea, keeping the mean temperature of the coldest month along the Skagerrak coast below 0°c. Even in areas of shallow relief, minor depressions can be subject to lower temperatures than the surrounding areas and form frost-prone hollows.

Extremely low temperatures are normally restricted in extent, because they are partly caused by local cooling and therefore depend on the nature of the local countryside. The lowest screen temperature recorded in Britain is −27·2°c at Braemar in the eastern Scottish Highlands, but −24°c has been observed at Buxton and −23°c in central Wales (Manley, 1970). Manley also considers that the local occurrence of −25°c in severe frost hollows in southern England is not impossible. Extreme low temperatures with similar values have been reported from most of northern Europe and indeed there is no part of Europe where the extreme minimum temperature is likely to be above 0°c.

## 10   Frost damage

To assess the importance of frost in economic terms, it is necessary to make some assumptions concerning the effects of low temperatures on particular crops and also on costs. Hogg (1968) has considered this with respect to black currant growing at Bradford-on-Avon, Somerset. Frost damage is most important in the spring, and so Hogg assumed a complete loss of the crop with a minimum temperature of −2·2°c (28°F) or below in April and May, and a partial loss (say 50 per cent) with a temperature of −1·1°c (30°F). At this particular site, climatological data suggest that during the critical period, 10 years out of 10 will have temperatures below 0°c, 7 in 10 below −1·1°c and 5 in 10 below −2·2°c. Thus no crop should be expected in 5 years, half a crop in 2 years, and a full crop in the other 3. If, apart from frost damage, the crop can be assumed to sell at £330, the gross return over 10 years will be £3,300. Allowing for reduced crops because of frost, the gross return over 10 years will be $(3 \times 330) + \frac{1}{2}(2 \times 330) = £1,320$.

Growers can minimize frost damage by installing frost-prevention equipment or by choosing sites which are relatively frost-free. Few sites in Europe are completely frost-free, but frost tends to be infrequent on slopes overlooking relatively warm seas, where there is free drainage for the air. An extreme example is provided by the garden at Inverewe, where a variety of subtropical evergreens flourish at nearly 58°N. Here, as in other parts of western Scotland, killing frosts are rare on the steep slopes facing the Atlantic. Similar phenomena are observed at other favourable sites along the Atlantic coasts of Europe and allow very early crops to be produced. Hogg (1967b) considers that the most favourable sites are found on southerly slopes with perhaps lights soils, where there is also strong advection of warm air-masses off a warm sea. Further discussion on the interaction of solar radiation with sloping terrain is found in chapter 1.

# Precipitation

*1 Distribution*

Precipitation amounts over land are greatly influenced by relief, while those measured over the ocean are presumably free from orographic influence. Unfortunately, it is difficult to measure precipitation over the open ocean directly from ships. Tucker (1961) has carefully estimated from meteorological 'present weather' observations the distribution of precipitation over the North Atlantic for the 5-year period 1953–1957, and his results are reproduced in figure 9.5. Tucker's average annual precipitation map suggests that precipitation increases from about 400 mm near the Azores anticyclone to over 1,200 mm in the seas between Greenland and Iceland. Precipitation amounts also increase towards the east, reaching a maximum just off the coast of western Europe.

It might be expected that the maximum precipitation over the North Atlantic would occur in the summer, since this is the season when the air is warmest and will therefore hold the most moisture. The seasonal charts in figure 9.5 clearly show that this is not so, and that the maximum precipitation occurs in winter over large areas of the ocean. Estimates of mean monthly precipitation at the various Atlantic Ocean weather ships are given in table 9.5, and all except Ship K report maximum monthly precipitation values during the winter

**Table 9.5** Mean monthly and mean annual precipitation (mm) over Atlantic Ocean weather ships during the 5-year period January 1953–December 1957

| Ocean Weather ship | Jan | Feb | Mar | Apr | May | Jun | Jul | Aug | Sept | Oct | Nov | Dec | Annual |
|---|---|---|---|---|---|---|---|---|---|---|---|---|---|
| I | 97 | 86 | 50 | 42 | 55 | 65 | 61 | 72 | 92 | 92 | 78 | 103 | 893 |
| J | 78 | 55 | 49 | 43 | 55 | 64 | 56 | 77 | 69 | 68 | 64 | 92 | 770 |
| K | 35 | 47 | 52 | 24 | 35 | 26 | 20 | 29 | 28 | 27 | 28 | 34 | 385 |
| M | 142 | 146 | 88 | 92 | 78 | 73 | 59 | 52 | 102 | 121 | 116 | 125 | 1,194 |

*After Tucker (1961).*

months of December, January, and February. Similarly, minimum monthly precipitation values occur either in July and August or April and May. These seasonal changes indicate that precipitation over the North Atlantic is very much controlled by the incidence of synoptic systems.

Since western Europe receives much of its weather from the North Atlantic, the distribution of precipitation over western Europe (figure 9.6) reflects the distribution over the North Atlantic. Cyclonic activity leads to much of the annual precipitation in Ireland, Scotland, and western Scandinavia, but is a less important factor in France, the Netherlands, and Germany. In the western areas it is often intensified by mountain ranges, and in Scotland annual precipitation locally exceeds 2,000 mm, while in western Norway there are regions of more than 3,000 mm that is to say, nearly three times as much as Tucker's estimate for the nearby North Atlantic. Large amounts are also experienced in the mountains of southern Europe, but in the relatively flat areas of western and central Europe the precipitation is rarely more than 750 mm per year. Areas of very low precipitation are generally observed on the leeward sides of mountain barriers, for example, as in south Norway, Sweden, and Finland east of the Scandinavian mountains, east and southeast of the Appenines, in eastern Greece, and eastern Romania, etc.

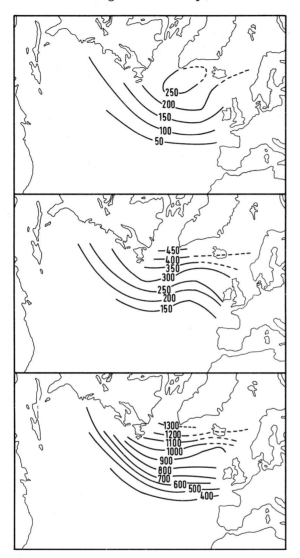

**Figure 9.5** Precipitation over the north Atlantic Ocean during the period 1953–1957 (mm). *Top:* average summer; *middle:* average winter; *foot:* average annual. (*After Tucker, 1961.*)

*2   Seasonal variation*

Seasonal variations of rainfall are partly related to the distance from the North Atlantic, for in the west the cyclonic type of precipitation is more dominant than further eastwards where the convective type of precipitation gradually becomes of greater significance. Over the North Atlantic, cyclonic activity leads to a winter maximum and a summer or spring minimum; it is reasonable, therefore, to expect to find a similar regime over the western parts of Europe. In the eastern, more continental parts, the dominantly convective precipitation gives rise, in contrast to the west, to a maximum in the summer. The southern parts of Europe tend to have a regime of winter rainfall and summer drought. associated with the movement of the subtropical anticyclones.

*3   Synoptic conditions*
The importance of cyclonic rainfall in western Europe is illustrated by the work of Beaver and Shaw (1970), who have analysed the rainfall at Keele, England, according to synoptic type. Their results, which are illustrated in table 9.6, show that for 9 months of the year, over half of the precipitation comes from frontal situations; only June, July, and August have totals of less than 50 per cent for warm front, cold front, and occlusion rainfall combined. The winter months receive much of their rain from warm fronts (29 per cent), but the most outstanding proportion is that of 31 per cent for occlusion rainfall in March. Thunderstorm rainfall is almost limited to the 6 summer months and reaches a maximum in July.

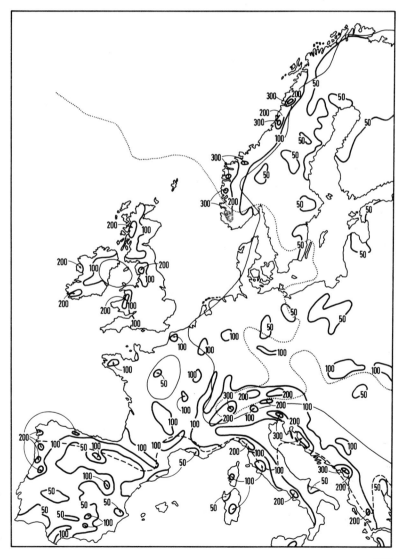

**Figure 9.6** Annual precipitation in Europe (cm). Thin continuous line shows western limit of region of predominant rainfall in summer and early summer; dashed line shows northern limit of region of predominant drought in summer; the line of dots shows southern limit of region with more than 40 days snowfall. (*After Blüthgen, 1958 and Wallén, 1970.*)

**Table 9.6** Percentage of monthly precipitation resulting from various synoptic weather situations at Keele (North Staffs) for 1952–69

| Month | Warm front | Warm sector | Cold front | Occlusion | Polar maritime | Polar low | Polar continental | Arctic | Thunder storm |
|-------|-----------|-------------|-----------|-----------|----------------|-----------|-------------------|--------|---------------|
| Jan   | 29 | 15 | 13 | 16 | 16 | 7  | 1 | 3 | 0  |
| Feb   | 23 | 10 | 10 | 25 | 14 | 9  | 3 | 5 | 0  |
| Mar   | 17 | 11 | 16 | 31 | 14 | 6  | 2 | 3 | 0  |
| Apr   | 28 | 12 | 11 | 16 | 22 | 8  | 1 | 1 | 2  |
| May   | 16 | 8  | 16 | 21 | 22 | 9  | 0 | 0 | 9  |
| Jun   | 22 | 15 | 11 | 16 | 18 | 12 | 0 | 1 | 6  |
| Jul   | 19 | 12 | 13 | 13 | 16 | 12 | 0 | 0 | 15 |
| Aug   | 19 | 12 | 11 | 13 | 26 | 14 | 0 | 0 | 4  |
| Sept  | 19 | 13 | 20 | 16 | 25 | 5  | 0 | 0 | 3  |
| Oct   | 22 | 13 | 23 | 16 | 19 | 7  | 0 | 0 | 0  |
| Nov   | 29 | 9  | 15 | 21 | 14 | 10 | 1 | 1 | 0  |
| Dec   | 29 | 15 | 11 | 16 | 19 | 8  | 0 | 2 | 0  |

*After Beaver and Shaw (1970).*

## 4 Snow

In the colder months some of the precipitation will fall as snow, which normally increases in quantity with distance inland and with altitude, since it is dependent on the mean temperature. Indeed, it has been found statistically that the monthly mean temperature must stay at about $-2°c$ in order that there be a permanent snow cover during the month. Because of the relatively mild winter temperatures, snowfall in the lowlands of Britain is rare enough to cause special comment if it lies for more than a few days. However, on the mountains, it often lies for long spells, the mean exceeding 100 days in the highest Scottish mountains and 50 days in the highest uplands of northern England. Manley (1969) has estimated the range of behaviour of the snow cover with altitude in Britain, and found that on average the duration of cover increases by 1 day for every 6·5 m; his results are sum-

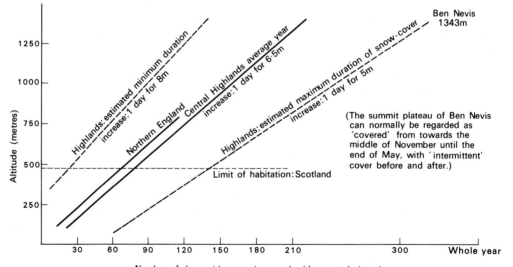

**Figure 9.7** Estimated range of behaviour of the snow cover in Britain: maximum and minimum expectation. (*After Manley, 1969.*)

254

marized in figure 9.7. The comments made about Britain apply to much of western Europe, for there too the lowlands normally have little snowfall. Paris, for instance, has about 10 days with snow lying, northwest Germany about 20, and southwest Norway from 10 to 50 days. Snow is also infrequent along the Mediterranean coasts of Europe, amounting on average to 2–6 days per year, but the frequency and amount increases rapidly in the mountains, reaching over 240 days at above 2,700 m in the Pyrenees.

Eastwards in Europe, the winter cold increases and much of the winter precipitation falls as snow. In the south of Germany, snow is most frequent in January, February, and March, the ground being snow-covered for about 20 days per year. The duration of snow cover again increases rapidly with altitude, reaching 167 days per year on the Feldberg, Black Forest (1,493 m).

Snow or sleet is likely to fall whenever a sufficient depth of cold air is present, and this usually means that the wet bulb temperature of the surface layer should not exceed 3°C (Manley ,1970). In practice, the height of the freezing-level above the ground and the thermal structure of the lowest layer of the atmosphere largely determine whether snow can reach the ground. Evaporation, sublimation, and melting are the processes which destroy falling snowflakes, and once a snowflake falls into air which is above the freezing-point, its rate of melting is probably measured by the time-integral of the temperature difference, air minus snowflake (Lamb, 1955). Murray (1952) has examined the occurrence of rain or snow with various freezing-level heights in the British Isles and found that frozen and unfrozen forms become equally likely when the freezing-level is at about 305 m; it also seems that this value is equally applicable to continental locations. Unfortunately, the problem is complex because long-continued rain and sleet falling through dry air may cool the atmosphere, cause the freezing-level to lower, and consequently turn the sleet eventually to snow.

Clarke (1969) has analysed the synoptic situations which have produced snowfall over southeast England and his results are illustrated in table 9.7. He considers that snow occurs

**Table 9.7**  Total number of daily occurrences of sleet or snow in association with different synoptic situations for southeast England during the years 1954–1969

| | |
|---|---|
| Minor troughs/polar lows, snow showers in northerly situations | 114 |
| Minor troughs/polar lows, snow showers in easterly situations | 163 |
| Frontal snow in association with cold fronts or cold occlusions | 32 |
| Frontal snow in association with warm fronts or warm occlusions | 113 |

*After Clarke (1969).*

basically in two types of situation: frontal, i.e., ahead of warm fronts and warm occlusions, and at or behind cold fronts or cold occlusions; and showery or unstable cold air-streams in which minor troughs or polar lows may be embedded. Table 9.7 suggests that warm fronts, particularly those approaching from the south and southwest, produce more frequent snowfalls than cold fronts, and that snow is more likely to occur in easterly than in northerly air-streams. Clarke's results do not apply to continental Europe, but they are useful in identifying snow-producing situations.

Snow and ice surfaces are important climatically because they have a very different effect upon the overlying air from that of either dry or moist land or deep water. The diurnal range of temperature of the actual snow surface is large, and of the order of that of dry ground, but the maximum temperature of the surface cannot rise above 0°c until the snow is removed, a process delayed by the requirement of much latent heat to be supplied from the air, from warm rain, from the soil, or by direct solar radiation. According to Lamb, the winter snow limit often advances rapidly, because of the passage of a single depression or the invasion of a cold air-mass and there then follows a period of stability or of more gradual retreat in which segments only of the snow margin are lost at any one time. The reason for the slow disappearance of the snow becomes clear from ablation studies which suggest that invasions of mild air over extensive snowfields in northern Europe in winter, typically accompanied by screen temperatures of about 3°c, normally thaw about 35 mm of snow in 24 hours. If there is appreciable rainfall, the ablation rate may increase to about 50 to 100 mm/day, and therefore the disappearance of 150 to 200 mm of snow may be expected to require either a continuous supply of mild air of several days duration, or a fall of 25 mm or more of rain. In practice, Lamb considers that in central and northern Europe, including the British Isles, warm air is·the most effective agent for removing snow cover in winter.

Perennial snows on temperate peaks are bounded by a lower line, the so-called snow-line, where as much snow melts in summer as falls in winter. Clearly it is the highest position of the 'temporary' snow-line which generally lies below it and fluctuates widely with the seasons. The snow-line coincides with neither contour nor isotherm but zigzags acutely up and down according to summer heat, radiation, precipitation, humidity, cloudiness, rock-structure, and slope aspects, etc. (Charlesworth, 1957). In the Alps, it is highest on Gran Paradiso (3,350 m) and sinks rapidly westwards (Mont Blanc group, 2,800 m), but less rapidly northwards to Monte Rosa (3,200 m) and Santis (2,504 m). The Norwegian snow-line reaches its highest point just south of Dovrefjeld and lies on Svartisen at 800–1100 m, on Folgefonn at 1,450–1,500 m, on Jostedalsbrae at 1,500–1,600 m, and on Jotunfjeld at 1,900 m. It falls to below 800 m in parts of Finnmark. Elsewhere in Europe it is above the highest summits (figure 9.8).

## Evapotranspiration and water balance

Mohrmann and Kessler (1955) have calculated the average annual potential evapotranspiration in Europe by Turc's formula, and a simplified version of their map is reproduced in figure 9.9. The chart only gives a generalized picture, and the values appear to be rather low in the north when compared with estimates made by Penman's method, but it nevertheless does show the main features of the distribution. Figure 9.10 illustrates the annual march of evapotranspiration values and water balance for a selection of western European sites which were mostly calculated by Penman's formula. Although the latter method underestimates winter values and exaggerates those for spring and early summer, it is considered to be the best method that actually exists for use with European climatological data. Solar radiation undergoes marked seasonal changes, being greatest in summer and least in winter, and since it is strongly correlated with potential evapotranspiration, this undergoes a similar variation. The result is that everywhere in western Europe precipitation exceeds evapotranspiration in winter, leading to saturated soils and surface runoff, while in summer the reverse is experienced with various degrees of soil moisture deficit. Extreme

256

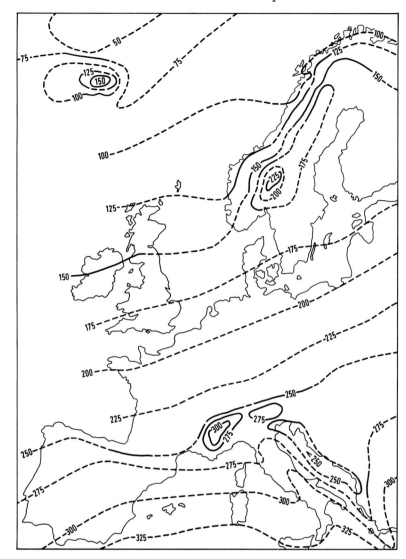

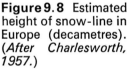

**Figure 9.8** Estimated height of snow-line in Europe (decametres). (*After Charlesworth, 1957.*)

examples are found in the Mediterranean lands, where the summers are hot and dry resulting in the soil moisture becoming completely exhausted for several months, but even in the north there is some drying of the soil in summer. In the Mediterranean lands the lack of soil moisture in the summer restricts plant growth, which mostly takes place in the comparatively cool winter. Further north, plant growth is restricted in winter by lack of radiation and low temperatures, so it takes place in summer when it may nevertheless be retarded by lack of water.

*1  Southern Europe*
Iberia may be taken as being typical of southern Europe, and water-balance diagrams for four stations are contained in figure 9.10. Potential evapotranspiration values are high in

the peninsula, reaching values of 1,300 mm per year along the coast of the Gulf of Cadiz, and at other points on the lower Ebro River (Escardo, 1970). In the south of Portugal, most of Andalucia, Estremadura, Alicante and in wide zones along the Ebro, it exceeds 1,000 mm per year. The lowest values occur in the northwest and on high ground, reflecting decreasing temperatures and radiation values. Escardo considers that in the average year the soil moisture reserves in the dry regions of the peninsula are exhausted by the end of May. This is confirmed by the water-balance diagrams for Lisbon, Madrid, and Sevilla, which indicate that soil moisture is adequate up to the end of April, but by the end of May it becomes almost exhausted, and in June the soil is dry with very low values of actual evapotranspiration. Madrid, in the centre of the peninsula, is somewhat drier than the other two stations. Almeria, in the southeast corner of Spain, has an almost desert climate,

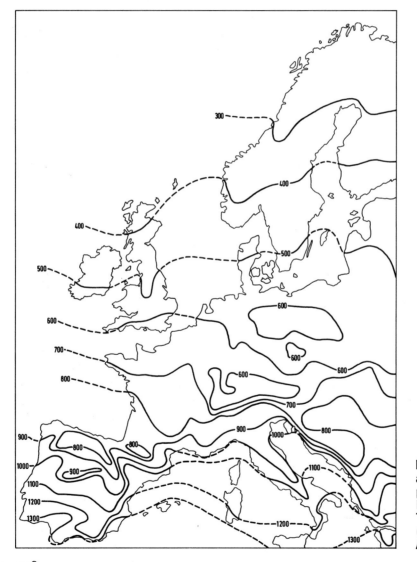

**Figure 9.9** Average annual potential evapotranspiration in Europe calculated by Turc's formula (mm). (*After Mohrmann and Kessler, 1955.*)

with water deficits throughout the year. Its climate really belongs to North Africa rather than Europe. Though plant growth is limited by a desert type climate in summer, the winter moist period is long and lasts 6–7 months at the more favourable stations. It is interesting to note that the net radiation available during the winter growing season is less than that available during summer at stations further north, for instance in Belgium. The water balance diagram for Ajaccio, Corsica, is included in figure 9.10 as an example of a station further east in the Mediterranean, but it is very similar to the Iberian stations and needs no special comment.

### 2 Northern Europe

Potential evapotranspiration decreases northwards with decreasing radiation values, so the extreme summer dry periods of the Mediterranean lands are not found in central or northern Europe, but all areas do experience a period of soil moisture deficit in summer. This is clearly seen in the example of water balance for eastern England which was discussed in chapter 1 and illustrated in figure 1.18. If Iberia is taken as representative of the near-desert southern margins of southern Europe, Scandinavia may be taken as an example of the more northerly limits.

Mean annual potential evapotranspiration over Scandinavia is illustrated in figure 9.11, which is based on the researches of Johannessen (1970). Since in winter the radiant energy available is small, evapotranspiration is negligible and most of it takes place in the summer half of the year. The highest values are found in the southeast, where the incoming radiation is greatest, and the lowest values in the north. A selection of water-balance diagrams for Swedish stations is contained in figure 9.10, which shows that all except the wettest stations experience a period of soil moisture deficit during the summer months; some even appear to be semi-arid.

Wallén (1966) has calculated the annual potential water balance (the difference between annual precipitation and annual potential evapotranspiration) for Sweden. Practically all Sweden has a positive balance, as has the rest of Scandinavia. A small annual deficit exists in the extreme southeast, of which Malmo is a typical station where the soil appears to be almost completely dry during the three summer months. The summer situation is not quite so extreme if a field capacity above 100 mm is assumed, but even so some period of moisture stress must occur in the average year.

### 3 Western Europe

A period when potential evapotranspiration exceeds rainfall appears in most of the water-balance budgets for lowland western Europe. Provided that the available moisture content of the soil is not exhausted, this is not necessarily detrimental to plant growth, since if the reverse were true the soil would be completely water-logged throughout the year. Under irrigation, for example, it is normal to maintain a small moisture deficit of perhaps about 25 mm. Even so, in drier years considerable water stresses must exist and therefore even in northern Europe lack of water can restrict plant growth over wide areas.

The low altitudinal limit to cultivation in the extreme west of Europe has been partly explained in terms of temperature, but it must also be a result of water balance. Precipitation increases rapidly with altitude on hills and mountains near the Atlantic, while potential evapotranspiration at best only decreases slightly. The result is that at quite low levels there is, on average, a water excess in every month of the year. In the average year the soil

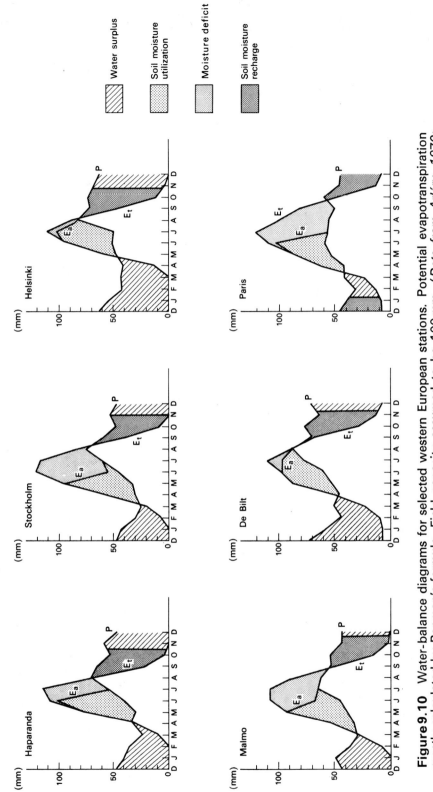

**Figure 9.10** Water-balance diagrams for selected western European stations. Potential evapotranspiration mostly calculated by Penman's formula. Field capacity assumed to be 100 mm. (*Data from Arléry, 1970; Escardo, 1970; Johannessen, 1970; and Mohrmann and Kessler, 1955.*)

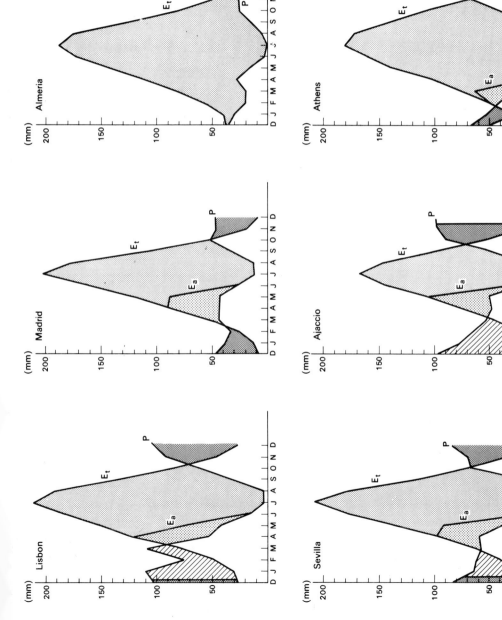

**Figure 9.10** *See opposite page.*

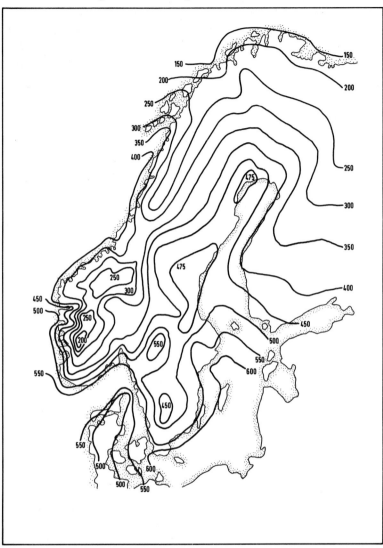

**Figure 9.11** Mean annual potential evapotranspiration from a grass surface for Scandinavia. Penman's formula used for Swedish and Norwegian stations, Turc's formula for Finnish and Danish stations. (*After Johannessen, 1970.*)

will be water-logged throughout the year, producing bogs and giving rise to mountain streams. This occurs above about 350–400 m in the Pennines. Obviously in drought years, the moors will become dry, but this is a relatively rare occurrence. Further east, where precipitation amounts are less, the level is raised; for example in southern Sweden it is at least above 500 m. In upland Britain, the average line of permanent water-logging is lower than the altitude limit to agriculture set by temperature, and consequently field drainage may improve yields on the lower slopes.

### 4 Plant growth

The importance of soil moisture and evapotranspiration to temperate agriculture is illustrated by the work of Smith (1967, 1968) on the growth of grass in England and Wales. He

262

considers that the two main meteorological factors in the growth of grass are the incident energy and the extent to which the plant is subject to soil moisture deficits, and since both influence the actual evapotranspiration, it is reasonable to use these as a meteorological parameter when analysing crop yields. Provided there is no major check to growth when the calculated soil moisture deficit is less than 50 mm in the main root zone, Smith discovered that for grass, the following relationship holds:

$$Y = 7 \cdot 07 + 1 \cdot 47 \, E_E + 0 \cdot 32 \, N$$

where Y is the yield of hay (cwt per acre) in England; $E_E$ is the effective evapotranspiration (inches) from start of growth to start of harvest, calculated assuming a field capacity of 50 mm (2 inches); and N is the nitrogen applied (cwt per 100 acres). As grass is consumed by cattle, Smith has found further correlations between the distribution in England and Wales of both cattle and dairy herds with the effective evapotranspiration.

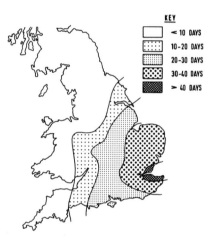

KEY

☐ < 10 DAYS
▨ 10–20 DAYS
▨ 20–30 DAYS
▨ 30–40 DAYS
▨ > 40 DAYS

**Figure 9.12** Average number of days lost to grass growth in England and Wales between April and September. A grass growing day is defined as a day when the soil moisture deficit does not exceed 50 mm. (*After Hurst and Smith, 1967.*)

Hurst and Smith (1967) have also proposed the use of a parameter known as 'grass growing days', defined as the number of days between April and September (inclusive) when the soil moisture deficit does not exceed 50 mm. The average number of days lost to grass growth from April to September in England and Wales are illustrated in figure 9.12. It is interesting that the pattern reflects that of the distribution of arable and permanent pasture.

For a given value of the net radiation, Budyko (1956) suggests that plant productivity will be greatest when the radiational index of dryness, $R_N/PL$, approaches unity. In Europe this index varies from below 0·3 in the north to over 2 in the south, but over large areas of central western Europe, it approaches unity. On this basis, the area covered by central and northern France, the southern British Isles, and southern Sweden to the Alps is a region of the world which is unusually favourable for agriculture. The areas in the south of this region probably have a slight advantage over those in the north because of higher temperatures and radiation values. This is illustrated by the distribution of regions with high yields of winter wheat which are found in Denmark, the Netherlands, eastern England, Germany, Belgium, and northern France. Yields are poor in the north, and also in the south except in the Po plain and near Valladolid.

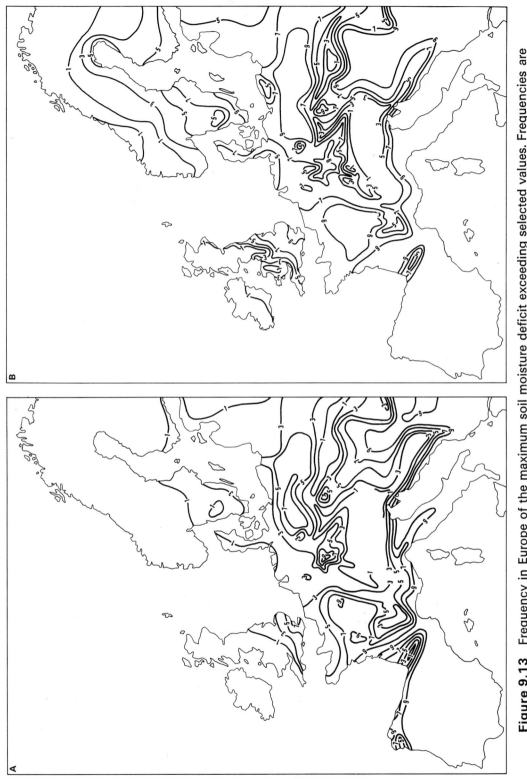

**Figure 9.13** Frequency in Europe of the maximum soil moisture deficit exceeding selected values. Frequencies are given in terms of number of years in 10 that the stated values are exceeded. **A**: Frequency of a maximum soil moisture deficit of 200 mm or more. **B**: Frequency of a maximum soil moisture deficit of 100 mm or more. (*After Mohrmann and Kessler, 1955.*)

**Table 9.8** Type of soil, depth of root zone, soil-root value

| Type of soil | Distribution | Max % available water | Depth of the profile | Soil root value (50% of the available water in the root zone) | | |
|---|---|---|---|---|---|---|
| | | | | Class I | Class II | Class III |
| Glacial deposits | | | | | | |
| loamy (including boulder clay) | Approximately the entire region north of latitude 52 and the Alpine area + margin, (regions covered by glaciers during the ice age) | 15% | 50–> 200 cm | 10–40 mm | 40–75 mm | 40–150 mm |
| sandy | | 10% | 50–> 200 cm | 5–25 mm | 25–50 mm | 25–100 mm |
| Cover sand | Small strip south of the boulder clay area, approx. between latitudes 51 and 52 | 15% | 50–> 200 cm | 10–40 mm | 40–75 mm | 40–150 mm |
| Loess (Chernozems) | Strip varying in width south of the boulder clay and cover sand region up to latitudes 45 to 50; wider in eastern Europe than in western Europe. | 20% | 50–> 200 cm | 10–50 mm | 50–100 mm | 50–200 mm |
| Mediterranean soils (e.g., terra rossa) | South of latitude 45, along the Mediterranean Sea | 20% | 25–100 cm | 10–50 mm | 25–100 mm | 25–100 mm |
| Shallow soils (including mountain ranges and degraded soils) | Coastal strip in western Scandinavia; central European mountain ranges; Alps. Apennines. Pyrenees, etc. | 17·5% | 0–45 cm | 0–45 mm | 0–45 mm | 0–45 mm |
| Alluvial and marine soils | Scattered along rivers and coasts | | | | | |
| clayey | | 20% | 50–> 200 cm | 10–50 mm | 50–100 mm | 50–200 mm |
| sandy | | 10% | 50–> 200 cm | 5–25 mm | 25–50 mm | 25–100 mm |
| Peat soils | Scattered | 25% | 50–> 150 cm | 12–60 mm | 60–125 mm | 60–180 mm |

Depth of the root zone in a deep, homogeneous and permeable soil:

Class I: shallow-rooted market-garden crops, or seedlings    10– 50 cm
Class II: grassland and moderately deep-rooted crops    50–100 cm
Class III: deep-rooted crops, including trees    100–200 cm

*After Mohrmann and Kessler (1955).*

## 5 Drought

The success of agriculture in Europe depends partly on the frequency of drought, and this in turn will depend on the frequency of large soil moisture deficits. Mohrmann and Kessler have investigated the frequency in Europe of maximum precipitation deficits of 100 mm and 200 mm or more, and simplified versions of their charts are shown in figure 9.13. Without some information on field capacities these maps are not particularly meaningful, but Mohrmann and Kessler have produced suitable estimates which are contained in table 9.8. from which it is seen that 200 mm represents the upper limit. Over most of Europe south of 45°N the soil moisture deficit exceeds 200 mm in nearly 10 years out of 10, and this partly explains the poor agricultural performance of the Mediterranean regions. Other particularly drought-prone areas emerge in central France, central Germany, and the lower part of the Thames Valley in eastern England. In contrast, the soil moisture deficit exceeds 100 mm less than once in 10 years over large parts of Ireland, Scotland, and Norway, while over much of the rest of northern Europe the corresponding values are between 3 and 7 years out of 10. Some of the best agricultural districts suffer from frequent water shortages, and Mohrmann and Kessler consider that supplemental irrigation is capable of increasing the crop yield throughout lowland Europe with the possible exception of the extreme north.

## References

ANONYMOUS. 1962: The calculation of irrigation need. 6th Ed. Great Britain, Ministry of Agriculture, *Fisheries and Food Technical Bulletin* 4.

ARLÉRY, R. 1970: The climate of France, Belgium, the Netherlands and Luxembourg. In Wallén, C. C. (editor), *Climates of northern and western Europe*. Amsterdam: Elsevier Publishing Co. p. 135.

BARRY, R. G. and CHORLEY, R. J. 1968: *Atmosphere, weather and climate*. London: Methuen.

BAUR, F. 1947a: *Einführung in die Grosswetterkunde*. Wiesbaden: Dietrich Verlag.

BAUR, F. 1947b: *Musterbeispiele europäisher Grosswetterlagen*. Wiesbaden: Dietrich Verlag.

BEAVER, S. H. and SHAW, E. M. 1970: The climate of Keele. *Keele University Library Occasional Publication* 7.

BELASCO, J. E. 1952: Characteristics of air-masses in the British Isles. *Geophysical Memoirs* 87. London: HMSO.

BERG, H. 1940: Die Kontinentalität Europas und ihre Änderung 1928/37 gegen 1888/97. *Annals Hydrogr.*, Berlin, 124.

BIERHUIZEN, J. F. 1958: Some observations on the relation between transpiration and soil moisture. *Netherlands Journal of Agricultural Science* 6, 94.

BLÜTHGEN, J. 1958: Annual precipitation map of Europe. In *Grosser Herder Atlas*. Freiburg: Troll.

BUDYKO, M. I. 1956. *The heat balance of the earth's surface*. Translated by N. I. Stepanova. Washington: US Weather Bureau.

BUDYKO, M. I. 1968: Solar radiation and the use of it by plants. In *Agroclimatological methods. Proceedings of the Reading Symposium*. Paris: UNESCO, p. 39.

CHARLESWORTH, J. K. 1957: *The quarternary era*. London: Edward Arnold.

CLARKE, P. C. 1969: Snowfalls over southeast England, 1954–1969. *Weather* 24, 438.

ESCARDO, A. L. 1970: The climate of the Iberian Peninsula. In Wallén, C. C. (editor), *Climates of northern and western Europe*. Amsterdam: Elsevier Publishing Co., p. 195.

FLOHN, H. 1954: *Witterung und Klima in Mitteleuropa*, Stuttgart: S. Hirzel.

HOGG, W. H. 1967a: The analysis of data with particular reference to frost surveys: suggestions on the expression of results in economic terms. *Proceedings Regional Training Seminar on Agrometeorology*, 13–25 May 1968, Wageningen, p. 235.

HOGG, W. H. 1967b: Meteorological factors in early crop production. *Weather* **22**, 84.

HOGG, W. H. 1968: Climate and surveys of agricultural land use. In *Agroclimatological methods. Proceedings of the Reading Symposium*. Paris: UNESCO, p. 281.

HURST, G. W. and SMITH, L. P. 1967: Grass growing days. In Taylor, J. A. (editor), *Weather and agriculture*. Oxford: Pergamon Press, p. 147.

JOHANNESSEN, T. W. 1970: The climate of Scandinavia. In Wallén, C. C. (editor), *Climates of northern and western Europe*. Amsterdam: Elsevier Publishing Co., p. 23.

LAMB, H. H. 1950: Types and spells of weather around the year in the British Isles: annual trends, seasonal structure of the year, singularities. *Quarterly Journal Royal Meteorological Society* **76**, 393.

LAMB, H. H. 1955: Two-way relationships between the snow or ice limit and 1,000–500 mb thickness in the overlying atmosphere. *Quarterly Journal Royal Meteorological Society* **81**, 172.

MANLEY, G. 1969: Snowfall in Britain over the past 300 years. *Weather* **24**, 428.

MANLEY, G. 1970: The climate of the British Isles. In Wallén, C. C. (editor), *Climates of northern and western Europe*. Amsterdam: Elsevier Publishing Co.

MOHRMANN, J. C. J. and KESSLER, J. 1955: Water deficiencies in European agriculture—A climatological survey. *International Institute for Land Reclamation and Improvement Publication* **5**, Wageningen.

MURRAY, R. 1952: Rain and snow in relation to the 1,000–700 mb and 1,000–500 mb thickness and the freezing level. *Meteorological Magazine* **81**, 5.

PUTNINS, P. 1963: An outlook on some problems of the atmospheric circulation around the southern part of Greenland. In *Studies on the Meteorology of Greenland—Second Interim Report*, 15 June 1962–15 March 1963. Sponsored by US Army Electronics Material Agency, Fort Monmouth, N.J. Washington: US Weather Bureau, p. 75.

SCHMAUSS, A. 1938: Synoptische Singularitäten. *Meteorologische Zeitschrift*, **55**, 385.

SMITH, L. P. 1960: The relation between weather and meadow-hay yields in England, 1939–1956. *Journal British Grassland Society* **15**, 203.

SMITH, L. P. 1962: Meadow-hay yields. *Outlook on Agriculture* **3**, 219.

SMITH, L. P. 1967. Meteorology and the pattern of British grassland farming. *Agricultural Meteorology* **4**, 321.

SMITH, L. P. 1968: Effective transpiration: a meteorological parameter for grassland. In Eckardt, F. E. (editor), *Functioning of terrestrial ecosystems at the primary production level. Proceedings of the Copenhagen Symposium*. Paris: UNESCO, p. 429.

TAYLOR, J. A. 1967: *Weather and agriculture*. Oxford: Pergamon Press.

TUCKER, G. B. 1961: Precipitation over the north Atlantic Ocean. *Quarterly Journal Royal Meteorological Society* **87**, 147.

TURC, L. 1953: The soil water balance: inter-relations of rainfall, evaporation and runoff. *African Soils* **III**, 139.

TURC, L. 1954: Le bilan d'eau des sols. Relations entre les précipitations, l'évaporation et l'écoulement. *Annales Agronomiques* **5**, 491 and **6**, 5.

WALLÉN, C. C. 1966: Global solar radiation and potential evapotranspiration in Sweden. *Tellus* **18**, 786.

WALLÉN, C. C. 1970: *Climates of northern and western Europe. World survey of climatology.* Vol. **5**. Amsterdam: Elsevier Publishing Co.

# 10
# Climates of the continental interiors and eastern coasts

The interiors of the continents are far removed from oceanic influence and therefore in middle latitudes experience extreme continentality of climate. The major effects of continentality, which were discussed in chapter 8, are to produce extreme seasonal temperature variations and to depress the mean annual temperature below the latitudinal average. Since the oceans are the main atmospheric moisture sources, the continental interiors also tend to be dry, allowing the subtropical deserts to extend into temperate latitudes. Because of the prevailing eastward circulation in middle latitudes, the east coast regions of the major continents experience a climate which is only a slightly modified version of that in the interior, in that the temperatures are slightly less extreme; since precipitation increases from the interior eastwards, they are also rather more moist, but can be discussed together with the interior climates.

## Atmospheric circulation

Since the main examples of extreme continental climates only occur in North America and Asia, discussion is restricted to these continents, although a minor area with a similar climate does exist to the east of the Andes in southern South America. The particular climate of central Asia arises because of the extreme distance from the sea, for according to Borisov (1965), maritime air from Europe frequently reaches the 40th meridian in the USSR but very rarely the 60th. In North America, the mountain ranges along the west coast curtail the penetration of moist Pacific air-masses inland and have the same drying effect as the whole of Europe has on air-masses entering Asia.

### *1  Seasonal variation*
In many ways the climates of interior and eastern North America are similar to those found in interior and eastern Asia, but there are important differences. In winter, the interiors of both continents are dominated by large anticyclones, which are normally of the cold or polar continental type. The most intense is over Asia and is located on average to the east of the 85°E meridian, that is in an area over 95° longitude from the Atlantic Ocean. The anticyclone over North America is a weak and unstable feature which is to be interpreted as a statistical average rather than as a semi-permanent and quasi-stationary system. Both systems act as the sources of continental polar air, which is intensely cold and dry. In summer, the continental anticyclones vanish and are replaced by shallow lows.

Intense thermal troughs are located in winter along the eastern coasts of both North America and Asia, while thermal ridges are found over western North America and western Europe (see chapter 8). Although the ridge over western North America directly influences the weather over the interior of that continent, a similar relationship does not exist between the ridge over western Europe and central Asia. Examination of upper air charts reveal in winter a further trough-ridge pair of small amplitude located between the main ridge over western Europe and the east Asia trough. The secondary trough runs approximately from the White Sea to the Black Sea while the ridge has its axis about 85°E, just to the west of the surface anticyclone. The east coast troughs may be correlated with the extreme winter cold of these climates, and it is interesting that in Siberia the lowest winter temperatures are to the east of the weak upper ridge.

The monsoon wind system of southern Asia introduces further differences between the two continents. In winter the subtropical branch of the jet stream flows around the southern slopes of the Himalayas, while the polar-front jet occurs to the north of the plateau of Tibet, and they both meet in an upper convergence zone over China. This upper convergence zone leads to subsidence over inner China and therefore to generally fine weather in winter. The two jet streams tend to reinforce each other over Japan, creating extremely high wind speeds which can reach 100 m/sec at the 200 mb level.

Summer circulation patterns are less complex because the upper winds are lighter and the upper troughs are not so pronounced. In Asia, the subtropical jet stream flows entirely to the north of Tibet and the upper convergence over China has disappeared.

## 2 Contrasting air-masses

The interiors of both continents are open to invasion by currents from arctic and subtropical regions, which can travel far without much modification, thus providing strongly contrasted air-masses. Canadian meteorologists (Anderson, Boville and McClellan, 1955; Penner, 1955) divide the atmosphere over North America into four climatological zones, separated by three climatological fronts:

1  the polar front separating tropical maritime air from polar maritime;
2  the maritime arctic front separating polar maritime air from arctic maritime;
3  the continental arctic front separating arctic maritime air from arctic continental.

In winter the continental arctic front is located in the mean over northwest Canada extending eastward to Maine, while the polar front is off the east coast of the United States. In summer the continental arctic front is usually located along the north coast of Alaska, from where it dips southeastward across Keewatin. It is suggested by Barry (1967) and Bryson (1966) that these frontal positions may have some ecological implications in that the summer and winter positions of the arctic front correspond respectively with the northern and southern limits of boreal forest in central Canada. Russian meteorologists report similar correlations (Borisov), because larch and fir do not apparently spread westwards beyond the mean track of ridges of high pressure which travel southeastward from the North Cape in summer. The northern boundary of forest-steppe also coincides with the extension of ridges from Siberia. In these examples the dynamic processes in the atmosphere are reflected by geobotanical boundaries because of the effect that these processes have on thermal and precipitation conditions.

## 3    Tornadoes

Decaying tropical storms, which were discussed in chapter 5, cross the eastern coasts of both North America and Asia and cause heavy rainfall. Intense tropical storms rarely penetrate far into the continental interiors and need not be considered further. The interiors of the continents are, however, subject to extreme and destructive storms, the most common of which is the tornado. The tornado is a small vortex of about 100 m in diameter with winds revolving tightly around the core at up to 50–80 m/sec; it is associated with destructive hail and the pressure drop in the centre amounts to 0·1 to 0·3 atmospheres. Evidence from damage suggests that the destruction of buildings arises as much from the explosion, when the pressure outside drops suddenly, as from the violent wind. Tornadoes normally travel along paths with lengths ranging from very short distances to 80 km, but the width of the area sustaining damage may only be 100 to 200 m, giving the impression that an enormous vacuum cleaner has swept the ground clean of vegetation, loose soil, and all other movable objects.

The midwestern and western plains' states of the USA are visited by more severe tornadoes than any other area in the world, and they are most frequent during the spring and summer months. Since the physical dimensions of the storms are so small, the actual probability of a tornado passing over any particular point is exceedingly small (Thom, 1963). Tornadoes can form under a variety of conditions, but they seem to require a shallow layer of conditionally unstable moist air separated from dry air above by an inversion layer at about 1,500–2,000 m. In addition, a crucial aspect of the wind structure is the occurrence of a low-level southerly 'jet stream' in the humid air, overlain by a westerly jet stream above the inversion. It is suggested that, under these conditions, local convective columns penetrate the inversion layer allowing converging air beneath to funnel violently up through the break.

The climates discussed in this chapter are full of extremes and probably exhibit more violent variations than any other climate in the world. In winter great blizzards blow causing the temperature to fall to arctic levels while in summer temperatures can be subtropical with the added hazard of violent thunderstorms. Some of the more important environmental aspects are developed later in this chapter, when the full potential for catastrophe, particularly in the interior, will become more clear.

## Radiation and temperature

### *1    Radiation*

The distribution of radiation over the middle latitude continents is illustrated in figures 2.2 to 2.7, and the seasonal variation is summarized in table 2.1. Incoming radiation values can vary widely between winter and summer even in temperate latitudes, and in the continental interiors where temperature is more closely correlated with radiation, they produce great seasonal variations of temperature which can exceed 45°C.

In both North America and central Asia, the smallest values for the intensity of global radiation are in December or early January, the areas north of about 67°N receiving no solar radiation at this time. The time of maximum global radiation depends partly on the seasonal variation of the dust and water vapour content of the atmosphere; for instance, it is reached in March in Tashkent, May in Moscow, and in July in the Bay of Tikhaya (Borisov, 1965).

*Plate 12* A sequence of photographs showing the development of a tornado in the central USA. *Reproduced by permission of the United States Department of Commerce, National Oceanic and Atmospheric Administration.*

*Albedo* Surface albedo is of importance in that it determines the amount of global radiation which is absorbed by the vegetation and soil, and in middle latitudes it is strongly dependent on the extent of the snow cover. Kung, Bryson, and Lenschow (1964) have investigated the seasonal distribution of the surface albedo over the North American continent and their results are summarized in table 10.1. In the subtropics, south of 35°N, it changes only slightly from winter to summer but in higher latitudes it can fluctuate from about 50 per cent in winter to 15 per cent in summer. The winter surface albedo is related to the extent of snow cover, and between 40° and 45°N the energy absorbed by the land surface in a mild winter may exceed that absorbed in a cold winter by more than 60 per cent. Similar conditions apply to central Asia, where the albedo varies from 70 per cent in winter to less than 20 per cent in summer, but becomes more seasonally uniform over the deserts to the south.

The average annual radiation balance of the surface is positive in both North America

**Table 10.1**  Zonal and continental means of surface albedo over North America

| Latitudinal zone (°N) | Continental surface albedo | | | | |
| --- | --- | --- | --- | --- | --- |
| | Winter of mean snow depth | Winter of max snow depth | Winter of min snow depth | Transitional seasons | Summer |
| 70–65 | 82·8 | 82·8 | 82·7 | 82·8 | 16·1 |
| 65–60 | 67·3 | 67·3 | 58·2 | 67·3 | 15·6 |
| 60–55 | 59·1 | 59·1 | 54·8 | 57·7 | 16·5 |
| 55–50 | 50·3 | 50·3 | 45·8 | 48·0 | 14·6 |
| 50–45 | 46·4 | 48·9 | 28·4 | 37·6 | 14·8 |
| 45–40 | 37·9 | 50·4 | 19·0 | 30·5 | 15·8 |
| 40–35 | 28·5 | 40·8 | 16·0 | 21·1 | 16·5 |
| 35–30 | 19·1 | 26·2 | 16·9 | 17·4 | 17·2 |
| 30–25 | 17·8 | 17·8 | 17·8 | 17·9 | 17·9 |
| 25–20 | 15·8 | 15·8 | 15·8 | 15·8 | 15·8 |

*After Kung, Bryson, and Lenschow (1964).*

and Asia, except in the extreme north (figure 2.4). In both cases, the values are less than over the oceans at similar latitudes, and north of 50°N show a slight tendency to decrease towards the east. In general, values over interior North America are similar to those over Asiatic Russia, while the southeastern half of the USA corresponds to China.

*2  Winter temperatures*
Temperatures are closely correlated with the radiation climate, and the seasonal changes in albedo together with the marked continentality produce extreme temperatures in both Asia and North America. In January, almost the whole of Canada, the north interior, and northeast of the United States have mean temperatures below 0°C, the 10°C isotherm not being reached until just short of the Gulf of Mexico. Similarly, all Asia north of about 40°N has mean January temperatures below 0°C, and in China the 0°C isotherm reaches 33°N, nearer to the equator than anywhere else in the world. The lowest January temperatures in Asia are observed in the vicinity of Verkhoyansk and Oimyakon, in eastern Siberia, where they average −50°C, but North American temperatures are not so extreme and only reach about −35°C on the northern coasts of Canada. Winter surface radiation balances are largely negative over both continents, so unless temperatures are to fall to very low values the energy loss to space must be replaced by atmospheric heat advection. Borisov states that air-masses from the Atlantic import large quantities of heat to the USSR and he considers that their influence is appreciable as far as the Urals and still noticeable on the River Yenisei. As a result, the winter isotherms in the USSR have a meridional rather than a zonal trend, with the lowest temperatures in the east. Therefore, the very low temperatures in eastern Siberia can be related to the mean ridge of high pressure over the area which causes stagnation of the air together with intense cooling under cloudless skies. In North America, the passage of warm air-masses inland is restricted by the Rockies and other coastal ranges, so that extreme continental conditions are found in the plains just to the east of the mountains. The gradient of mean January temperature is over 10 times steeper in western Canada than that in western Europe.

*Cold winds* The interiors of the two continents act as the winter source regions of polar continental air which flows over neighbouring land areas as an intensely cold air current. The air outflowing from the continental anticyclones is very shallow and usually only extends 1,000–2,000 m above ground; its advance is greatly affected by topography and it is rapidly heated over warm ocean surfaces. Cold continental air is steered away from the western coasts of North America by the mountain ranges but on average 3 or 4 outbreaks a year will reach the southern coasts causing temperatures to fall as low as −12°C at Galveston and New Orleans. The drop in temperature with the arrival of the cold northern air is often dramatic; for instance at Houston the temperature has fallen from 24°C to −5°C in 9 hours. Because of dynamic considerations (see chapter 3) air flow towards the equator tends to subside and diverge; the cold air decreases in depth as it moves south, and this makes it more liable to surface warming. Over the cold winter land the amount of warming is only slight, but over the warm waters of the Gulf of Mexico it is rapid, and the off-shore islands rarely experience temperatures below 0°C; Key West, for example, has never recorded a frost, its lowest temperature being 5°C. In a similar manner, cold waves sweep across north China and Korea at intervals of about one week throughout the winter. The Chinling and Nanling Hills frequently delay the advance of the cold air and in places it is restricted to the valleys. As in the USA, the arrival of the cold air can cause sudden falls in temperature of up to 10° or 20°C, bringing surprisingly low temperatures; −14°C has been quoted at Nanking, to give one instance. The colder water off the northeastern coasts of the continents is less effective in warming easterly winds; January mean temperatures in Newfoundland range from −1°C to −7°C, while those in Japan are between 8°C and −8°C.

## 3   Summer temperatures

July is a month when the surface radiation balance is positive over nearly the whole of the northern hemisphere and this is reflected by a more nearly zonal trend in the isotherms over central Asia and North America, where the mean temperature is above 10°C everywhere except in the Arctic and mountainous regions. The positive surface radiation balance also leads to relatively uniform temperatures over the interiors of the two continents. Cool water along the coasts of Labrador and Newfoundland cause the July mean isotherms to dip southward, and a similar trend is also observed along the eastern coast of Asia. Extreme summer temperatures of 30°C, are recorded even in the north of the USSR and similar temperatures are observed in Canada, where the extreme maximum temperature at Winnipeg has exceeded 40°C.

Extreme continentality within the cores of the great land-masses causes the air temperature to follow very closely the annual march of solar radiation. Thus in both central and eastern Asia and continental North America, the transitions between the winter and the summer temperature distributions take place rapidly in April and September. As in western Europe, the temperature curves are not smooth but contain evidence of singularities. Wahl (1954) has studied a temperature singularity common to central North America occurring about 15 October, when the frequency of cold days and snow probability suddenly increase. Similarly, Wahl (1952) has found a marked increase in mean temperature for the 3-day period 21–23 January along the east coast of the USA between Boston and Atlanta.

## 4  Land and water modifications

Temperatures are further modified by local topographical features such as areas of open water and mountain ranges. The influence of the Great Lakes in North America is well known: they cause a slight northward swing in the isotherms in October, which becomes more pronounced until January when it diminishes, and in March they cease to provide warmth. Similarly, Hudson Bay remains unfrozen until about January, so that its eastern land margins are distinctly less cold in autumn and early winter than are the opposite western margins. Lake Baikal, which is 1,500 m deep, remains ice-free until the end of December and then freezes until June. In summer, its temperature is 10°c cooler and in the early winter months, 14°c warmer than that of its surroundings. Unusual winter warmth also exists at times to the east of the Rockies, and is the result of warm dry winds with foehn characteristics which are known locally as chinooks. They are most pronounced along the eastern edge of the Rockies, where they melt the winter snow and make some grazing possible throughout most winters. The North American chinook has been described in some detail by Brinkmann (1971) and Brinkmann and Ashwell (1968). Foehn winds are also reported from the USSR where they are known by various local names. In many ways their influence on the climate is similar to that of the chinook.

## 5  Permafrost

The intense winter cold has a marked influence upon soil temperatures in the northern parts of the continents. In January and February the temperature at a depth of 20 cm is below 0°c throughout the USSR apart from the southern Crimea, west Transcaucasia, and locally in forests and swampy tundra areas (Borisov, 1965). At this depth the minimum temperature is reached in February over large parts of the USSR and is close to the air minimum temperature; similarly, the maximum temperature is reached in July and over large areas varies between 15°c and 20°c. The intense winter cold can result in the soil becoming cemented by frozen water, and this is known as permafrost if the soil layers remain in this state throughout the year. The distribution of permafrost (figure 10.1) shows that it occurs in Alaska and over much of Canada, but its greatest extent is in Siberia where it covers about 75 per cent of the Asiatic USSR. In winter the upper limit is usually the soil surface, but in summer the upper layers of the soil thaw forming a moist layer which overlies the permanently frozen ground below. Charlesworth (1957) has suggested that the depth to which permanfrost extends is related to the mean annual air temperature as follows: 1–2°c, 20 m; 0–1°c, 50 m; below 0°c, more than 80 m. However, experience in Asia indicates that it occurs mainly where the snow cover does not reach great depths. There are two theories on its origin: the first, that it is a relic of the glacial period, and the second, that it is the result of the present climate. Buried blocks of ice and very deep permafrost layers which are found in places are obviously not the result of the present climate, but nevertheless, Charlesworth (1957) considers that most permafrost is stable and corresponds to present conditions. Permafrost must be considered when constructing foundations for buildings and roads, and also when installing service pipes in the ground. It affects the vegetation by its influence on drainage, by repressing all deep-rooted species, and by limiting growth to only those which have shallow roots.

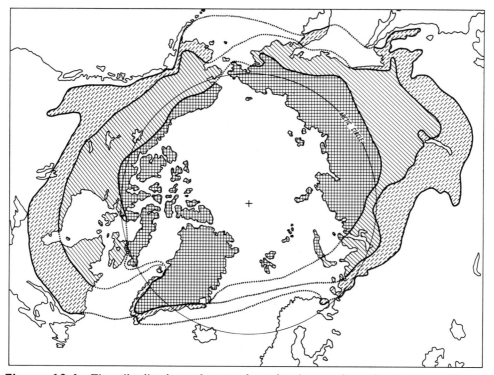

**Figure 10.1** The distribution of permafrost in the northern hemisphere. Cross-hatching, continuous permafrost; oblique shading, discontinuous permafrost; broken lines, zone with islands of permafrost. (*After Charlesworth, 1957.*)

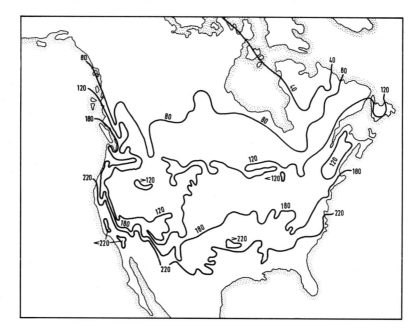

**Figure 10.2** Mean annual number of frost-free days in North America. (*After Kendrew, 1961.*)

## 6 Frost

Agriculture is not only checked by the extreme winter cold but also by a general liability to severe frosts in late spring and early autumn. The mean annual number of days without frost in North America is shown in figure 10.2. The most favoured areas are in the south, but even here severe killing frosts occur every winter. The frost-free period decreases rapidly towards the north, being between 120 and 80 days in the Canadian prairies and only 40 days in the extreme north. North of 50°N in the prairies, frost may appear even in summer, but the severity decreases at higher latitudes because of the longer days. A comparable situation exists in Asia where over most of Siberia the number of frost-free days ranges between 95 and 115, but they increase to about 245 just north of the Caspian Sea.

## 7 Plant growth

Since plants only grow in the summer months, the radiation and temperature of these months are particularly important. The radiation in the wavelength range 0·4–0·7 $\mu$m is photosynthetically active in plants, and values for the vegetation growth season have been charted over the USSR by Efimova (1965); a simplified version of her chart is shown in figure 10.3. Except in the east, the distribution shown is basically zonal, and this is reflected in the zonal distribution of vegetation in northern Asia. In western Asia a situation exists which is similar to that in Europe, because in the north plant growth is restricted by low temperatures and lack of radiant energy, while further south it is restrained by a shortage of water. The sequence of natural vegetation in both western Asia and central North America is therefore from tundra in the extreme north, through coniferous forests (taiga), deciduous mixed forest, grasslands (prairie, steppe), to semi-desert. Further east where moisture is more readily available the grasslands and semi-deserts are missing.

*a  Tundra vegetation*  Tundra vegetation is found in the extreme north of both Asia and North America in regions where there are long periods of winter darkness and the photo-synthetically active radiation is of the order 15 k cal/cm. The long summer days provide an almost continuous flow of solar energy which allows rapid development of the tundra vegetation in the short growing season of 2–3 months; at this time the intensity of solar radiation can be greater than in areas further south. Average monthly summer temperatures reach 10°–12°c, but in winter they can fall to −11° and lower (to −30°c). The poor climate is reflected in the net primary production of the vegetation (annual production of leaves and organs of grass, as well as of wood and roots), which is estimated by Rodin and Basilevič (1968) to be only 1–2·5 tons per hectare.

*b  Forests*  Coniferous forest occupies large areas of both the USSR and Canada, and in an average growing season receives 20–25 k cal/cm of photosynthetically active radiation, which is just under twice that received in the tundra. The growing season is short, for there are only 120 to 130 days which have an average daily temperature above 5°c, but mean monthly temperatures in the summer can be high ranging from 15° to 20°c. The climate tends to be extremely continental, with winter temperatures often below those further north but these are compensated by the higher summer temperatures. The extreme cold at Oimyakon, in eastern Siberia, has already been mentioned, for here temperatures can fall

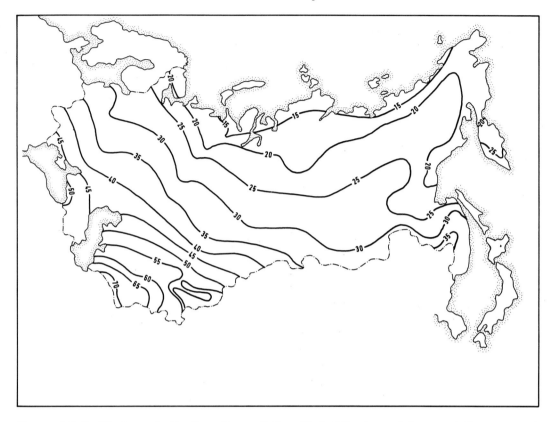

**Figure 10.3** Photosynthetically active radiation (0·4–0·7 microns) for vegetation season in the USSR (k cal cm⁻²). (*After Efimova, 1965 and Budyko, 1968.*)

to −70°c, while at Yakutsk the temperature may reach 39°c in summer. Rodin and Basilevič estimate that the net primary production in coniferous forests range from 4·5–8·5 tons per hectare, which is twice to four times that in the tundra.

The climate of the mixed deciduous forests differs from that of the coniferous forests by being less continental, for here winters are less severe, the summers are warmer, and the incoming radiation is slightly greater than in the coniferous zone. Since monthly mean temperatures vary between −4° and −20°c in the winter, this climatic type is often found along the eastern edges of the continents where the ocean has a slight modifying influence. Net primary production in these forests is only slightly greater than in the southern fir forests of the taiga.

*c  Grasslands*  The hearts of both Asia and North America contain extensive grasslands known as steppes or prairies, where the growth is restricted by a lack of moisture. Incoming radiation values in these areas are higher than in the forest regions to the north, and the resulting increased evapotranspiration partly explains the water shortage. In the Russian steppes the photosynthetically active radiation during the growing season varies between

40 and 45 k cal/cm. The climate is very continental with winter monthly mean temperatures between 0°c and −20°c and summer monthly mean temperatures exceeding 20°c, the temperature changes in spring being particularly rapid. Extreme winter temperatures rarely fall below −50°c and in summer the temperature can reach 40°c. Both the steppes and the prairies are exposed to outbreaks of very cold air from the north, but nevertheless winter temperatures can be very mild, particularly in the south, leading to very changeable conditions and an unstable snow cover. In the south the grasslands gradually merge into areas of semi-desert where the winter months may still be below 0°c but the summer temperatures may rival those of the subtropical deserts. Because of moisture limitations the biomass is restricted in steppe and prairie regions, but the growing season is long—often exceeding 6 months. The incoming radiation is high, and therefore the net primary production is surprisingly large, often exceeding 11 tons per hectare which is greater than that reached in coniferous and some broad-leaved forests.

*d Theoretical yields* Ničiporovič (1968) has estimated, for optimal conditions (i.e., appropriate plant varieties, optimal supply of mineral nutrients and water), the potential photosynthetic production in terms of dry biomass at various latitudes assuming the appropriate photosynthetically active radiation during the possible growing season. His

**Table 10.2** The length of the vegetative period, amounts of incoming PAR solar energy (C), calculated theoretically possible maximal (B) and real record yields (D) of various plants (E), grown in different north latitudes

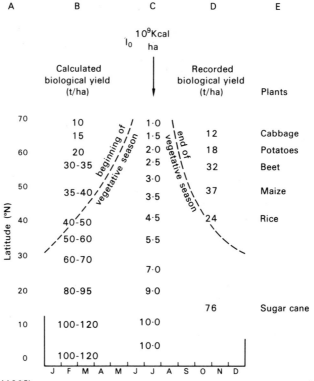

*After A. A. Ničiporovič (1968).*

data are presented in table 10.2, which also contains data on maximum biological yields recorded. It can be seen that the latter are close to the theoretical yields, especially in high latitudes. However, Ničiporovič comments that average yields are considerably lower than those given as theoretically possible, and he considers that this fact clearly shows the vast possibilities of increasing the photosynthetic productivity of plants. The differences can also be accounted for in terms of insufficient supply of mineral nutrients and water to the crops, but nevertheless it does appear that some widely grown crops do not make good use of the radiation available during the growing season. For example, Ničiporovič has found that some wheat crops make good use of radiation in the first half of the growing season, but not in the second. In contrast, he found that maize makes poor use of the first half of the period, and in the vicinity of Moscow does not finish its growth cycle before the end of the growing season. Obviously, in these cases the yields could be improved by breeding new varieties, or by growing more suitable crops.

## Precipitation

A generalized picture of mean annual precipitation over North America and Asia is contained in figure 2.16. In both the continental interiors precipitation totals are low, mainly because of distance from the oceanic moisture sources (see chapter 8 for greater detail). In the United States the region between the Rockies and 100°w has annual means between 300 and 500 mm, while vast areas of central Asia receive less than 500 mm, but in both cases amounts increase towards the east coasts where they can exceed 1,000 mm. In central Asia precipitation falls from a maximum of 400 to 500 mm per year in the central regions, to below 300 mm in both the arctic north and the semi-deserts of the south. A similar distribution is observed in North America, where the Canadian arctic receives only about 300 mm per year and totals decline from the Great Plains towards the arid southwest of the United States. It is interesting that precipitation amounts are particularly low in Siberia near Verkhoyansk and Oimyakon (which have already been mentioned in relation to extremely low winter temperatures). This is because of the general lack of moist air advection from all directions (which is part of the explanation for the low winter temperatures), and also because of the frequent inversions and intense cold in winter.

### *1  Seasonal variation*

The seasonal variation of precipitation in the interior and eastern parts of both North America and Asia can be complex, but for a variety of reasons there is normally a summer maximum. In a large area of interior North America, extending from the south of the USA to the forests of the north of Canada, the early summer months have the most precipitation, with a pronounced maximum in June. In the central area, including Alberta, Saskatchewan, most of Manitoba, and in the United States the plains west of the line Duluth (Minnesota)–El Paso (Texas), more than half is in the months of May to August, according to Kendrew (1961). In the south of Alberta, and in North and South Dakota, 70 per cent or more of the annual rainfall is received in the summer half-year. Further north in the north and northeast of Canada, the summer maximum is delayed until August or later, autumn having much more precipitation than spring. Much of the summer rainfall in these interior regions is caused by thunderstorms and instability showers. Most of the eastern half of the USA experiences a late summer maximum, due to hurricanes or thunderstorms, while eastern Canada has a more uniform annual rainfall or even a winter maximum in places. Similar precipitation patterns, with summer maximums, are found over extensive areas of central and eastern Asia.

Since northern China and Japan occupy similar positions in Asia to the eastern United States in North America, it is instructive to compare their respective precipitation patterns. In the eastern United States south of about Jacksonville (Florida), two maximums of precipitation appear: the first in July is due to local thunderstorms, and the second one in September and October is due to hurricanes. Further north, in the Wilmington (North Carolina) area, only one maximum is observed in July, and this arises from thunderstorm activity. In higher latitudes near Washington or Boston the precipitation is brought mostly by frontal activity and is evenly distributed without clear maximums. Over north China, the seasonal variation of precipitation is characterized by a pronounced summer maximum and a comparatively dry winter. Westward of the Tasueh Shan mountains 75 to 95 per cent of the annual total is recorded during May to September, and in Hopei 90 per cent of the annual total occurs within this period (Watts, 1969). The seasonal variation of precipitation in north China is similar to that over a large part of Japan, where according to Arakawa and Taga (1969) six seasons can be recognized: winter, spring, Bai-u, summer, shurin, and autumn. In much of the Yangtze Valley and over Japan, there is a season of prolonged rainfall between spring and summer, which is known as the Mai-Yu in China and the Bai-u in Japan. It is caused by shallow depressions which advance slowly northeast to Japan from central China (Kendrew, 1961). These are associated with the movement northwards of the subtropical jet stream in spring, and this movement in turn is connected with large-scale atmospheric changes taking place over north India and the Himalayas. Reiter (1961) has observed that the onset of the Mai-Yu rains coincides with the start of the southwest monsoon in India. The causes of the autumn rainy season, known as Shurin in Japanese, are complicated; it results partly from decaying typhoons and partly from the disturbances connected with the southward passage of the subtropical jet stream. The seasonal precipitation patterns over eastern Asia are therefore more complex than those in eastern North America because of the influence of large-scale circulation changes associated with the Asian monsoon.

## 2 Snow

Both interior North America and Asia receive much of their winter precipitation in the form of snow. This is illustrated for the USSR by figure 10.4 which indicates the mean depth of snow cover together with its duration. For the European USSR, the maximum depth is observed in the central Urals where it reaches 90 to 100 cm. In western Siberia the snow depth varies from 40 to 90 cm, but further east in Kamchatka it reaches 100 cm. The depth of snow is not only related to the distance from oceanic moisture sources but also to the air temperature, because in the warm regions some of the winter precipitation will fall as rain. Chebotarev (1962) states that the average maximum water equivalent of the snow in the European USSR reaches the following magnitudes at the end of winter: in the northeast, 180 to 200 mm, in the central regions, 160 mm, and in the south, 30–60 mm.

Usually the establishment of a firm snow cover in the USSR is preceded by a pre-winter period during which it is unstable and appears only to disappear at a later date. Borisov considers that the pre-winter season becomes shorter with increasing continentality. Thus in the west of the USSR the number of temporary snow covers increases from north to south, but towards the centre of the country the reverse occurs, and there is an increase from south to north. The snow cover is unstable throughout the southern part of the USSR and can almost completely disappear in winter thaws. The steppes and deserts are characterized by a snow cover duration varying from 20 to 120 days. This compares with a snow cover of 160 to 260 days in the forested areas of the Asiatic USSR and 200 to 260 days in the tundra. The final melting of the snows in spring takes up to 50 days, depending on the state of the snow and the local geographical conditions. Winter snow conditions are very similar in North America, where apart from large falls in the western mountains, snow depths increase from the interior towards the east coast.

### 3   Intensity

The intensity of precipitation varies with latitude in both North America and Asia, reflecting the temperature and moisture content of the atmosphere. In the USSR the mean intensity varies from 2–5 mm per hour in the north to 8–10 mm per hour at 51°N, while daily precipitation intensities also show a similar trend. In the USA extreme precipitation values decrease northwards from the Gulf of Mexico (figure 10.5). In both interior USA and Asia, heavy showers and thunderstorms are frequent in the summer months, often causing flooding or damage from hail. In both countries these storms decrease in frequency from north to south, but hail causes significant damage as far north as the Canadian prairies.

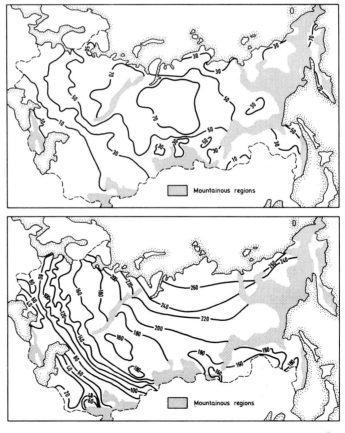

**Figure 10.4** Snow in the USSR. *Upper:* mean depth of snow cover in cm. *Lower:* duration of snow cover in days. (*After Rikhter, 1948 and Borisov, 1965.*)

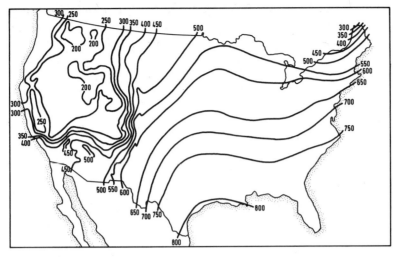

**Figure 10.5** Probable maximum precipitation for a duration of 6 hrs over an area of 10 sq miles in the USA (mm). (*After US Weather Bureau, 1956, 1960.*)

*4    Hailfall*

Hailfall in central Alberta has been studied by Summers and Paul (1967). The hail season in Alberta lasts from May to September, and as shown in a selection of statistics in table 10.3, the frequency of hail increases rapidly at the start of the season reaching a peak in

**Table 10.3    Hail days in central Alberta**

|  | May | June | July | August | September |
|---|---|---|---|---|---|
| Average number of hail days | 5·3 | 20·3 | 20·3 | 15·7 | 4·9 |
| Probability of a hail day (per cent) | 17 | 68 | 66 | 51 | 16 |
| Mean hail onset time (MST) | 1527 | 1607 | 1655 | 1612 | 1517 |

*After Summers and Paul (1967).*

June. It then gradually falls off through July and August and is effectively over by mid-September. Cumulonimbus clouds show distinct diurnal variations and in Alberta, hail is most frequent between 1100 and 2000 local time. The reason for hail damage is clear from table 10.4 which lists the percentage frequency of maximum hail size. In 6 per cent of the

**Table 10.4    Percentage frequency of maximum hail size in central Alberta (1957–66)**

| 1–4 mm | 4–12 mm | 12–20 mm | 20–32 mm | 32–50 mm | 50 mm |
|---|---|---|---|---|---|
| 9·4 | 38·7 | 33·7 | 12·2 | 4·7 | 1·3 |

*After Summers and Paul (1967).*

storms the maximum hail size was greater than 32 mm. Fortunately the duration of hailfall is short (table 10.5), and most storms last for less than 12 minutes.

**Table 10.5    Hailfall duration in central Alberta**

| Duration (minutes | ⩽6 | 7–12 | 13–18 | 19–24 | ⩾25 |
|---|---|---|---|---|---|
| Frequency (per cent) | 47 | 26 | 13 | 7 | 7 |

*After Summers and Paul (1967).*

*Plate 13*  Hail storm in the USA. Top photographs show typical cumulonimbus clouds while lower right illustrates a hail storm.  *Below left:* a cross-section through a hailstone. Hail can be extremely destructive in parts of Canada and the USA. *Reproduced by permission of the United States Department of Commerce, National Oceanic and Atmospheric Administration.*

## Evapotranspiration and Water Balance

*1 Distribution*

Charts showing the distribution of potential evapotranspiration in central USA, Canada, the USSR, and China, are contained in figures 10.6 to 10.9. The chart for the central and eastern USA is based on the classical study by Thornthwaite (1948) on the rational classification of climate. In the USA annual values range from about 1,350 mm along the Gulf of Mexico to 500 mm along the Canadian border, while in Canada they fall to only 250 mm in the far north. A similar range of values is observed in China, whereas in the arctic regions of the USSR they fall to below 125 mm per year. Everywhere the distribution pattern of potential evapotranspiration tends towards the zonal, illustrating the important correlation with summer net radiation.

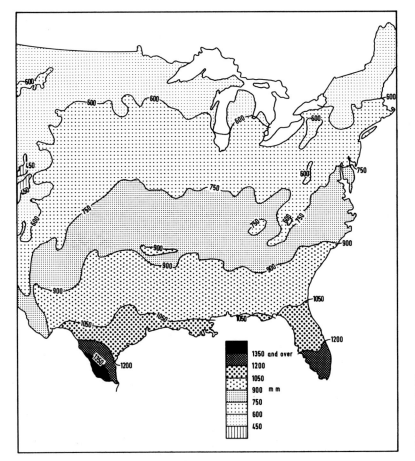

**Figure 10.6** Mean annual potential evapotranspiration for central and eastern USA (mm). (*After Thornthwaite, 1948.*)

*2 Seasonal variation*

The annual march of potential evapotranspiration is similar in most localities, since it is negligible in the winter months and reaches a maximum in June or July. The interesting environmental parameter is the water balance, which is the result of the interaction of

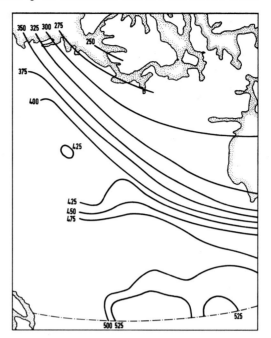

**Figure 10.7** Mean annual potential evapo-transpiration for central Canada (mm). Calculations by Thornthwaite's method. (*After Sanderson, 1948.*)

potential evapotranspiration with precipitation, and this is illustrated for North America. Edmonton (Alberta) and Manhattan (Kansas) can be taken as typical of the continental interior, since they both illustrate widely-occurring features. At both localities there is a summer maximum in precipitation, but this is exceeded by the potential evapotranspiration, leading to a soil water deficit which normally has become excessive by the end of the summer. Since winter precipitation is slight, the soil moisture is often not replaced until

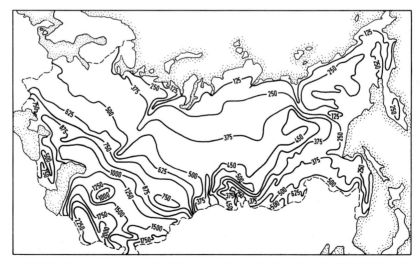

**Figure 10.8** Mean annual potential evapotranspiration for the USSR (mm). (*After Borisov, 1965.*)

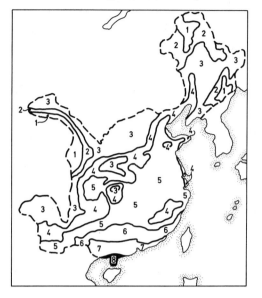

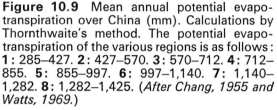

**Figure 10.9** Mean annual potential evapotranspiration over China (mm). Calculations by Thornthwaite's method. The potential evapotranspiration of the various regions is as follows: **1**: 285–427. **2**: 427–570. **3**: 570–712. **4**: 712–855. **5**: 855–997. **6**: 997–1,140. **7**: 1,140–1,282. **8**: 1,282–1,425. (*After Chang, 1955 and Watts, 1969.*)

early spring. Further east, the climate is less arid and at Willard (North Carolina) the potential evapotranspiration and the precipitation are nearly the same in all the growing-season months, while there is a massive water surplus in the winter. As the potential evapotranspiration values at Manhattan and Willard are not very different, in this case it is the increase in precipitation towards the east coast which makes the climate more humid. A period when evapotranspiration exceeds precipitation is normal at most stations, and is even found as far north as Fort Simpson (North West Territories).

Rasmusson (1971) has recently studied the hydrology of eastern North America using atmospheric water vapour flux data. He finds that the Thornthwaite's climatic water-balance data appear to overestimate the difference between precipitation and evapotranspiration during winter and underestimate it during summer. This paper should be consulted for further details of water-balance studies in the eastern USA.

### 3 *Moisture regions*

In the United States, Thornthwaite distinguishes between the dry climates of the interior, where there is a prevailing water deficit, and the moist climates of the east, where there is a water surplus. They are divided by his moisture index line of value 0 (figure 10.10) which extends from western Minnesota southward to the Gulf of Mexico, following the 96th meridian almost exactly. In Canada this particular isopleth runs from near Winnipeg northeastwards keeping just to the west of the Great Slave and Great Bear Lakes. Thornthwaite's moisture regions show some general relationship to natural vegetation regions; for example, in southern Canada the zero isopleth separates coniferous forest from deciduous mixed forest, and there is some correspondence in eastern North America.

Borisov has suggested average values for both the actual and potential evapotranspiration in the various physico-geographical zones of the USSR; these are reproduced in table 10.6, while some sample water balances for selected forests are shown in table 10.7. Potential exceeds actual evapotranspiration by 3–4 times in the steppes, by 2–3 times in the

**Table 10.6** Actual and potential evapotranspiration values for the different physico-geographical zones of the USSR (units : mm per year)

| Zone | Actual evapotranspiration | Potential evapotranspiration |
|---|---|---|
| Tundra | 70–120 | 50–200 |
| Taiga | 200–300 | 150–500 |
| Mixed forests | 250–430 | 400–800 |
| Steppes | 240–550 | 600–1,000 |
| Semi-deserts | 180–200 | 900–1,400 |
| Deserts | 50–100 | 1,000–2,400 |

*After Borisov (1965).*

mixed-forest zone and by 1·5–2 times in the taiga and in the region of broad-leaved forests in the east. In contrast, in the tundra regions actual evapotranspiration is 1·5–2 times less than the potential value. The ratios of precipitation to potential evapotranspiration for the various zones of the USSR are listed in table 10.8, where a value of unity corresponds to zero on Thornthwaite's moisture index scale. It is evident that there is a marked water shortage in the steppe areas, and it must therefore be concluded that natural grassland vegetation in the interiors of the continents is a response to a severe water shortage in summer.

## 4 Drought

Drought is reported from all areas, including the tundra, but is most obvious in the marginal agricultural lands of the prairies and steppes. It is a comparatively frequent phenomenon in the Russian steppe, and on average one in every 3 or 4 years is a drought year. In

**Table 10.7** Mean annual water balance of selected forest types in USSR (units : mm per year)

| Type | Precipitation | Total evapo-transpiration | Water retained in tree canopies | Evaporation from soil and grass cover | Transpiration from stand | Surface runoff | Infiltration |
|---|---|---|---|---|---|---|---|
| Mixed pine and spruce forests in northern taiga | 525 | 340 | 108 | 56 | 176 | 95 | 90 |
| Spruce and exploited deciduous forests in middle taiga | 600 | 323 | 158 | 65 | 101 | 13 | 264 |
| Spruce forest in southern taiga | 730 | 464 | 128 | 80 | 256 | 175 | 91 |
| Long-stem moss pinewoods of mixed forests | 550 | 445 | 90 | 190 | 165 | 15 | 90 |
| Oak and ash forest growing near forest-steppe boundary | 513 | 447 | 70 | 113 | 264 | 21 | 45 |

*After Molchanov (1971).*

the interior of North America, particularly dry regions are located in the Canadian prairies, Montana, South Dakota, and Wyoming. The largest, in southern Alberta, Saskatchewan, and Manitoba, has been studied by several climatologists, including Villimow (1956) and Laycock (1960). Laycock calculates that the average soil moisture deficit (figure 10.11) at the end of the summer ranges from 300 mm in the driest areas to between 100 and 150 mm

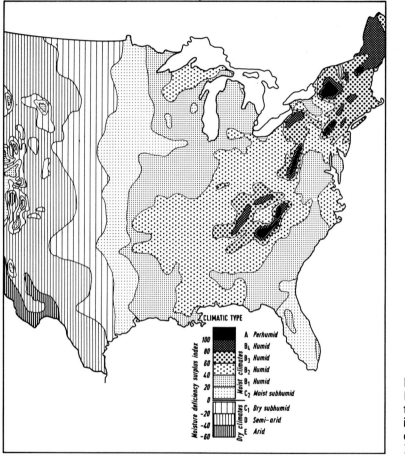

**Figure 10.10**
Distribution of Thorn-thwaite's moisture index over central and eastern USA. (*After Thornthwaite, 1948.*)

$$\text{Moisture index} = \frac{100 \times \text{annual water surplus} - 60 \times \text{annual water deficit}}{E_t}.$$

**Table 10.8**  Ratio (K) of annual precipitation to annual potential evapotranspiration

| K | Vegetation when there is sufficient heat |
|---|---|
| 1·50 | Humid forest |
| 1·49–1·00 | Moist forest |
| 0·99–0·60 | Forest-steppe |
| 0·59–0·30 | Steppe |
| 0·29–0·13 | Semi-deserts |
| 0·12–0·00 | Deserts |

*After Borisov (1965).*

288

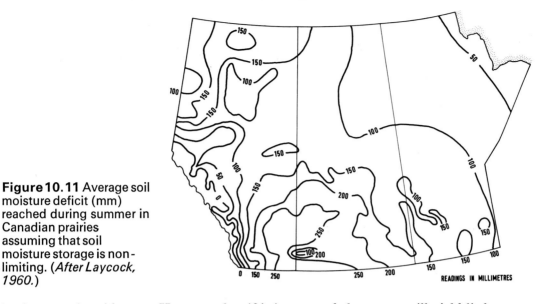

**Figure 10.11** Average soil moisture deficit (mm) reached during summer in Canadian prairies assuming that soil moisture storage is non-limiting. (*After Laycock, 1960.*)

in the more humid areas. He states that if it is assumed that crops will yield little or no return to the farmer when the soil moisture deficit exceeds 200 mm, then almost all the agricultural areas of the Canadian prairies will have experienced at least one failure during any 30-year period. Villimow has found that in the driest summers, parts of the prairies may not receive any significant precipitation for up to 75 days. Soil moisture patterns are not necessarily the same in successive dry years. Laycock discovered that, in 1927, most of the southern Canadian regions that are normally dry, were relatively moist, and the Peace River District in the north was dry. In 1936, a large part of the southern plains were very dry, while the Peace River District was relatively moist. Again, in 1950 the major drought area was centred well to the west in Alberta and the northern and eastern areas were moist.

*a Determination of drought* The water balances and vegetation covers of the continental interiors vary widely, and obviously a water shortage in one area would pass unnoticed in another. Palmer (1965) considers that in spite of the differences which exist, people in humid climates seem to mean much the same thing when they refer to drought as do the people in a semi-arid region; *viz.*, that the moisture shortage has seriously affected the established economy of their region. He therefore defines drought as a prolonged and abnormal moisture deficiency, and suggests that the problem be approached by the use of hydrologic accounting and the use of indexes of moisture anomaly.

Palmer summarizes his procedure for drought determination as follows:

I   undertake a detailed water-balance study for each month of a long series of years;
2   summarize the results to obtain coefficients which are dependent on the climate of the area being analysed;
3   re-analyse the series using the derived coefficients to determine the amount of moisture required for 'normal' weather during each month;
4   convert the departures to indices of moisture anomaly;
5   analyse the index series for the start and finish of drought periods and also the severity.

*The middle and high latitude atmosphere*

In dry climates it is usual for the actual evapotranspiration to fall below the potential values, and this can be expressed in the form

$$a = \frac{\overline{E}_a}{\overline{E}_t},$$

where $a$ is the coefficient of evapotranspiration, and $\overline{E}_a$ and $\overline{E}_t$ are the monthly averages of the actual and potential evapotranspiration respectively.

Similarly, coefficients of recharge ($\beta$), runoff ($\gamma$), and loss ($\delta$) can be derived:

$$\beta = \frac{\overline{R}}{\overline{PR}},$$

where $\overline{R}$ is the average monthly recharge and $\overline{PR}$ is the average monthly potential recharge defined as the amount of moisture required to bring the soil to field capacity;

$$\gamma = \frac{\overline{RO}}{\overline{PRO}},$$

where $\overline{RO}$ and $\overline{PRO}$ are the actual and potential runoff values, and potential runoff is the difference between the field capacity of the soil and the potential recharge (equal to the available moisture at the start of the month).

$$\delta = \frac{\overline{L}}{\overline{PL}},$$

where the actual and potential values are as above, potential loss being defined as the amount of moisture that could be lost from the soil provided the precipitation during the period was zero, and the potential evaporation and initial soil moisture conditions were as observed.

Palmer considers that these coefficients, when used in conjunction with the potential value for a particular month, enable an estimate to be made of the value that is 'climatically appropriate for existing conditions' (CAFEC). It is therefore possible to calculate for any particular month the CAFEC values (denoted by a circumflex) for evapotranspiration, recharge, runoff, and loss:

$$\widehat{E}_a = a\,E_t,$$
$$\widehat{R} = \beta\,PR,$$
$$\widehat{RO} = \gamma\,PRO,$$
$$\widehat{L} = \delta\,PL,$$

The CAFEC precipitation is given by

$$\widehat{P} = E_a + \widehat{R} + \widehat{RO} - \widehat{L},$$

and the difference (d) between this and the actual precipitation (P) for each month

$$d = P - \widehat{P},$$

provides a meaningful measure of the departure of the precipitation from normal. It is apparent that the significance of the departure d will depend on the locality and the seasons. So Palmer took two areas, one in central Iowa and the other in western Kansas, assumed that the driest year in each would have similar consequences for the inhabitants, and compared the values of d. On this assumption, it is possible to obtain a weighting factor, K, which can be applied to the values of d to make them of equal significance:

$$K = \bar{d}_{KAN}/\bar{d}_{IOWA}$$

Further research led Palmer to believe that the climatic characteristic K can be estimated for each calendar month as follows:

$$K = (\bar{E}_t + \bar{R})/(\bar{P} + \bar{L}).$$

It is now possible to define a moisture anomaly index, z, such that

$$z = dK$$

Each value of z expresses on a monthly basis the departure of the moisture climate from the average for the month.

Both the US Department of Agriculture and the US Weather Bureau have used the terms mild, moderate, severe, and extreme to describe droughts. The severity of a drought is a function of both the value of the index z, and also of the duration of the spell, because clearly the situation will tend to deteriorate with time. Drought severity X is therefore given by the expression

$$X_i = \sum_{t=1}^{i} z_t/(0{\cdot}309\, t + 2{\cdot}691)$$

where t is the time in months. Palmer has assigned values of X to various categories of drought, which are listed in table 10.9; these values may be used to determine the start and finish of a drought and its severity at any time during its occurrence.

Analysis by Palmer of a period of 76 years in western Kansas shows that drought occurred in 37 per cent of the months, wet spells in 37 per cent and near-normal conditions in 12 per cent. This indicates that in the interior grasslands the weather tends towards extremes and so-called average conditions are rare. He further found that in western

**Table 10.9**  Classes for wet and dry periods

| X | Class |
|---|---|
| $\geqslant 4{\cdot}00$ | Extremely wet |
| $3{\cdot}00$ to $3{\cdot}99$ | Very wet |
| $2{\cdot}00$ to $2{\cdot}99$ | Moderately wet |
| $1{\cdot}00$ to $1{\cdot}99$ | Slightly wet |
| $0{\cdot}50$ to $0{\cdot}99$ | Incipient wet spell |
| $0{\cdot}49$ to $-0{\cdot}49$ | Near normal |
| $-0{\cdot}50$ to $-0{\cdot}99$ | Incipient drought |
| $-1{\cdot}00$ to $-1{\cdot}99$ | Mild drought |
| $-2{\cdot}00$ to $-2{\cdot}99$ | Moderate drought |
| $-3{\cdot}00$ to $-3{\cdot}99$ | Severe drought |
| $\leqslant -4{\cdot}00$ | Extreme drought |

*After Palmer (1965).*

Kansas the drought was mild during 11 per cent of the months, moderate in 11 per cent, severe in 8 per cent, and extreme in 6 per cent. The drought during the 1930s was the longest and most serious so far recorded in western Kansas, for between August 1932 and October 1940 there were 38 months of extreme drought. These years were further characterized by unusually strong winds which created terrible duststorms and combined with the drought to produce really disastrous conditions. Other serious droughts occurred in 1894, 1913, and from 1952 until 1956.

*b Effects of drought*   The influence of drought on the natural vegetation of the prairies is indicated by the researches of Weaver and Albertson (1957), who made basal cover measurements each autumn in an ungrazed area near Hays, Kansas. Their results are contained in table 10.10, which clearly shows the changes in the drought years of the 1930s.

**Table 10.10**   Total basal cover of short grasses in an ungrazed prairie near Hays, Kansas

| Year | Per cent cover |
|------|----------------|
| 1932 | 89 |
| 1933 | 86 |
| 1934 | 85 |
| 1935 | 65 |
| 1936 | 58 |
| 1937 | 26 |
| 1938 | 30 |
| 1939 | 22 |
| 1940 | 20 |
| 1941 | 56 |
| 1942 | 94 |
| 1943 | 90 |
| 1944 | 95 |
| 1945 | 93 |
| 1946 | 89 |
| 1947 | 88 |
| 1948 | 93 |
| 1949 | 92 |
| 1950 | 91 |
| 1951 | 90 |
| 1952 | 93 |
| 1953 | 38 |
| 1954 | 20 |

*After Weaver and Albertson (1957).*

They found even greater changes on overgrazed areas, where the cover was reduced from 80 per cent in 1932 to 30 per cent in 1934.

The steppes of the USSR are subject to droughts which are similar in nature to those experienced in Kansas. Borisov states that it is a comparatively frequent phenomenon in the steppe and that in some areas there were 7 years of severe drought in a period of 20 years. Very severe droughts occurred in the years 1889–92, 1930–31, 1934–39, and 1946.

In both North America and central Asia there is a tendency towards persistence of abnormal drought or wet weather, since drought years often follow each other or are separated by lengthy periods of sufficient moisture. So far it has not been possible to detect any periodicity in the occurrence of wet and dry years. The danger of steppe and prairie climates is that during moist periods agriculture is allowed to develop and intensify, and since these periods last several years, a sense of security is produced in the farming community. Eventually a drought period begins and the agricultural activity can no longer be maintained, leading to a collapse of the agricultural community and a rapid retreat from the land. Once the cycle of events is known, some allowance can be made for bad years, but the first experience of this cycle can be traumatic for the farmers involved.

# References

ANDERSON, R., BOVILLE, B. W. and MCCLELLAN, D. E. 1955: An operational frontal contour-analysis model. *Quarterly Journal Royal Meteorological Society* **81**, 588.

ARAKAWA, H. and TAGA, S. 1969: Climate of Japan. In Arakawa, H. (editor), *Climates of northern and eastern Asia.* Amsterdam: Elsevier Publishing Co., p. 119.

BARRY, R. G. 1967: Seasonal location of the Arctic front over North America. *Geographical Bulletin* **9**, 79.

BORISOV, A. A. 1965: *Climates of the USSR.* 2nd edition, 1959. Translated from Russian by R. A. Ledward. London: Oliver and Boyd.

BRINKMANN, W. A. R. 1971: What is Foehn? *Weather* **26**, 230.

BRINKMANN, W. A. R. and ASHWELL, I. Y. 1968: The structure and movement of the chinook in Alberta. *Atmosphere* **6**, 1.

BRYSON, R. A. 1966: Air-masses, streamlines and the boreal forest. *Geographical Bulletin* **8**, 228.

BUDYKO, M. I. 1968: Solar radiation and the use of it by plants. In *Agroclimatological methods. Proceedings of the Reading Symposium.* Paris: UNESCO, p. 39.

CHANG, JEN-HU. 1955: Climate of China according to the new Thornthwaite classification. *Annals Association American Geographers* **45**, 393.

CHARLESWORTH, J. K. 1957: *The quarternary era.* London: Edward Arnold.

CHEBOTAREV, N. P. 1962. *Theory of stream runoff.* Jerusalem: Israel Program for Scientific Translations, 1966.

EFIMOVA, N. A. 1965: Distribution of photosynthetically active radiation over the territory of the Soviet Union. *Trudy Glavnaia Geofizicheskaia Observatoriia,* **179**, 118.

KENDREW, W. G. 1961: *The climates of the continents.* London: Oxford University Press.

KUNG, E. C., BRYSON, R. A. and LENSCHOW, D. H. 1964: Study of a continental surface albedo on the basis of flight measurements and structure of the earth's surface cover over North America. *Monthly Weather Review* **92**, 543.

LAYCOCK, A. H. 1960: Drought patterns in the Canadian prairies. *Publication* **51**, *Symposium International Association Scientific Hydrology,* Helsinki, p. 34.

MOLCHANOV, A. A. 1971: In Duvigneaud, P. (editor), *Productivity of forest ecosystems. Proceedings of the Brussels Symposium.* Paris: UNESCO, p. 49.

NIČIPOROVIČ, A. A. 1968: Evaluation of productivity by study of photosynthesis as a function of illumination. In Eckardt, F. E. (editor), *Functioning of terrestrial ecosystems at the primary production level. Proceedings of Copenhagen Symposium.* Paris: UNESCO, p. 261.

PALMER, W. C. 1965: Meteorological drought. *US Department of Commerce Research Paper No.* **45**, Washington.

PENNER, C. M. 1955: A three-front model for synoptic analyses. *Quarterly Journal Royal Meteorological Society* **81**, 89.

RASMUSSON, E. M. 1967: Atmospheric water vapour transport and the water balance of North America: Part 1. Characteristics of the water vapour flux field. *Monthly Weather Review* **95**, 403.

RASMUSSON, E. M. 1968: Atmospheric water vapour transport and water balance of North America: Part II. Large-scale water balance investigations. *Monthly Weather Review* **96**, 720.

RASMUSSON, E. M. 1971: A study of the hydrology of eastern North America using atmospheric vapour flux data. *Monthly Weather Review* **99**, 119.

REITER, E. R. 1961: *Meteorologie der Strahlströme*. Vienna: Springer Verlag. English Translation, 1963. *Jet stream meteorology*. Chicago: University of Chicago Press.

RIKHTER, G. D. 1948: Vliianie snezhnogo pokrova na klimat. *Akademiia Nauk SSSR. Institut Geografii* **40**, 10.

RODIN, L. E. and BASILEVIČ, N. I. 1968: World distribution of plant biomass. In Eckardt, F. E. (editor), *Functioning of terrestrial ecosystems at the primary production level. Proceedings of Copenhagen Symposium*. Paris: UNESCO, p. 45.

SANDERSON, M. 1948: Drought in the Canadian northwest. *Geographical Review* **38**, 289.

SUMMERS, P. W. and PAUL, A. H. 1967: Some climatological characteristics of hailfall in central Alberta. *Proceedings Fifth Conference on Severe Local Storms*, St. Louis, Mo., p. 313.

THOM, H. C. S. 1963: Tornado probabilities. *Monthly Weather Review* **91**, 730.

THORNTHWAITE, C. W. 1948: An approach toward a rational classification of climate. *Geographical Review* **38**, 55.

US WEATHER BUREAU. 1956: Seasonal variation of the PMP east of the 105th meridian for areas from 10–1,000 square miles and durations of 6, 12, 24, and 48 hours. *Hydrometeorological Report No.* **33**.

US WEATHER BUREAU. 1960: Generalized estimates of probable maximum precipitation for the United States west of the 105th meridian for areas to 400 sq miles and durations to 24 hours. *Technical Note No.* **38**.

VILLIMOW, J. R. 1956: The nature and origin of the Canadian dry belt. *Annals Association of American Geographers* **46**, 211.

WAHL, E. 1952: The January thaw in New England. *Bulletin American Meteorological Society* **33**, 380.

WAHL, E. 1954: A weather singularity over the US in October. *Bulletin American Meteorological Society* **35**, 351.

WATTS, I. E. M. 1969: Climates of China and Korea. In Arakawa, H. (editor), *Climates of northern and eastern Asia*. Amsterdam: Elsevier Publishing Co., p. 1.

WEAVER, J. E. and ALBERTSON, F. W. 1957: Grasslands of the great plains. Lincoln, Nebr.: Johnson Publishing Co.

# 11
# Polar climates

Polar climates are the least promising environmentally, because they lack both warmth and water. Desert climates can be exploited when water is supplied because they have ample warmth, but both the essentials of life are lacking in the polar climates. In winter there is continuous darkness and though the ice-sheets may give the impression of a plentiful snow-fall the actual precipitation is very low.

## Distribution

The climatic type occurs in two contrasting areas, one of which is largely an elevated plateau and the other an ocean. As the Antarctic continent has an area of about 14,000,000 km², of which less than 3 per cent is estimated to be free from a permanent ice-sheet, the prospects for settlement there are poor. The environment is even less inviting because of the elevation: 55 per cent of the surface is above 2,000 m and about 25 per cent is above 3,000 m. The highest mountains in Antarctica are not found in the area of the great icy plateau of east Antarctica, but belong to the Sentinel Range, Ellsworth Mountains, where the summit of the Vinson Massif (78·6°s, 85·4°w) rises to 5,140 m. In contrast, the Arctic Ocean is almost completely landlocked except for one main access point to the warmer waters of the Atlantic between Greenland and Norway. The central Arctic basin is covered by a thin but permanent ice-pack which extends to the continents during the winter. The one major permanent ice-sheet, which occurs on the edge of the Arctic Ocean in Greenland, covers an area of 1,726,000 km² or approximately 4/5 of the island, although several other islands do have minor ice-caps.

Both polar regions are located in regions of general atmospheric subsidence, though the climate is not particularly anticyclonic and the winds are not necessarily easterly. The moisture content of the air is low because of the intense cold, and horizontal thermal gradients are normally weak, with the result that energy sources do not exist for major atmospheric disturbances, which are rarely observed. The climate of both regions is very much controlled by the radiation balance, and the interaction of radiation with the surface, together with its modification by atmospheric heat advection. Vowinckel and Orvig (1970) suggest that the arctic atmosphere can be defined as the hemisphere cap of fairly low kinetic energy circulation lying north of the main course of the planetary westerlies, which places it roughly north of 70°N. The situation over the south polar regions is more complex and the boundary of the Antarctic is not so clear.

## Radiation and temperature

At high latitudes the total daily solar radiation depends largely on day-length which in turn varies widely with the season of the year. Values of day-length are summarized in table 11.1

**Table 11.1**   Daily duration of possible sunshine (in hrs and min) on the fifth day of each month at various latitudes

|       | 60      | 65      | 70      | 75      | 80      | 85      | 90      |
|-------|---------|---------|---------|---------|---------|---------|---------|
| Dec   | 6 h 43  | 5 h 02  | —       | —       | —       | —       | —       |
| Mar   | 11 h 44 | 11 h 40 | 11 h 33 | 11 h 23 | 10 h 50 | 9 h 50  | —       |
| Jun   | 18 h 49 | 21 h 53 | 24 h 00 | 24 h 00 | 24 h 00 | 24 h 00 | 24 h 00 |
| Sept  | 12 h 55 | 13 h 07 | 13 h 26 | 13 h 57 | 15 h 10 | 18 h 15 | 24 h 00 |

*After Gavrilova (1963).*

which shows that they vary from 24 hours at 70°N and above in June to zero above the same latitude in December. The radiation climate produced with continuous darkness in winter and continuous daylight in summer is distinctly different from that found at lower latitudes.

The duration of the polar night, which is defined as the period when the solar altitude is less than 0°50′, depends on the latitude, and increases towards the pole. By astronomical calculations which disregard refraction of the sun's rays and twilight, the polar night should last from 179 days at 90° to 24 hours at the Arctic Circle. Solar refraction in practice reduces the duration of the polar night to 175 days at the pole and zero on the Arctic Circle. In actual conditions the polar night is shortened still further by the phenomenon of twilight, which is defined as the period when the solar altitude is between −6° and 0° 50′. At the equator twilight lasts for only a few minutes, whereas near the pole it may continue for several days in succession, but it is only important for illumination, since the influx of short-wave radiation is negligible. Dates of the beginning and end of the polar day and night are given for various latitudes in table 11.2.

**Table 11.2**   Dates of the beginning and end of the polar day and polar night (taking refraction into account)

| Latitude | Polar day Beginning | End      | Polar night Beginning | End    |
|----------|---------------------|----------|-----------------------|--------|
| 70 N     | 17 V                | 27 VII   | 25 XI                 | 17 I   |
| 80 N     | 14 IV               | 30 VIII  | 22 X                  | 21 II  |
| 90 N     | 19 III              | 25 IX    | 25 IX                 | 19 III |

*After Gavrilova (1963).*

*1   Cloud effect*

In fact the distribution of global radiation is controlled not only by astronomical considerations, but also by the distribution of cloud, which is closely related to the character

*Plate 14*   Orographic cloud (multi-layered wave clouds) formed in airflow over the mountains of Graham Land. Taken from Argentine Islands (65°S, 65°W) looking northeast. *Reproduced by permission of W H Townsend and the Meteorological Office, Bracknell.*

*Plate 15* Fohn Wall and wave clouds over Signy Island, South Orkney. *Reproduced by permission of P A Richards and the Meteorological Office, Bracknell.*

of the atmospheric circulation processes. According to Gavrilova (1963) the most cloudy regions of the Arctic are the North Atlantic and European sectors where, because of the intense cyclonic activity, the frequencies of overcast skies are very high. The least cloudy regions in the Arctic are the northeastern parts of the Canadian Archipelago and North Greenland which are subject to the prolonged influence of the polar and Greenland anti-cyclones. A second minimum is observed in eastern Siberia, where it is low along the coast and decreases inland. Throughout much of the Arctic the greatest number of clear days occur at the end of the winter and the beginning of spring, while the maximum number of overcast days are observed in the autumn, before the beginning of the polar night. In the North Atlantic, Norwegian, and Greenland Seas there is almost no variation in cloudiness during the year.

A high frequency of low-level clouds is observed during almost the whole year in the Arctic, there being a small increase from winter to summer. These are mainly stratus clouds, for cumulus clouds are rarely encountered, except during the warm period of the year over the continents. The main features of Arctic clouds are their low water content and small depth. Dergach, Zabrodskiĭ, and Morachevskiĭ (1960) discovered that the water content of the clouds most often encountered in the Arctic is only measured in hundredths or tenths of a g/m³, whereas the water content of clouds at lower latitudes reaches several g/m³. Similarly, it has been established by Zavarina and Romasheva (1957) that the mean thickness of Arctic low-level clouds averages 350–550 m, while the corresponding value of medium level clouds is 400–450 m. The clouds are particularly thin in the cold season of the year (100–150 m), and thickest in the warm season (up to 1,000 m); however, in 76 per cent of the cases the cloud thickness in the Arctic is less than 700 m. The corresponding mean values for clouds at lower latitudes are as follows: stratus and stratocumulus 700–800 m; altostratus and altocumulus 500–800 m.

Information on cloudiness in the Antarctic is less complete, partly because of the complications introduced by the elevation of the plateau. Schwerdtfeger (1970) states that the circumpolar band of maximum cloud amount is located north of the belt of lowest pressure at sea level and south of the belt of strongest westerly winds. The situation over the continent is not clear, except that cumulus clouds are occasionally reported along convergence lines.

## 2   Distribution of radiation

Details of the distribution of global radiation in the Arctic are contained in figure 11.1. Minimums of global radiation are found during all months over the Norwegian Sea, and Vowinckel and Orvig (1970) suggest that this is a result of the high cloud amount, high cloud density, and the high atmospheric water vapour content. Maximums, observed mostly over eastern Siberia between 140° and 160°E, probably result from the extreme remoteness from open oceans. Secondary maximums, observed over Canada and the Canadian Archipelago, may be caused by the northward extension of the clear skies of the prairies. The thin clouds make the diffuse sky radiation a more important component of the global radiation than is usual in lower latitudes.

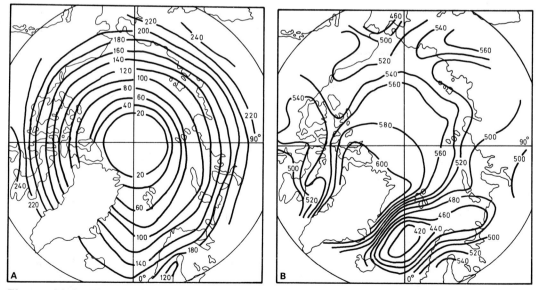

**Figure 11.1**   The distribution of global radiation in the Arctic (cal cm⁻² day⁻¹). **A:** March. **B:** June. (*After Vowinckel and Orvig, 1970.*)

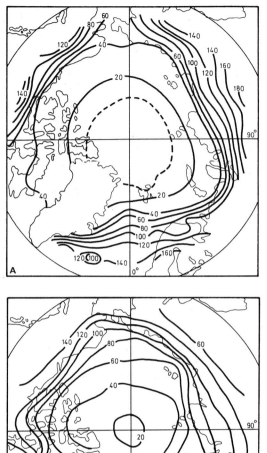

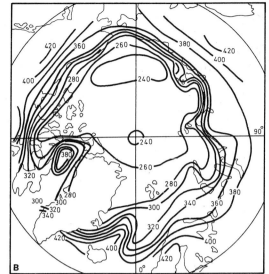

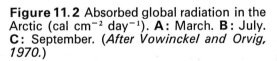

**Figure 11.2** Absorbed global radiation in the Arctic (cal cm⁻² day⁻¹). **A**: March. **B**: July. **C**: September. (*After Vowinckel and Orvig, 1970.*)

The short-wave radiation actually absorbed by the surface (figure 11.2) is obtained by multiplying the global radiation by (1 — albedo). Over most parts of the world the albedo values are relatively small and uniform over large areas, but as the albedo of snow and ice is very high, in regions of seasonal snow cover there will be wide variations in albedo. Vowinckel and Orvig (1970) suggest that the smallest Arctic albedos are found in July, August, and September along 65°N, where the absorbed radiation is only 5–6 per cent smaller than the global radiation; but to the north and in winter the albedo becomes a factor of equal or greater importance than all other depletion factors. The charts of absorbed radiation indicate some extremely steep gradients, which are apparent in all months, but especially in spring and early summer. Over the Atlantic Ocean the steep

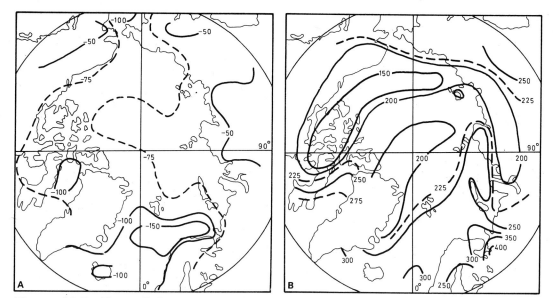

**Figure 11.3** Net radiation balance of the Arctic surface (cal cm$^{-2}$ day$^{-1}$). A: January. B: July. (*After Vowinckel and Orvig, 1970.*)

gradients follow the ice limit, but conditions are more complex over the continents because coniferous forests with low albedos extend far northwards beyond the snow-line.

Long-wave atmospheric counter radiation is particularly important at high latitudes, where the short-wave radiation becomes negligible during the winter. Atmospheric long-wave radiation is the only radiactive heat source for the surface during the polar night and even in midsummer the long-wave contribution is higher than that of the solar radiation (table 11.3).

**Table 11.3** Per cent contribution by insolation and atmospheric counter-radiation to total surface radiation income in June

|            | 65 | 70 | 75 | 80 | 85 | 90 |
|------------|----|----|----|----|----|----|
| Long wave  | 60 | 64 | 68 | 69 | 69 | 69 |
| Short wave | 40 | 36 | 32 | 31 | 31 | 31 |

*After Vowinckel and Orvig (1970).*

The net radiation balances of the Arctic surface during January and July are shown in figure 11.3. During winter, the maximum radiation loss is not in the polar basin proper but just to the south, where the ice-free ocean begins. In particular, a major area of loss appears over the Norwegian Sea and the adjoining oceans, forming one of the major radiational heat sinks of the earth's surface. Towards spring, steep gradients arise because of variations in albedo, but in summer the net radiation over the Arctic becomes positive and relatively uniform.

As shown in figure 11.4, total radiation balances for the earth–atmosphere system during February and August indicate that at all seasons of the year the balance is negative, signifying a continual heat loss to space which must be made up by energy advection from low latitudes. In February, the non-radiative processes must represent 100 per cent of the energy expenditure (that is, disregarding any heat stored in the ground), and 100 per cent

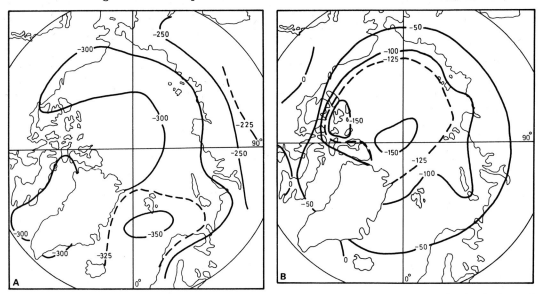

**Figure 11.4**  Net radiation balance of the earth–atmosphere system in the Arctic (cal cm⁻²
day). **A**: February. **B**: August. (*After Vowinckel and Orvig, 1970.*)

of the radiative energy turnover, but the importance of non-radiative processes becomes
less towards summer. Vowinckel and Orvig (1970) have calculated the percentage contribution of non-radiative processes to the energy turnover in the Arctic, and some of their
results are contained in table 11.4, which shows that the winter heat advection is extremely
powerful.

**Table 11.4**  The percentage contribution of non-radiative processes
to the turnover of energy at the North Pole

| | |
|---|---|
| Jan | 100 |
| Feb | 100 |
| Mar | 100 |
| Apr | 48 |
| May | 10 |
| Jun | 9 |
| Jul | 22 |
| Aug | 25 |
| Sept | 79 |
| Oct | 100 |
| Nov | 100 |
| Dec | 100 |

*After Vowinckel and Orvig (1970).*

Radiation studies in the Antarctic are not as advanced as those in the Arctic, but when
allowance is made for the ice-covered high plateau surface, the same general principles hold
there as in the Arctic. Very large amounts of solar energy reach the surface of the plateau
in summer, and Rusin (1961) has made the interesting comment that 'at the latitudes of
the 80s where the coldest point on earth is situated, the point of maximum monthly

amounts of solar energy is also situated'. The high surface albedo and the thin clouds enhance the importance of diffuse sky radiation, so that the global radiation on cloudy days differs little from that on clear days.

The seasonal variation of surface radiation balance over the continent and the adjacent ocean has been described by Zillman (1967). The entire region experiences a surface net radiation gain in January, although values approaching zero are found over the greater part of the plateau. By April, all the continent and the ocean southward of 55°s experience a net radiation loss, which is particularly large over the water oceans surrounding Antarctica. In July, the region of net radiative loss from the surface has expanded northward to 45°s, with the largest values in the 60–65°s latitude band. The return of the sun in October brings a net radiative gain to parts of the Ross Sea and the eastern escarpment of the plateau extending as far south as 85°, but over the remainder of the region the net radiation is still negative.

## 3 Inversions

There are two main types of semi-permanent inversions in the world: the first type is found in the subtropics and the second in the polar regions. While the inversions of the subtropics are normally dynamic in origin, the polar inversions are complex in origin and are maintained not only by the subsidence of warm air but also by intense surface cooling. Vowinckel and Orvig (1967) consider that temperature inversions dominate the Polar Ocean practically throughout the year, since no month averages less than 59 per cent inversion conditions, and in late winter the frequency reaches 100 per cent. In winter over 80 per cent of the inversions start at the surface and are not destroyed by the occasional cyclones, for the advection of warm air seems to further stabilize the surface conditions. A similar phenomenon is found in the Antarctic where, with the exception of two summer months, an inversion is an ever-present feature over the high plateau, and a frequent one over the rest of the continent including the coastal areas.

## 4 Surface temperatures

During most of the year the Polar Ocean is covered by a relatively thin layer of cold air and the air temperature near the surface is primarily dependent on the temperature of the ice surface. In summer, the prevailing melting of snow and ice holds the surface temperature close to 0°C, but positive temperatures are usually observed near the pole in the second half of July. The winter temperatures over the Arctic pack-ice remain nearly constant for a considerable time and this represents a balance between heat loss from the ice and snow by radiation, heat conducted through the ice from the underlying water, and also the heat transported into the Arctic by intense warm air advection in cyclones. Winter minimum air temperatures occur when the net radiative loss is only balanced by the transfer of heat through the ice from the underlying water, and according to Vowinckel and Orvig (1970), this gives temperatures of about −40°C, or even down to −50°C over thick ice. Winter maximum temperatures are nearly a linear function of wind speed, because of the transport of heat downwards from above.

Antarctic surface temperatures are complicated by the relief of the continent, large areas of which are elevated above sea level. Taljaard, Van Loon, Crutcher, and Jeune (1969) have found that in January average temperatures vary from −30°C near the centre of the continent to −2°C along the coasts, while in July the coldest areas of the plateau average

—70°c, and the coasts reach about —28°c. The surface temperature in Antarctica and particularly over the Antarctic Plateau decreases very little during the winter night, from April to August, presenting the so-called coreless winter phenomenon. Wexler (1959) considers that when the sun sets, the temperature drops rapidly over the continent, but less so over the surrounding oceans, thus creating a strong meridional temperature gradient. This initiates the formation of numerous intense cyclones which move vast quantities of warm marine air southward and in this way ventilate large portions of Antarctica above the surface inversion, so slowing the decline of surface temperature.

## Snow cover, precipitation, and water balance

During the greater part of the year the Arctic surface is snow-covered since only in July and the first half of August is the snow in a state of rapid melting. The snow cover is less stable on the margins of the Arctic Ocean and vanishes during the summer. According to Gavrilova (1963), on the northern islands of the Soviet Arctic seas the snow cover is established in the first half of September, but at the end of September or the beginning of October in the northern part of the Asian continent, on the coasts of the Laptev, East Siberian and Beaufort Seas, on the southern islands of the Canadian Archipelago, and on the coasts of the Karsk and Chukotsk Seas. In northern Europe the snow cover only appears at the middle or the end of October.

The destruction of the Arctic snow cover begins in the middle of May in the continental areas and on the coast of the Barents Sea, in the first half of June on the coast of the remaining part of the Arctic, and in the second half of June on the islands and most northern parts of the continents.

Annual precipitation over the Polar Ocean is meagre, and falls mainly in the form of snow during autumn and late spring. This precipitation is largely frontal in nature and therefore decreases poleward from 250 mm per year on the margins to about 135 mm in the centre (Vowinckel and Orvig, 1970). The distribution of precipitation over the Antarctic continent is not well known, but the annual accumulation of snow appears to vary from about 60 g/cm along the coast to about 5 g/cm in the centre of the plateau (Giovinetto, 1964, 1968).

It is possible to compute annual water balances for individual ice-sheets and seas, but the type of soil moisture balance calculation used in previous chapters is of little use in polar regions, because soil and vegetation are mostly non-existent in these extremely hostile environments. Since evaporation rates are normally very small, the meagre precipitation can still lead to the accumulation of substantial ice-sheets, as for instance those found in Antarctica.

## References

DERGACH, A. L., ZABRODSKIĬ, G. M. and MORACHEVSKIĬ, V. G. 1960: Opyt kompleksnogo issledovaniya oblakov tipa st-sc i tumanov v Arktike. (Experience in the complex investigation of St-Sc type clouds and of fogs in the Arctic.) *Akademiia Nauk SSSR, Izvestiia, Ser. Geofiz.* **1**, 107.

GAVRILOVA, M. K. 1963: *Radiation climate of the Arctic.* Jerusalem: Israel Program for Scientific Translations, 1966.

GIOVINETTO, M. B. 1964: The drainage systems of Antarctica: accumulation. In Mellor, M. (editor), Antarctic snow and ice studies. *American Geophysical Union, Antarctic Research Series* **2**, 127.

GIOVINETTO, M. B. 1968: Glacier landforms of the Antarctic coast, and the regimen of the inland ice. Thesis, University of Wisconsin, Madison.

ORVIG, S. (editor) 1970: *Climates of the polar regions. World survey of climatology.* Vol. **14**. Amsterdam: Elsevier Publishing Co.

RUSIN, N. P. 1961: *Meteorological and radiational regime of Antarctica.* Jerusalem: Israel Program for Scientific Translations, 1964.

SCHWERDTFEGER, W. 1970: The climate of the Antarctic. In Orvig, S. (editor), *Climates of the polar regions.* Amsterdam: Elsevier Publishing Co.

TALJAARD, J. J., VAN LOON, H., CRUTCHER, H. L. and JEUNE, R. L. 1969: *Climate of the upper air.* Part I. *Southern Hemisphere.* **1**. *Sea level pressures and selected heights, temperatures and dew-points.* NAVAIR 50–16–15. Washington: US Government Printing Office.

VOWINCKEL, E. and ORVIG, S. 1967: The inversion over the polar ocean. In Orvig, S. (editor), *WMO–SCAR–ICPM. Proceedings of Symposium on Polar Meteorology. WMO Technical Note* **87**, 39.

VOWINCKEL, E. and ORVIG, S. 1970: The climate of the north polar basin. In Orvig, S. (editor), *Climates of the polar regions.* Amsterdam: Elsevier Publishing Co., p. 129.

WEXLER, H. 1959: Seasonal and other temperature changes in the Antarctic atmosphere. *Quarterly Journal Royal Meteorological Society* **85**, 196.

ZAVARINA, M. V., and ROMASHEVA, M. K., 1957. Moshchnost' oblakov nad arkticheskimi moryami i Tsentral'noi Arktikoi. (Thickness of clouds over Arctic seas and over the central Arctic.) *Problemy Arktiki* **2**, 127.

ZILLMAN, J. W. 1967: The surface radiation balance in high southern latitudes. In Orvig, S. (editor), *WMO–SCAR–ICPM Proceedings of Symposium on Polar Meteorology,* Geneva 1966. *WMO Technical Note* **87**, 142.

# Glossary

Included in the following glossary are a few technical terms which may be less familiar to some readers. For further details, readers should consult one of the published meteorological glossaries, the best of which is *Meteorological Glossary* published by Her Majesty's Stationery Office, London, ·1972, and compiled by Dr D. H. McIntosh for the Meteorological Office. Two other useful sources are published by the World Meteorological Organization, Geneva: *International Meteorological Vocabulary* (1966) and *International Glossary of Hydrology* (1969).

**Accumulated Temperature:** The integrated excess or deficiency of temperature measured with reference to a fixed datum over an extended period of time. If on a given day the temperature is above the datum value for n hours and the mean temperature during that period exceeds the datum line by m degrees, the accumulated temperature for the day above the datum is nm degree-hours or nm/24 degree-days. By adding up the daily entries arrived at in this way, the accumulated temperature above or below the datum value may be evaluated for periods such as a week, a month, a season, or a year.

**Adiabatic:** An adiabatic process (thermodynamic) is one in which heat does not enter or leave the system.

**Advection:** The process of transfer (of an air-mass property) by virtue of motion.

**Air-Mass:** A large body of air in which horizontal gradients of temperature and humidity are relatively slight and which is separated from an adjacent body of air by a more or less sharply defined transition zone (front) in which these gradients are relatively large.

**Albedo:** A measure of the reflecting power of a surface, being that fraction of the incident radiation which is reflected by a surface.

**Angular Momentum:** The angular momentum (or moment of momentum) per unit mass of a body rotating about a fixed axis is the product of the linear velocity of the body and the perpendicular distance of the body from the axis of rotation. The dimensions are $L^2 T^{-1}$.

**Angular Velocity:** The angular velocity of a moving point about a fixed axis is the rate of change of the angle between a plane drawn through the axis and the moving point, and a fixed plane through the axis. The angular velocity of a solid body about an axis is the angular velocity of any point of the solid body about that axis. Angular velocity is a vector quantity which is normally measured either in revolutions per unit time or in radians per unit time. The angular velocity of the earth may be represented by a vector parallel to the axis of the earth and directed northwards, with a magnitude of $2\pi$ rad/sidereal day = $7 \cdot 292 \; 10^{-5}$ rad/s.

**Baroclinic:** A baroclinic atmosphere is one in which surfaces of pressure and density intersect at some level or levels. The atmosphere is always, to some extent, baroclinic. Strong baroclinicity implies the presence of large horizontal temperature gradients and thus of strong thermal winds.

**Barotropic:** A barotropic atmosphere is that hypothetical atmosphere in which surfaces of pressure and density coincide at all levels.

**Blocking:** The term applied in middle latitude synoptic meteorology to the situation in which there is interruption of the normal eastward movement of depressions, troughs, anticyclones, and ridges for at least a few days.

**Boundary Layer:** That layer of a fluid adjacent to a physical boundary in which the fluid motion is much affected by the boundary and has a mean velocity less than the free-stream value.

**Bowen Ratio:** The ratio of the amount of sensible heat to that of latent heat lost by a surface to the atmosphere by the processes of conduction and turbulence.

**Capillary Potential:** A concept used in soil moisture studies, being the force of attraction exerted by soils on contained water, or the equivalent force required to extract the water from the soil against the capillary forces (surface tension) acting in the soil pores. It is generally expressed in the pressure unit of atmospheres.

**Convection:** A mode of heat transfer within a fluid, involving the movement of substantial volumes of the substance concerned. Two distinct, though not mutually exclusive types of convection occur in the atmosphere. In 'forced' convection the vertical motion is produced by mechanical forces, as in the passage of air over rough or high ground, and the vertical transport of properties is effected by 'eddies'. In 'free' or 'natural' convection buoyancy forces operate to effect vertical mixing through the agency of convection cells or 'bubbles'.

**Coriolis Acceleration:** An apparent acceleration which air possesses by virtue of the earth's rotation, with respect to axes fixed in the earth.

**Coriolis Parameter:** A quantity, denoted f, defined by the equation $f = 2\,\Omega \sin \phi$, where $\Omega$ is the magnitude of the earth's angular velocity and $\phi$ the latitude.

**Development:** In synoptic meteorology, the intensification of circulation, cyclonic or anticyclonic.

**Diabatic:** A diabatic thermodynamic process is one in which heat enters or leaves the system. The established equivalent term 'non-adiabatic' is generally preferred.

**Divergence:** The divergence of the flux of a quantity expresses the time-rate of depletion of the quantity per unit volume. Negative divergence is termed 'convergence' and relates to the rate of accumulation.

In meteorology, divergence (or convergence) is mostly used in relation to the velocity vector and so refers to the flux of air particles themselves. The 'divergence of velocity' is a three-dimensional property which expresses the time-rate of expansion of the air per unit volume. The following relation holds:

$$\text{div V (or } \nabla \cdot \text{V)} = \frac{\partial u}{\partial x} + \frac{\partial v}{\partial y} + \frac{\partial w}{\partial z},$$

where u, v, w are the components of the velocity vector V in the Cartesian x, y, z directions, respectively.

In the atmosphere, div V is small and of little direct interest, and therefore the main concern is with the 'horizontal divergence of velocity' ($\text{div}_H$ V), which expresses the time rate of horizontal expansion of the air per unit area.

Since div V $\approx$ o,

$$\text{div}_H \text{ V} \equiv \frac{\partial u}{\partial x} + \frac{\partial v}{\partial y} = -\frac{\partial w}{\partial z},$$

i.e., horizontal divergence of air is closely associated with vertical contraction of an air column centred at the level concerned.

**Direct Circulation:** A circulation in which the potential energy represented by the juxtaposition of relatively dense and light air-masses is converted into kinetic energy as the lighter air rises and the denser air sinks. In an 'indirect circulation' the converse holds and the potential energy increases as kinetic energy decreases.

**Evapotranspiration:** The combined process of evaporation from the earth's surface and transpiration from vegetation. 'Potential evapotranspiration' is the addition of water vapour to the atmosphere which would take place by these processes from a surface covered by green vegetation if there were no lack of available water.

**Field Capacity:** The mass of water (per cent of dry soil) retained by a previously saturated soil when free drainage has ceased, is known as the soil's field capacity or water-holding capacity.

**Geostrophic Wind:** That horizontal equilibrium wind ($V_G$), blowing parallel to the isobars and representing an exact balance between the horizontal pressure gradient force and the horizontal component of the Coriolis force ($fV_G$). Low pressure is to the left of the wind vector in the northern hemisphere, to the right in the southern hemisphere.

**Hadley Cell:** A simple thermal circulation first suggested by George Hadley in the eighteenth century as a partial explanation of the trade winds and still thought to be observed in the troposphere between latitudes o° and 30°.

**Hydrostatic Equation:** In an atmosphere at rest with respect to the earth, the variation of pressure (p) with height (z) is given by

$$\frac{\partial p}{\partial z} = -\rho g$$

where g is gravitational acceleration and $\rho$ is air density.

*Glossary*

**Index Cycle:** A term often applied to alternating periods of predominantly zonal and meridional flow.

**Isentropic:** Without change of entropy, generally equivalent in meaning to adiabatic. Isentropic surfaces in the atmosphere are surfaces of constant potential temperature.

**Potential Temperature:** That temperature $(\theta)$, which a given sample of air would attain if transferred at the dry adiabatic lapse rate to the standard pressure of 1,000 mb.

**Potential Vorticity:** In adiabatic motion of a column of air the quotient of the absolute vorticity of the air column $(\zeta + f)$ to the pressure difference between the top and bottom of the column $(\Delta p)$ is constant, i.e.,

$$\frac{\zeta + f}{\Delta p} = \text{constant}$$

where $\zeta$ is the relative vorticity of the column and f is the Coriolis parameter. The value of the absolute vorticity of a column which corresponds to a standard value $\Delta p$, is termed the potential vorticity.

**Precipitable Water:** The precipitable water of a column of air is the depth of water that would be obtained if all the water vapour in the column, of unit area cross-section, were condensed on to a horizontal plane of unit area.

**Rossby Number:** A dimensionless parameter found to be important in studies of the convection chamber type.

**Rossby Parameter:** The northward variation of the Coriolis parameter, arising from the spherical shape of the earth.

**Roughness Length:** A quantity $(z_0)$, also called the 'roughness coefficient' or 'roughness parameter', which enters as a constant of integration into the form of the logarithmic velocity profile appropriate to fully rough flow near a surface. See chapter 1 for equations.

**Soil Moisture Deficit:** The amount of rainfall or irrigation required to restore soil to its field capacity.

**Solenoids:** The intersection in a baroclinic atmosphere of surfaces of constant pressure with surfaces of constant density. The existence of such solenoids in the atmosphere tends to produce a direct circulation.

**Stomatal Diffusion Resistance:** The resistance to moisture flux through plant leaves. The dimensions are $TL^{-1}$.

**Thermal Wind:** The thermal wind in a specified atmospheric layer, at a given time and place, is the vertical geostrophic wind shear in the layer concerned. The term 'thermal wind' was adopted because wind shear is determined by the distribution of mean temperature in the layer concerned. The direction of the wind is such that it blows parallel to the mean isotherms of the layer with low mean temperature on the left in the northern hemisphere and on the right in the southern hemisphere.

**Vorticity:** The vorticity at a point in a fluid is a vector which is twice the local rate of rotation of a fluid element. The component of the vorticity in any direction is the circulation per unit area of the fluid in a plane normal to that direction. The dimensions are $T^{-1}$.

**Vorticity Equation:** The vorticity equation as used in meteorology relates the rate of change of the vertical component of vorticity to the horizontal divergence. In pressure co-ordinates the equation can be approximately written:

$$\frac{d}{dt}(\zeta + f) = -(\zeta + f)\ \text{div}_p\ V$$

where $\zeta$ is the vertical component of vorticity, and f is the Coriolis parameter.

**Wilting Point:** The point at which the soil contains so little water that it is unable to supply it at a rate sufficient to prevent permanent wilting of plants.

# Symbols

The most important mathematical symbols are defined below:

| | |
|---|---|
| A | Reciprocal of the mechanical equivalent of heat; average annual temperature range; constant |
| a | Constant |
| B, b | Constants |
| C | Energy gained or lost by convection; atmospheric runoff (local evapotranspiration which does not form local rainfall); speed of propagation of a Rossby wave; frequency of continental air-masses; fractional cloudiness |
| $C_p$ | Specific heat of air at constant pressure |
| D | Characteristic dimension |
| $\Delta D$ | Horizontal distance |
| d | Zero-plane displacement; average depth |
| E | Vertical transfer of water vapour (evaporation) |
| $E_a$ | Actual evapotranspiration; aerodynamic component of evapotranspiration |
| $E_t$ | Evapotranspiration, potential evapotranspiration |
| e | Screen vapour pressure |
| $e_a$ | Vapour pressure at some height above surface |
| $e_s$ | Vapour pressure at the apparent evaporating surface |
| F | Flux of radiation |
| f | Coriolis parameter; number of tropical storm passages per year |
| $f_R$ | Annual runoff |
| G | Storage of heat in a crop |
| $G_V$ | Annual extreme wind population |
| g | Acceleration due to gravity |
| H | Vertical flux of sensible heat |
| h | Height |
| I | Intensity of hydrologic cycle |
| i | Rainfall intensity |
| K | Sensible heat transfer to the atmosphere; index of continentality |
| $K_H$ | Eddy thermal diffusivity |
| $K_h$ | Thermal diffusivity |
| $K_M$ | Kinematic eddy viscosity |

## Symbols

| | |
|---|---|
| $K_v$ | Water vapour eddy diffusivity |
| k | Karman's constant, coefficient of the hydrologic cycle |
| L | Latent heat of vaporization |
| $L_o$ | Net radiation as $R_s \to 0$ |
| $\overline{L}$ | Actual monthly loss |
| l | Wave length; vertical Coriolis parameter $(2\,\Omega \cos \phi)$ |
| M | Absolute angular momentum per unit mass about earth's axis |
| N | Frequency of air-masses other than continental; nitrogen applied |
| O | Runoff |
| P | Mean annual precipitation, mean monthly precipitation |
| $\overline{PL}$ | Average monthly potential loss |
| $\overline{PR}$ | Average monthly potential recharge |
| $\overline{PRO}$ | Average monthly potential runoff |
| p | Pressure, mean monthly precipitation |
| $p_E$, $p_T$ | Probabilities |
| $\Delta p$ | Difference in height of sloping surface; pressure difference |
| Q | Total horizontal transport of water vapour, total energy content of one gram of moist air |
| q | Specific humidity |
| R | Gas constant for dry air; thermal Rossby number |
| $R_a$ | Radiation absorbed by leaf surface |
| $R_L \uparrow$ | Upwards flux of long-wave radiation |
| $R_L \downarrow$ | Downwards flux of long-wave radiation |
| $R_N$ | Net radiation |
| $R_{NL}$ | Net long-wave radiation |
| $R_s$ | Global radiation |
| $R_s \uparrow$ | Upwards flux of short-wave radiation |
| $R_s \downarrow$ DIR | Downwards flux of short-wave direct radiation |
| $R_s \downarrow$ DIF | Downwards flux of short-wave diffuse radiation |
| $\overline{R}$ | Average monthly recharge |
| $\overline{RO}$ | Actual monthly runoff |
| r | Albedo; radius; rainfall |
| $r_A$ | Component of rainfall formed by moisture advected by winds |
| $r_E$ | Component of rainfall formed by moisture from local evapotranspiration |
| S | Sensible heat transfer to the soil; amount of a given quantity in unit volume of air |
| T | Absolute temperature; mean annual temperature |
| $T_a$ | Air temperature |
| $T_e$ | Surface temperature of leaf |
| $T_m$ | Mean temperature of layer |
| $T_s$ | Surface temperature |
| $\overline{T}$ | Mean temperature |
| t | Mean monthly temperature |
| u | Eastward component of wind velocity; deep drainage of water |
| $u_1$, $u_2$ | Wind speeds at heights $z_1$ and $z_2$ respectively |
| $u_z$ | Wind speed at height z |

314

| | |
|---|---|
| u | Average wind velocity |
| V | Wind speed |
| $V_a$ | Actual vapour pressure |
| $V_g$ | Geostrophic wind |
| $V_s$ | Saturated vapour pressure |
| $V_z$ | Net downward flux of water at depth z |
| $V_o$ | Velocity at bottom of layer, inflow velocity |
| $V_1$ | Velocity at top of layer, outflow velocity |
| $V'$ | Thermal wind |
| v | Northward component of wind velocity |
| $W, W_o, W_1, W_e$ | Precipitable water content |
| $\Delta W$ | Change in soil water storage |
| w | Vertical component of wind velocity, sustained on-shore component of the wind |
| X | Drought severity index |
| Y | Yield of hay |
| y | Shape parameter |
| z | Height; moisture anomaly index |
| $z_o$ | Roughness length |
| $z_o, z_1, z_2$ | Height of levels 0, 1, and 2 respectively |
| $\alpha$ | Short-wave albedo; coefficient of evapotranspiration; constant |
| $\beta$ | Heating coefficient; Bowen ratio; scale parameter; Rossby parameter; coefficient of recharge; constant |
| $\gamma$ | Psychrometer constant; coefficient of runoff |
| $\Delta$ | Slope of temperature-vapour pressure curve |
| $\delta$ | Coefficient of loss |
| $\epsilon$ | Infra-red emissivity; surface emissivity |
| $\zeta$ | Relative vorticity |
| $\zeta'$ | Vorticity of thermal wind |
| $\eta$ | Ratio of heating of air column by condensation to cooling by adiabatic expansion in the rising air |
| $\theta$ | Potential temperature, volumetric soil water content |
| $\lambda$ | Constant dependant on cloud height, wavelength |
| $\lambda_m$ | Wavelength of maximum energy |
| $\rho$ | Air density |
| $\sigma$ | Constant |
| $\phi$ | Latitude |
| $\Omega$ | Angular velocity of earth |

# Author Index

# Geographical Index

*In this Index place-names are arranged alphabetically under the following headings:*
*Africa, Asia, Australasia, Europe, North America, Oceans, Polar Regions, South America.*

# Subject Index